MANUFACTURING ENGINEERING AND INTELLIGENT MATERIALS

PROCEEDINGS OF THE INTERNATIONAL CONFERENCE ICMEIM 2015, GUANGZHOU, CHINA, 30–31 JANUARY 2015

Manufacturing Engineering and Intelligent Materials

Editors

Li Lu
National University of Singapore, Singapore

Nooh Abu Bakar
Universiti Teknologi, Malaysia

CRC Press
Taylor & Francis Group
Boca Raton London New York

CRC Press is an imprint of the
Taylor & Francis Group, an **informa** business

A BALKEMA BOOK

Published by:
CRC Press/Balkema
P.O. Box 447, 2300 AK Leiden, The Netherlands
e-mail: Pub.NL@taylorandfrancis.com
www.crcpress.com – www.taylorandfrancis.com

First issued in paperback 2020

Typeset by V Publishing Solutions Pvt Ltd., Chennai, India

ISBN 13: 978-0-367-73798-6 (pbk)
ISBN 13: 978-1-138-02832-6 (hbk)

This book contains information obtained from authentic and highly regarded sources. Reasonable efforts have been made to publish reliable data and information, but the author and publisher cannot assume responsibility for the validity of all materials or the consequences of their use. The authors and publishers have attempted to trace the copyright holders of all material reproduced in this publication and apologize to copyright holders if permission to publish in this form has not been obtained. If any copyright material has not been acknowledged please write and let us know so we may rectify in any future reprint.

Visit the Taylor & Francis Web site at
http://www.taylorandfrancis.com

and the CRC Press Web site at
http://www.crcpress.com

Manufacturing Engineering and Intelligent Materials – Lu & Abu Bakar (Eds)
© 2015 Taylor & Francis Group, London, ISBN 978-1-138-02832-6

Table of contents

Preface

The 2015 International Conference on Manufacturing Engineering and Intelligent Materials (ICMEIM2015) provides an academic platform for leading experts and researchers in these fields for sharing and exchanging the latest research results and problem solving solutions. ICMEIM has collected advanced results and trends in the field of manufacturing engineering and intelligent materials.

This book is a collection of quality papers. All the papers accepted are reviewed and edited by 2-3 expert referees from our academic committee board. This book aims to collect the advanced and practical result of research developments on manufacturing system, control engineering, materials engineering, etc.

The Scientific Committee of the Conference has made sure that this book will provide the readers with the latest advanced knowledge in these fields, also providing a valuable summary and reference.

We express our sincere appreciations to the authors for their contribution to this book. We would also like to express our sincere gratitude to all the experts and referees for their valuable comments and the editing of the papers. Thanks also to CRC Press/Balkema (Taylor & Francis Group).

Manufacturing Engineering and Intelligent Materials – Lu & Abu Bakar (Eds)
© *2015 Taylor & Francis Group, London, ISBN 978-1-138-02832-6*

Organizing committee

ORGANIZED BY:

Hongkong Information Science and Engineering Research Center, Hongkong

CO-CHAIRMEN

Professor Li Lu, *National University of Singapore, Singapore*
Professor Nooh Abu Bakar, *Universiti Teknologi, Malaysia*
Professor Li Xiaoping, *National University of Singapore, Singapore*

INTERNATIONAL SCIENTIFIC COMMITTEE

Professor Mohd Sapuan Salit, *Universiti Putra, Malaysia*
Professor Dr. Mohamad Nor Berhan, *Universiti Teknologi MARA, Malaysia*
Professor Kyle Jiang, *University of Birmingham, UK*
Professor Huang Shiuh-Jer, *National Taiwan University of Science and Technology, Taiwan*
Professor Li Lu, *National University of Singapore, Singapore*
Professor Nooh Abu Bakar, *Universiti Teknologi, Malaysia*
Professor Li Xiaoping, *National University of Singapore, Singapore*

LOCAL ORGANIZING CHAIRMAN

Dr. Jason Lee, *Hongkong Information Science and Engineering Research Center, Hongkong*

Manufacturing Engineering and Intelligent Materials – Lu & Abu Bakar (Eds)
© 2015 Taylor & Francis Group, London, ISBN 978-1-138-02832-6

Experimental investigation into the pulsating degree of in-line pulse CVT

K.C. Liu & Z.X. Zheng
College of Mechanical Engineering and Automation, Fuzhou University, Fuzhou, Fujian, China

ABSTRACT: The performance of in-line pulse CVT and its effect factors are investigated by experimental method. An experimental device of in-line pulse CVT is designed and manufactured, a pulsating degree test bench of the CVT is established, and the pulsating degree experiment of the CVT is carried out. The experiment results show that the pulsating degree depends on the distance between the speed control point and the input shaft, the pulsating degree of in-line pulse CVT is less than 15%. Experimental results verify that the new structure of in-line pulse CVT has good work performance and usability.

Keywords: Continuously variable transmission; Pulse CVT; Test bench; Pulsating degree

1 INTRODUCTION

The pulse CVT (Continuously Variable Transmission) is a novel type of mechanical transmission that has been widely used in the textile, food and packaging industries. This device has many advantages, such as reliability, broad adjustable-speed range, compact structure and lighter weight, among others (Ruan 1999, Zhou 2001).

Currently, Germany, the United States and Japan have produced the most sophisticated types of pulse CVT. Of these, the German pulse CVT GUSA and the American ZERO-MAX CVT are examples that have mastered this relatively mature technology. Singh and Nair and Mangialardi. L, Mantriota. G (Singh & Nair 1992, Mangialardi & Mantriota 1999) have established the relationships among transmission efficiency, input torque, angular velocity of the drive shaft and transmission ratio. Kazerounian and Furu-Szekely Kazem & Zoltan (2005) proposed a new parallel disk continuously variable transmission, and provided analysis and testing of the dynamic characteristics of the device. Nobuyuki, Hiroki and Takeshi (Nobuyuki et al. 2010) presented alternative schemes of transmission element design using a finite element model in the vibration mode. Domestically, pulse CVT had been studied from the early 1970s, which improved and promoted the basic types of pulse CVT, the GUSA and ZERO-MAX, but, so far, there has been no systematic and integrated design theory nor effective measures for calculating CVT performance in practice (Zhu & Liu 2003).

In practice, pulse CVT has three main shortcomings. First, the inertia force produced by the movement of the connecting rod is difficult to balance. Vibration in the unit is caused by the unbalanced inertia force and inertia moment when the speed of the connecting rod is high. Second, the overrunning clutch of the output mechanism is the weak link in the transmission chain of the impulse CVT. Its bearing capacity and shock resistance are relatively poor and this limits the capacity of power transfer of the impulse CVT. Third, pulsation cannot be completely eliminated (Sun et al. 2012).

Pulsating degree is a primary parameter of motion and a major quality index for pulse CVT. In-line pulse CVT described in this paper presents a whole new design, full uniform motion output can be achieved by using cam-connected rod combination mechanism with a combination law of motion and one-way clutch. The pulsating degree of the CVT is investigated by experimental method, a pulsating degree test bench of the CVT is established. In order to improve its dynamic performance, the pulsating degree experiment of the CVT is carried out under the different transmission ratio.

2 MOTION PRINCIPLE OF THE IN-LINE PULSE CVT

2.1 *Mechanism diagram*

The new structure of in-line pulse CVT is depicted in Fig. 1. According to the typical pulse CVT, the new structure of in-line pulse CVT mainly includes the transmission mechanism, output mechanism (overrunning clutch) and the speed control mechanism. In Fig. 1, the No.1 is a cam which is connected with axis of input shaft at point A in the form of revolute joint. The No.2 is cam roller which is connected with the cam 1. The No.3 is lower clamp, the No.6 is upper clamp, the No.4 is spring, the No.5 is guide rods. These components are combined into compact mechanism which ensures that the cam 1 and the cam roller 2 keep proper contact throughout the whole working process. The No.7 is a connecting rod which is connected with the cam roller 2 in the form of hinge at the end of point B. The No.11 is rocker which is connected with the connecting rod 7 in the form of hinge at the end of point D. On the other end of point E, it connects with axis of output shaft in the form of revolute joint. The No.12 is speed-regulating handle, the No.10 is screw, the No.8 is speed nut, the No.9 is swing block. These components are combined into the speed control mechanism which can adjust the speed of output shaft.

2.2 *Principle of in-line pulse CVT*

When the input shaft is rotating under a certain transmission ratio, it drives the cam 1 with the cam roller 2 rotation synchronously. Then the cam roller 2 pushes the connecting rod 3 to rotate around the sliding bottom 9, while the connecting rod 3 slides along the sliding bottom 9 which drives rocker 11 to swing back and forth at the same time. One-way rotation property of overrunning clutch is installed on the rocker 11, which can make the rocker 11 do unidirectional intermittent rotation. No matter whether the input shaft is in motion or

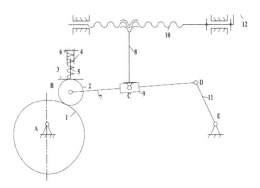

1-cam; 2-cam roller; 3; 4; 5; 6-compact mechanism; 7-connecting rod; 8-speed nut; 9-swing block; 10-screw; 11-rocker; 12-speed-regulating handle

Figure 1.   The structure diagram of in-line pulse CVT.

not, the screw 10 can be revolved. If the screw 10 revolves, the speed nut 8 must move along its axis, so the swing block 9 moves synchronously, then the ratio of the length of connecting rod 7 is changed. It thus changes the pendulum angle and angular velocity of the rocker 11 and the speed regulations of in-line pulse CVT can be achieved.

In order to achieve the rocker 11 continuously unidirectional rotation, this paper utilizes three cam-connected rod combination mechanisms which were installed evenly at the circumference along the axis of the drive shaft.

3 PULSATING DEGREE EXPERIMENT

3.1 *Experimental prototype and parameters*

The main structure parameters of experimental prototype are: the center distance LAE = 350mm, the rocker LDE = 150mm, the connecting rod LBD = 232.5mm, the roller radius r = 40mm, the distance between swing block and frame e = 85mm. The main experimental parameters are: the rated power input P1 = 1kw, the rated speed input n1 = 1500rpm, the range of transmission ratio i1H = 8.161~209.408.

3.2 *Experimental lines*

Experimental lines are shown in Fig. 2. Three-phase asynchronous motor 1 with rated power input P1 = 1kw, rated speed input n1 = 1500rpm, which provides driving force. Magnetic particle brake 12 of model CZ-20 is chosen as the loading device for machinery experiments. Digital torque and speed measuring instruments 5, 6 of model JN-338M-A are chosen to record torque, speed and power of input (or output) shaft with torque sensor 4, 9 of model JN-338. Current regulator 13 of model WIJ-3A meets the requirements for controlling the loading process of the experimental prototype 7. Coupling between parts are connected by the elastic pin coupling. Actually, the speed of input shaft needs to satisfy the regeneration rate of spring of compact mechanism, but this requirement can be realized through the wheel speed 2, and the really input speed of formal testing is 570rpm.

3.3 *Experimental conditions*

The values of the speed ratio and the speed of input shaft keep constant during the experiment, the input power and output power change with the output torque. According to the selection of input speed n1 = 570rpm, the output torque by M2 = 20Nm, 30Nm, ... , and gradually loaded to 150Nm.

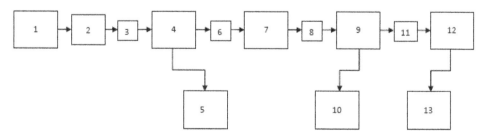

1-three-phase asynchronous motor; 2-wheel speed; 3, 6, 8, 11-coupling; 4, 9-torque sensor; 5, 10-digital torque and speed measuring instruments; 7-experimental prototype; 12-magnetic particle brake; 13-current regulator

Figure 2.   Test circuit diagram of the transmission efficiency measurement.

4 EXPERIMENTAL RESULTS AND ANALYSIS

4.1 Experimental results

Several groups of speed fluctuations are gained in the different speed control points by varying the horizontal distance between the swing block 9 with the input shaft. Each speed control point's measurement interval time is about 3s, 10 measurements data are recorded. This paper chooses a set of data which loaded M2 = 60Nm (as shown in Table 1) for analysis.

4.2 Experimental analysis

Pulsating degree is used to evaluate stationary of the angular velocity of output shaft, which can be demonstrated using uneven coefficient of velocity. By definition, uneven coefficient of velocity is expressed as follows:

$$\delta = \frac{n_{2\max} - n_{2\min}}{n_{2m}} \tag{1}$$

where $n_{2\max}$ is the minimal angular velocity of the output shaft, $n_{2\max}$ is the maximum angular velocity of the output shaft, n_{2m} is the average angular velocity of the output shaft.

With this formulation, uneven coefficient of velocity d is shown in Table 1. In Fig. 4, curve is the change curve between the speed control point and uneven coefficient of velocity. Obviously, the speed points are farther away from the input shaft, the greater the pulsating degree. Otherwise, the speed point closer to the input shaft, the smaller the pulsating degree, the vibration of the experimental prototype will become more intense.

It is much convenient to use MS Excel to solve the problem of one variable linear regression. So we can fit the red linear regression in Fig. 3, its function equation is

$$\delta = 0.0025x - 0.2353 \tag{2}$$

Table 1. The test results.

speed control point/mm	The number of measurements in 3 seconds										$n_{2\max}/r\cdot\min^{-1}$	$n_{2\min}/r\cdot\min^{-1}$	$n_{2m}/r\cdot\min^{-1}$	δ
	1	2	3	4	5	6	7	8	9	10				
105	41.32	41.55	40.83	40.32	41.53	40.91	41.37	41.56	41.95	41.37	41.95	40.32	41.27	0.0395
110	36.84	38.56	37.89	38.84	36.75	38.74	36.84	37.89	38.26	37.45	38.84	36.75	37.81	0.0553
115	32.63	33.68	32.63	33.68	34.28	33.68	34.28	33.63	34.73	33.56	34.73	32.63	33.68	0.0623
120	31.68	30.63	30.47	31.68	30.52	29.78	31.57	32.13	31.25	31.43	32.13	29.78	31.11	0.0755
125	28.42	29.47	28.52	29.42	28.73	27.85	27.52	29.67	28.42	27.63	29.67	27.52	28.57	0.0753
130	26.31	27.32	26.31	25.26	27.42	27.36	26.31	27.42	25.15	25.26	27.42	25.15	26.41	0.0859
135	25.26	23.07	25.26	23.15	25.26	24.1	25.26	24.31	25.31	24.85	25.31	23.07	24.58	0.0911
140	24.26	21.83	23.15	23.23	22.1	23.15	24.38	23.16	22.15	24.21	24.38	21.83	23.16	0.1101
145	19.25	21.05	21.75	21.05	20	21.05	20	19.62	18.94	20	21.75	18.94	20.27	0.1386

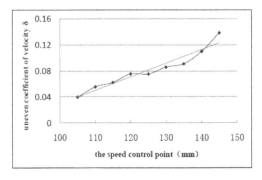

Figure 3. The change curve between the speed control point and uneven coefficient of velocity.

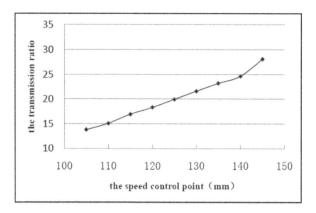

Figure 4. The change curve between the speed control point and the transmission ratio.

The correlation coefficient of the red linear regression $\hat{\rho}$ is 0.9609. According to review, the related correlation coefficient test table (Wu & Luo1999), $\hat{\rho}_{0.01}(8) = 0.789 < \rho$, which shows that significantly linear correlations between the speed point displacement and the pulsating degree in this section speed adjustment range.

Average transmission ratio of pulse CVT is expressed by

$$i_m = n_1 / n_{2m} \tag{3}$$

where n1 is rotational velocity of the input shaft, n2m is average rotational velocity of the output shaft.

With the formulation (3), the change curve between the speed control point and the transmission ratio as shown in Fig. 4. From the change curves of the transmission ratio, we can see as the increase of speed control point displacement, the machine transmission ratio becomes higher gradually.

In order for the experimental prototype to work properly, all the data from these experiments are only measured by the transmission ratio range from 13.811 to 28.12. Compared with the traditional connecting rod transmission mechanism such as GUSA type and Zero-Max type, it can be seen that the pulsating degree of in-line, pulse CVT is less than 15%[1]. Experimental results verified that the new structure of in-line, pulse CVT has good work performance and usability.

5 CONCLUSIONS

(1) The speed points are farther away from the input shaft, the greater the pulsating degree. Otherwise, the speed point closer to the input shaft, the smaller the pulsating degree, the vibration of the experimental prototype will become more intense.
(2) As the increase of speed control point displacement, the machine transmission ratio becomes higher gradually.
(3) The pulsating degree of in-line pulse CVT is less than 15%.

ACKNOWLEDGEMENTS

This work was financially supported by the Natural Science Foundation of Fujian Province, China (grant no.2013J01184).

REFERENCES

Kazem Kazerounian, Zoltan Furu-Szekely. 2005. Parallel disk continuously variable transmission (PDCVT) [J]. Mech and Mach Theory, 40 (9):1–30.

Mangialardi L, Mantriota G. 1999. Power flows and efficiency in infinitely variable transmissions [J]. Mech and Mach Theory, 34(8):973–994.

Nobuyuki Okubo, Hiroki Aikawa and Takeshi Toi. 2010. Noise reduction of continuously variable transmission for automobile. Society for Experimental Mechanics Inc, 2(1-4):763–770.

Ruan Zhongtang. 1999. Design and selecting guidelines of mechanical continuous variable transmission[M]. Beijing: Chemical Industry Press. (in Chinese)

Singh T, Nair S S. 1992. A mathematical review and comparison of continuously variable transmissions [C]. Seattle, Washington: SAE: 1–10.

Sun Jiandong, Fu Wenyu, Lei Hong, Tian E, and Liu Ziping. 2012. Rotational swashplate pulse continuously variable transmission based on helical gear axial meshing transmission[J]. Chinese Journal of Mechanical Engineering, 25(6):1138–1143. (in Chinese)

Wu Zongze, Luo Shengguo. 1999. Mechanical design course design manual[M]. Beijing: Higher Education Press. (in Chinese)

Zhou Youqiang. 2001. Mechanical continuous variable transmission[M]. Beijing: China Machine Press. (in Chinese)

Zhu Yu, Liu Kaichang. 2003. The present situation of research and development of impulse stepless speed variator[J]. Packaging and Food Machinery, 21(5): 11–14. (in Chinese)

Manufacturing Engineering and Intelligent Materials – Lu & Abu Bakar (Eds)
© 2015 Taylor & Francis Group, London, ISBN 978-1-138-02832-6

A study of migration patterns of hot spots formed in the wet-disc brake

T.H. Luo, Y.L. Wang, X.J. Zheng, B. Li & S.J. Dong
*College of Mechatronics and Automobile Engineering, Chongqing Jiaotong University,
Chongqing, China*

ABSTRACT: As multi-interfaced constrained hot spots migrate in wet brake, the migrating mechanism of hot spots was studied in this paper. Also, this paper analyzed the migration patterns of friction hot spots and established the mathematical model of hot spots migration. The regular pattern of hot spots was studied based on the established mathematical model. Also, this paper has researched the influence to the migration pattern of hot spots based on the friction speed, the wavelength, the growth rate of contact pressure, the migration speed and the number of hot spots from two aspects, which are size parameters and material properties of multi-interfaced constrained friction. The results show that changes in every parameter could influence the migration of hot spots of multi-interfaced constrained friction and the migration of hot spots can be controlled effectively by reducing the elastic modulus of friction material, reducing the thickness of the steel and increasing the thickness of friction disc, thus the problems of hot spots could be alleviated.

Keywords: multi-interfaced constrained friction; hot spots; migration pattern

1 INTRODUCTION

Hot spots are the area where high temperature and high pressure formed when the multi-interfaced constrained friction works, which can damage the surface of the material and change the properties of material of friction interface, such as the composition and organization. Hot spots usually emerge on the friction disc which is carbon steel Liu & Liu (2004). As shown in Figure 1, District A is the normal wearing area and District B is the hot spot produced area. Hot spots usually migrate at relatively low speed along the sliding direction, which is called the migration of hot spots. The migration of hot spots makes friction material contact with the area of high temperature and pressure. This is the main reason that causes friction and wear on the multi-interfaced constrained friction (Lin et al. 2006)

Researches on hot spot mainly focused on two aspects (Wang et al. 2009, Emery 2003) 1. Experimental study and simulation analysis of multi-interfaced constrained friction;

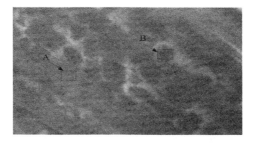

Figure 1. Hot spot on friction disc.

2. Previous research was confined to formation mechanism and distribution of hot spots. However, the pattern of migration of hot spots hasn't been studied. Thus, research on the migration of hot spots is very important for alleviating the hot spot problem.

In this paper, the migrating mechanism of hot spots was inquired and the migration patterns of friction hot spots were analyzed. Also, the mathematical model of hot spots migration was established by focusing on the regularity of migration of hot spots. According to the phenomenon of hot spots migrating in wet brake, this paper has researched the influence of the friction speed, the wavelength, the growth rate of contact pressure, the migration speed and the number of hot spots on the migration pattern of hot spots through the digital modeling and simulation tests on migration pattern. Therefore, the purpose to alleviate the hop spots problem could be achieved.

2 ANALYSIS ON THE MATHEMATICAL MODEL OF MIGRATION MECHANISM OF HOT SPOT

Assuming that all frictional work is converted into heat and the material shows no wear and tear, if we regard the input of thermal flux between frictional interface as the input of thermal flux of boundary, so the density of input of thermal flux on frictional surface is:

$$q(x,y,t) = \mu p(x,y,t) v(x,y,t) \tag{1}$$

where, $p(x,y,t)$ = the pressure on the frictional contact surface. μ = friction factor, $v(x,y,t)$ = relatively moving speed between the friction interface.

During braking, the direction of rotation of friction disc is set as counterclockwise and steel plate is fixed, so the equation of thermal conduction is:

$$\rho_i c_i \frac{\partial T_i}{\partial t} = \lambda_i \left[\frac{\partial}{\partial x}\left(\frac{\partial T_i}{\partial x}\right) + \frac{\partial}{\partial y}\left(\frac{\partial T_i}{\partial y}\right) + \frac{\partial}{\partial z}\left(\frac{\partial T_i}{\partial z}\right) \right] i = 1,2 \tag{2}$$

where, ρ_1, ρ_2 = material density of paper-based materials and steel. c_1, c_2 = specific heat capacity of paper-based materials and steel., λ_1, λ_2 = the thermal conductivity of paper-based materials and steel. t = braking time.

The density (q_1, q_2) of thermal flux between the upper and lower surfaces of friction interface is:

$$\begin{cases} q_1 = w_1 w_{II} F_f v(x,y,t) \\ q_2 = (1 - w_1) w_{II} F_f v(x,y,t) \end{cases} \tag{3}$$

where, F_t = tangential friction, $F_t = \mu p(x,y,t)$; w_1 = weights of density of steel's thermal flux of friction determined by the conditions of frictional thermal coupling. $0 \le w_1 \le 1$; w_{II} = the weights of frictional power transforming into density of thermal flux. It is assumed that all frictional power is converted into frictional heat, which means that $w_{II} = 1$.

The boundary conditions on the working surface of a paper-based material are that frictional surface which is contacted with steel has the input of free thermal conduction and friction heat flow. The effect of thermal radiation and convective heat of brake oil exists on non-contact surface:

$$\lambda_1 \frac{\partial T_1}{\partial z} = -[1 - g(m)] h_{a1}(T_1 - T_f) - [1 - g(m)] \times \sigma A(T_1^4 - T_f^4) + g(m)\lambda_c(T_2 - T_1) + g(m)q_1 \tag{4}$$

where, h_{a1} = heat transfer coefficient of steel plate working surface. T_f = temperature of brake oil. A = radiation area of the steel plate. σ = Steffen Boltzmann constant. λ_c = thermal conductivity of paper-based materials between contacted interface. When the point is in the heat source, $g(m) = 1$, when the point is outside the heat source, $g(m) = 0$.

Assuming that frictional thermal density is freely allocated between the friction interface according to the ideal thermal conductivity of interface and thermal physical properties of the disc, the above equation can be expressed as:

$$\lambda_1 \frac{\partial T_1}{\partial z} = -[1-g(m)]h_a(T_1 - T_f) - [1-g(m)] \times \sigma A(T_1^4 - T_f^4) + g(m)q_1. \tag{5}$$

Only lubricated media or convective heat transfer of air exists on the outer side of the friction plate and steel plate, so:

$$\begin{cases} \lambda_1 \dfrac{\partial T_1}{\partial x} n_x + \lambda_1 \dfrac{\partial T_1}{\partial y} n_y = -h_{1c}(T_1 - T_f) \\ \lambda_2 \dfrac{\partial T_2}{\partial x} n_x + \lambda_2 \dfrac{\partial T_2}{\partial y} n_y = -h_{2b}(T_2 - T_f) \end{cases} \tag{6}$$

where, n_x = normal direction outside the boundary surface x, n_y = normal direction outside the boundary surface y, h_{s2} = heat transfer coefficient on excircle flank of steel plate, h_{a2} = heat transfer coefficient on excircle flank of friction plate.

Equation 7 is the boundary conditions of materials of friction surface which are steel. The contact interface has the input of free thermal conductivity and flow of friction heat.

$$\lambda_2 \frac{\partial T_2}{\partial z} = \lambda_c(T_1 - T_2) + q_2. \tag{7}$$

$$\lambda_2 \frac{\partial T_2}{\partial z} = q_2. \tag{8}$$

Usually hot spots appear on the surface of the steel plate, the heat flux could be expressed as the thermal conduction equation:

$$k_s \nabla^2 T_2 = \frac{\partial T_2(x,y,t)}{\partial t}. \tag{9}$$

Thermal transfer in friction materials is expressed by the equation of thermal transfer with convection:

$$k_f \nabla^2 T_i = \frac{\partial T_i}{\partial t} + V \frac{\partial T_i}{\partial x}, \ i = 1,3. \tag{10}$$

where, $T_2(x,y,t)$ = the temperature of steel plate. ∇ = the differential sign. k_s = the thermal diffusive coefficient on steel plate. h = the coefficient of convective thermal transfer. k_f = the thermal diffusive coefficient on friction plate. V = the speed of heat flux on boundary.

Heat flux generated on the interface of friction:

$$q_j = fVp_j. \tag{11}$$

$$q = \rho c \left[\frac{\partial T}{\partial t} + \{V\}^T \{L\} T \right] + \{L\}^T \{q\}. \tag{12}$$

where, f = the friction coefficient. P_j = the contacted pressure. L = the wavelength.

Contacted pressure is expressed through the following equation:

$$p(x,t) = p_m + p_{a0} \cos \left[\frac{2\pi(x+ct)}{L} \right] \exp(bt) \tag{13}$$

where, P_m = the average exerted pressure. P_{a0} = the initial amplitude of variation of pressure. c = migration velocity of hot spot compared to steel plate's. b = the rate of contacted pressure.

According to the contacted theory of HERTZ, the boundary condition of thermal equilibrium of hot spots at the contacted interface is:

$$k_f \frac{\partial T_1}{\partial y} - k_s \frac{\partial T_2}{\partial y} = q_1$$

when $y = y_1$; $k_s (\partial T_2/\partial y) - k_f (\partial T_3/\partial y) = q_2$ when $y = y_2$

The equation of thermal convection on the contacted interface is:

$$fV\left\{P_{m1} + P_{a0} \cos\left[\frac{2\pi(x+ct)}{L}\right]\exp(bt)\right\} = k_f \frac{\partial T_1}{\partial y} - k_s \frac{\partial T_2}{\partial y}. \tag{14}$$

We then put above convective equations with every boundary condition into the form of weak integral:

$$\int_{vol}\left(fV\left\{P_{m1} + P_{a0} \cos\left[\frac{2\pi(x+ct)}{L}\right]\exp(bt)\right\}\right)d(vol)$$

$$= \int_{S1} k_f \frac{\partial T_1}{\partial y}d(S_1) - \int_{S2} k_s \frac{\partial T_2}{\partial y}d(S_2) + \int_{vol} qd(vol) \tag{15}$$

where, Vol = unit volume, S_i = the contacted interface, $i = 1, 2$.

In order to establish the equation controlled by finite element of temperature field on multi-interfaced constrained fiction, we firstly disperse the spatial domain Vol into finite element and the function of temperature field in the spatial domain and time domain is (Li et al. 2005).

$$T = \{N\}^T \{T_e\} \tag{16}$$

where, $\{N\}^T = \{N_1(x,y,z)N_2(x,y,z)\dots N_{n_g}(x,y,z)\}$ is the unit interpolation equation. $\{T_e\} = \{T_1(t)T_2(t)\dots T_{n_g}(t)\}$ = the temperature vector of node. n_g is the quantities of unit nodes.

3 EXPLORING THE FACTORS ABOUT THE INFLUENCE ON THE MIGRATION PATTERNS OF HOT SPOTS

Migration patterns of hot spots in the multi-interfaced constrained friction mainly reflect that changes in the rate of migration along the circumferential and the number of hot spots during braking. Changes in the migration velocity along the circumferential and the number of hot spots are also influenced by the relative sliding speed of friction element, the growth rate of contacted pressure and the wavelength. Friction characteristics and the size of friction determine the sliding speed, rate and wavelength. So the research on the migration patterns of hot spots is to explore the relationships among above parameters.

3.1 Determined the geometric parameters and material properties of multi-interfaced constrained friction

Geometry and physical properties of the steel disc and the friction disc of multi-interfaced constrained friction used simulation have been given in Tawble 1, the parameters given in brackets have modified for comparison. A and B represents two different friction materials.

The tangential velocity at the outer edge of friction plate respectively is 7 m/s, 15 m/s, 22 m/s, 30 m/s in the multi-interfaced constrained friction, which concludes the regularity

Table 1. Physical properties of material fiction and geometric parameters of friction disc and steel disc.

Name	Copper-based friction material A	Copper-based friction material B
Thickness of friction plate (*mm*)	0.70 (1.20)	0.70 (1.20)
Thickness of steel plate (*mm*)	2.70 (1.70)	2.70 (1.70)
Modulus of elasticity E (*Mpa*)	270	152.5
Poisson's ratio v	0.12	0.12
Thermal expansion coefficient α (1/K)	6.3e-5	6.3e-5
Coefficient of thermal conductivity K ($W \cdot m^{-1}K^{-1}$)	0.241	0.22
Specific heat capacity c_p ($J \cdot kg^{-1}K^{-1}$)	1610	1783
Density ρ ($kg \cdot m^{-3}$)	1125	1008

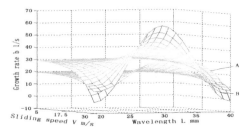

Figure 2. Dimensional graph of growth rate—sliding speed—wavelength.

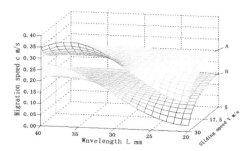

Figure 3. Dimensional graph of migration speed—sliding speed—wavelength.

changed by different growth rate of contacted pressure, speed of migration and wavelength. Table 1 shows the volume ratio of the original geometry of friction is 0.9333.

3.2 *Numerical simulation*

As shown in Figures 2, 3, 4, we simulate the established equations by software called MATLAB (respectively simulating material A, B and different size) and establish the relationships of each parameter.

The relationship of numerical simulation between each parameter has been clearly shown in the above figures. The "a" is the original friction size, b is the change after the increasing thickness of the frictional material, c is the change after the decreasing thickness of steel plate. The stability of the thermo elasticity in hot spot is marked by the changes of the growth rate of contacted pressure b. The heat produced by friction at the contacted interface is proportional to the contacted pressure b; the temperature in the region with more pressure will rise faster, which leads to partial pressure increase faster. The increasing of partial pressure and temperature has an enormous impact on the migration patterns of hot spots.

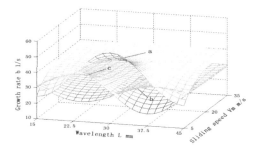

Figure 4.   Dimensional graph of growth rate—sliding speed—wavelength.

4   CONCLUSION

1. The heat generated by friction on the multi-interfaced constrained friction is proportional to the contacted pressure b. The temperature of area with higher pressure will rise faster, and the higher the temperature, the faster the local pressure increase. Local pressure and the increase of temperature have an enormous impact on the migration patterns of hot spots.
2. Changes in the elastic modules of the friction material can effectively influence the migration patterns and the mechanism of the hot spots on the multi-interfaced constrained friction. The growth rate of contacted pressure is increased by the increasing of sliding speed, and the migration velocity c is increased by the increasing of sliding speed. The growth rate of contacted pressure and the migration velocity of hot spots can be reduced by the decreasing of the elastic modules of the friction material.
3. Changes in geometry of the multi-interfaced constrained friction can effectively control the migration patterns of hot spots and ease the wear problems caused by hot spots. Increasing the thickness of friction plate and decreasing the thickness of steel plate can reduce the growth rate of contacted pressure, which enhancing the critical velocity with appearing hot spots, thus easing the problem of hot spots.

REFERENCES

Ding Qun, Xie Jilong. 2002. The temperature filed and stress field calculation of rake disc based on 3-dimension model[J]. *Journal of the china railway society*, 6:34–38.
Dufrenoy P, Weichert D. 2003. A thermo mechanical model for the analysis of disc brake fracture mechanisms[J]. *Journal of Thermal Stress*, 26:815–828.
Emery A.F. 2003. Measured and predicted temperature of automotive brakes under heavy or continuous Braking. SAE Paper 2003-01-2712.
Lloyd F.A, W.O. II. 2002. The effect of modulus and thermal diffusivity on sintered metal performance, SAE Paper 2002-01-1438.
Liu Gaowen, Liu Songling. 2004. Numerical Simulation of Unsteady Hot Streak Migration in a 1-1/2 Stage Turbine[J]. *Journal of Aerospace Power*, 12(30):84–88.
Lin Xiezhao, Gao Chenghui, Huang Jianmeng. 2006. Effects of operating condition parameters on distribution of friction temperature field on brake disc[J]. *Journal of Engineering Design*, 13(1):45–48.
Li Liang, Song Jian, Li Yong, etc. 2005. Study on fast finite element simulation model of thermal analysis of vehicle brake[J]. *Journal of System Simulation*, 12(17):2869–2872.
Lee Sangkook, Yeo Taein. 2007. Temperature and coning analysis of brake rotor using an axisymmetric finite element technique. *International Journal of Mechanical Sciences* 49(2007):129–137.
Wang P.H, Wu X, Jeon Y.B. 2009. Thermal-mechanical coupled simulation of a solid brake disc in repeated braking cycles[J]. *Journal of Engineering Tribology*, 11:1041–1048.
Wang Wenjing, Xie Jilong, Liu Zhiming, etc. 2002. Simulation of three-dimensional transient temperature field based on the cyclic symmetrical brake disc[J]. *Chinese Journal of Mechanical Engineering*, (12).
Zhao Qingjun. 2007. Investigation on flow characteristic and migration mechanisms of inlet hot streaks in a Vaneless counter-roating turbine. Institute of engineering thermophysics Chinese Academy of Science.

Manufacturing Engineering and Intelligent Materials – Lu & Abu Bakar (Eds)
© 2015 Taylor & Francis Group, London, ISBN 978-1-138-02832-6

Research of hydraulic buffer system on speed in overloading manipulator

T.H. Luo & X.Y. Ma
Country College of Mechatronics and Automobile Engineering, Chongqing Jiaotong University, Chongqing, China

ABSTRACT: According to the characteristics of speed regulating valve, the direction of the control valve and the accumulator which can absorb the storage pressure and reduce the pressure impact, the hydraulic buffer system of overloading manipulator speed is put forward to achieve the accuracy control of the robot and inhibit the unstable phenomenon of hydraulic drive. The hydraulic transmission components of manipulator and the model of auxiliary components accumulator were inferred in this papers. Combining with the working principle of hydraulic cylinder and accumulator and their impacts on the system pressure, the physical model of working stroke and idle motion were solved. Then the models were respectively simulated by the software called Fluid-SIM. Finally, results show that the acceleration of the speed in working stroke was reduced. And the buffer system was verified reasonably and effectively.

1   INTRODUCTION

Hydraulic impact often occurs in the process of hydraulic transmission, which could cause impact vibration, loosing packing element, increasing the leakage and sometimes making wrong actions Yao (2004). So the problems need to be solved, especially under the condition of overloading and high accuracy of manipulator. Aiming at the effects of hydraulic impact, the solutions mainly includes installing buffer device at the end of the hydraulic cylinder and the method of electrical control. However, electric control could reduce the response-speed. Therefore, the hydraulic circuit and hydraulic components need to be studied and the accumulator commonly was used.

In Japan, Norio Nakazawa and Yoichrio Kono used energy-saving technology of accumulator in the car and the utilization of the energy was increased about 50% Liu & Hand (1999). Nakazawa & kono (1987). In American, P.B guchward etc. verified energy-saving technology of accumulator by the same method Norio & Yoichrio (1987). Yokota S and Somada H et al. developed a new type of accumulator, which could reduce high frequency-hydraulic pulse caused by hydraulic pump effectively Liu & Hand (1999). HE Hai-yang etc. analysed the mechanize of hydraulic shock and presents a method of hydraulic shock cushion by accunulator with one-vay throttle valve and found that the hydraulic shock amplitude maybe reduced obviously (He et al. 2011). The mathematical model for analysis of accumulator was established by Zhao-di Chen et al., which proved that the device has the function of reducing the hydraulic shock (Chen et al. 1999). Yu Miao added the accumulator in the hydraulic system to ensure its performance with the braking system Miao & Shi (2011). Therefore, speed-buffer system of overloading manipulator is proposed in this paper and the buffer system of linear drive is solved. And the speed control loop lock was introduced in the loop to ensure safe.

2   DESIGN OF BUFFER SYSTEM SCHEME

According to some characteristics of hydraulic components, the accumulator not only could store and release the pressure of liquid components, but also could be used as a short-term supply and absorb the vibration of the system. To avoid hydraulic oil flowing back and the hydraulic pump being damaged, one-way valve is adopted in the buffer circuit. Then the flow valve is selected to achieve the stability of the speed and control the velocities. Therefore, the basic principle is as follows: When the components performing in the work schedule, parts of the oil is stored into the accumulator and the others flowing into a hydraulic cylinder through the reversing valve and the flow valve; and the oil together with the parts from accumulator flowing into the hydraulic cylinder when actuators are in a state of idle motion.

3   MATHEMATICAL MODEL OF BUFFER SYSTEM BASED ON LINEAR MOTION IN MANIPULATORS

When the oil with p (the pressure) and q (the flow) flowing into the rodless cavity, piston would move out to the right with v, and the oil flow out from the other side. The speed of piston to the right v is:

$$v = \frac{q\eta_v}{60A_1} = \frac{q\eta_v}{15\pi D^2} \tag{1}$$

$$q = q_1 + q_2 \tag{2}$$

where, A_1 describes the area of effective work for rodless cavity. q_1, q_2 represents the flow of hydraulic cylinder from accumulator and the flow valve respectively, and "+", "−" describes idle stroke and work schedule separately. And the accumulator volume was commonly calculated as flows:

$$V_0 = \frac{0.004 q p_2 (0.0164L - t)}{p_2 - p_1} \tag{3}$$

where, p_1 descried the biggest shock pressure of the system; p_2 was the work pressure before the valve opened or closed; q was the pipe flow before valve mouth shut down; t presented the duration of the valve from open to close; L described pipeline which had the impact. The flow of the buffer loop pipe was solved by the accumulator type (3):

$$q_1 = \frac{V_0 (p_2 - p_1)}{0.004 p_2 (0.0164L - t)} \tag{4}$$

The flow q through throttling could be described with the following formula, no matter what forms of the throttle:

$$q_2 = CA_2 \Delta p^m \tag{5}$$

where, C was the coefficient decided by the shape, size, and the liquid nature; A_2 was the area of cross-section from orifice; Δp described the pressure difference between before or after the orifice; m presents the index determined by aspect ratio. Mathematical model of the speed about mobile joint was calculated by Eqs. (4), (5), (2) and (1). So, the working stroke and idle stroke were as shown respectively:

$$v_1 = \frac{q\eta_v}{60A_1} = \frac{\left[\dfrac{V_0 (p_2 - p_1)}{0.004 p_2 (0.0164L - t)} + CA_2 \Delta p^m\right]}{60A_1} \tag{6}$$

$$v_1 = \frac{q\eta_v}{60A_1} = \frac{\left[\dfrac{V_0(p_2 - p_1)}{0.004p_2(0.0164L - t)} - CA_2\Delta p^m\right]}{60A_1} \qquad (7)$$

The parameter of type was the same as the aboved.

4   ANALYSIS OF SYSTEM BASED ON SPEED-CONTROL HYDRAULIC CIRCUIT

Actuators of hydraulic system mostly are hydraulic cylinder and hydraulic motor. Without considering the compressibility of hydraulic oil and leak case, the rate of hydraulic cylinder is:

$$v = \frac{q}{A_0} \qquad (8)$$

where, v is the output speed of the piston rod. From type (8), the change of speed could be reached by changing the input flow or the effective area. Commonly, the method of changing q was used. The impact of the movement part ΔF caused by inertia when the moving parts are braked was shown in formula (9) (Zhang et al. 2001):

$$\Delta F = \sqrt{\frac{ME}{V}}v_0 \qquad (9)$$

where, ΔF describes the impact pressure, M presents total quality of the moving parts, v_0 is the speed of the moving parts, E represents for bulk modulus of liquid, V is the volume of oil return cavity. Therefore, according to the type (8), (6) and (7), the impact of the work schedule and idle stroke were calculated as follows respectively:

$$F_1 = \sqrt{\frac{ME}{V}}v_1 = \sqrt{\frac{ME}{V} \times \frac{\left[\dfrac{V_0(p_2 - p_1)}{0.004p_2(0.0164L - t)} + CA_2\Delta p^m\right]}{A_1}} \times 10^{-3} \qquad (10)$$

$$F_2 = \sqrt{\frac{ME}{V}}v_2 = \sqrt{\frac{ME}{V} \times \frac{\left[\dfrac{V_0(p_2 - p_1)}{0.004p_2(0.0164L - t)} - CA_2\Delta p^m\right]}{A_1}} \times 10^{-3} \qquad (11)$$

From formula (9), we could conclude that the more quality of the moving parts and the faster the speed, the greater the impact was produced. Therefore, the way to reduce the impact of velocity is just to start with its speed and the principle of the system was solved and shown in Figure 2.

As is shown in Figure 1, the principle of system is composed of two parts: working stroke and idle stroke. Working stroke: it consists with speed regulating valve and the accumulator, and the function was achieved by on and off of reversing valve. Idle stroke: The accumulator was adopted in the loop to improve the efficiency. When the reversing valve was in the median, oil could be put into the accumulator by the hydraulic pump. The oil was transferred by the hydraulic pump and accumulator. At the same time, the locking circuit was also used in this system to ensure that executive component did not move caused by external force any more after a stop.

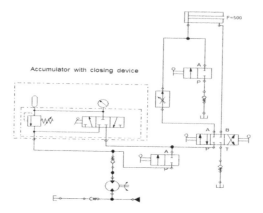

Figure 1.   The principle of the system in moving joint.

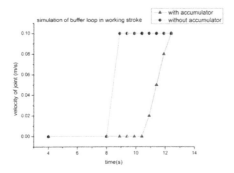

Figure 2.   Comparison with accumulator or not in common loop and buffer loop.

Table 1.   Parameters of elements in common loop and buffer loop.

Element	Set value
Check valve	1 Mpa
Accumulator	1 Mpa
Nominal pressure of relief valve	2 Mpa
Speed control valve	0.2641 gal/min
Load	500 N

Table 2.   Parameters of simulator elements about accumulator.

Element	Set value
Flow valve	2 L/min
Accumulator	2 Mpa
Relief valve	
Nominal pressure	2 Mpa
Flow	2 L/min
Load	500 N

5 SIMULATION OF THE BUFFER SYSTEM

To verify the proposed system above, buffer loop of moving joint and working loop with accumulator or not were simulated respectively by the software called FluidSim-h.

The results are shown in Figure 2, parameters of element were set in Tables 1 and 2 respectively. From Figure 2, the speed was decreased with the use of the overflow valve and accumulator, and the buffer system was verified effectively.

6 CONCLUSION

In hydraulic transmission, due to the moving parts with a larger kinetic energy, great impact or vibration would be produced when the hydraulic actuators with heavy load stop suddenly or reversing. To reduce the hydraulic impact, the accumulator, deceleration valve and other hydraulic components were adopted in the hydraulic system. And the models of hydraulic cylinder circuit and buffer circuit were solved. Then they were simulated and the simulation showed the effectiveness of the hydraulic circuit.

ACKNOWLEDGEMENTS

The research work was supported by National Natural Science Foundation of China (NO. 51375519).

REFERENCES

Chen Zhao-di, et al. 1999. Study of accumulator influence on the impact of pressure piping system [J]. *Hydraulics Pnenmatics & Seals*, (2): 2–5.

He Hai-yang, Yu Ke-yun, et al. 2011. Simulation and exp experimental research on hydraulic shock cushion of accumulator-pump system [J]. *Ship Science and Technology*, 33(7): 55–58.

Liu Y., Hand H. 1999. Technical note sliding mode control for a class of hydraulic position Servo [J]. *Mechatronics*, (1): 111–123.

Miao Yu, Shi Bo-qiang. 2011. Optimization design and robust analysis of accumulator in hydraulic brake system [J]. *Transactions of the CSAE*, 27(6): 132–136.

Norio Nakazawa, Yoichrio kono. 1987. Development of a Braking Energy Regeneration System for City Buses [J]. *SAE technical*, (11): 56.

Yao Huan-xin. 2004. Engineering machinery hydraulic transmission and control [M]. *Beijing: China Communications Press*.

Zhang Lu-jun et al. 2001. Summary of the types and Applications of Accumulator [J]. *Machine tool & hydraulics*, (6): 5–7.

Manufacturing Engineering and Intelligent Materials – Lu & Abu Bakar (Eds)
© *2015 Taylor & Francis Group, London, ISBN 978-1-138-02832-6*

Quantum mechanical calculations on absorption of COCl2 by Aluminum-Nitride Nanotube

A. Kazemi Babaheydari & Kh. Tavakoli
Department of Chemistry, University of Islamic Azad, Shahrekord Branch, Shahrekord, Iran

ABSTRACT: The results concerning an investigation employing the ab initio Molecular Orbital (MO) and B3LYP methods to calculate structural optimization and the major stabilizing orbital interactions on surface nanotube and phosgene were evaluated by Natural Bond Orbital (NBO) methodology. Based on the optimized ground state geometries using B3LYP/6-31G* method, the NBO analysis of donor-acceptor (bond-antibond) interactions revealed that the stabilization energies associated with the electronic delocalization.

Keywords: Molecular Orbital (MO); B3LYP; Natural Bond Oribital (NBO); Phosgene; Aluminum–Nitride

1 INTRODUCTION

Discovery of carbon nanotubes Iijima (1991) has sparked intense research activity within the past decade. These novel materials have a wide range of potential applications ranging from the. fields of nano electronics to nano-scale biotechnology. They may be used as molecular field-effect transistors (Mine et al. 2002), electron field emitters (Mine et al. 2002, Zhou et al. 2002), artificial muscles (Mine et al. 2002, Baughman et al. 1999), or even DNA sequencing agents (Gao et al. 2003). The adsorption behavior of single atoms or gas molecules on carbon nanotubes has been studied extensively in the past decade by experiments (Kong et al. 2000, Villalpando-Paez et al. 2004, ValenEni. 2004) and theoretical calculation (Peng & Cho 2000, Lu et al. 2003, Liu et al. 2005, Ding et al. 2006, Zhao et al. 2005). Aluminum nitride nanotubes (AlNNTs) are inorganic analog carbon nanotubes (CNTs). They are isoelectronic with CNTs, and have been synthesized successfully by different research groups (Baughman et al. 1999, Gao et al. 2003, Kong et al. 2000). Because of their high temperature stability, large energy gap, thermal conductivity, and low thermal expansion (Villalpando-Paez et al. 2004), AlNNTs and aluminum nitride nanomaterials are used widely in technological applications, mainly in micro and optoelectronics such as laserdiodes and solar-blind travioletphotodetectors and semiconductors (Villalpando-Paez et al. 2004). Unlike CNTs, AlNNTs exhibit electronic properties and semiconductor behavior independent of length, tubular diameter and chirality. Tuning the electronic structures of the semiconducting AlNNTs for specific applications is important in building specific electronic and mechanical devices. Phosgene is a major component of natural gas and its adsorption behavior in pores has been studied extensively. In this paper, B3LYP studies of the absorption behavior of phosgene gas on nanotube were performed in terms of absorption energy.

2 COMPUTATIONAL DETAILS

In our current study, extensive quantum mechanical calculations of structure of Aluminum-Nitride nanotube [zigzag (6, 0)] have been performed on a Pentium-4 based system using GAUSSIAN 03 program (Frisch et al. 1998. At first, we have modeled the nanotube

Table 1. Calculated energy values (kcal/mol) of AlN (Aluminum–Nitride) nanotube and phosgene in gas phase at the level of B3LYP/6-31G*.

Parameters	AlNNT	Phosgene
E (total)	−8925/918	−1033/714
EHOMO/ev	−0/236	−0/360
ELUMO/ev	−0/037	0/310
[I = −EHOMO]/ev	0/236	0/360
[A = −LUMO]/ev	0/037	−0/310
[η = (I−A)/2]/ev	0/099	0/335
[μ = −(I+A)/2]/ev	−0/136	−0/025
[s = 1/2η]/ev−1	0/049	0/167
[w = μ2/2η]ev	0/0009	0/0001

I = ionization potential, A = electron affinity, η = Global hardness, μ = chemical potential and w = electrophilicity.

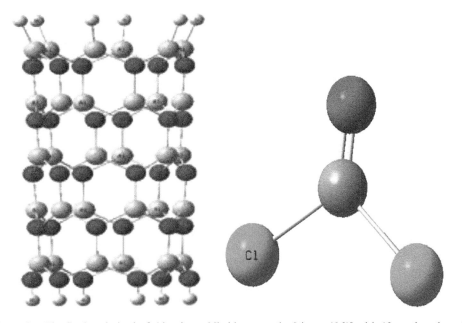

Figure 1. The final optimized of Aluminum-Nitride nanotube [zigzag (6,0)] with 12 nm length and phosgene gas, obtained through B3LYP (6-31G*) calculation.

with Nanotube Modeler package and then optimized at the B3LYP level of theory with 6-31G*basisset. After full optimization of nanotube, we have calculated adsorption of phosgene at the level of 6-31G*theoryon outside (external) of carbon nanotube and have been reported in Table 1, and finally we calculate Natural Bond Orbital (NBO) parameters for this structure, Table 2.

3   RESULTS AND DISCUSSION

In this paper, B3LYP method with 6-31G*basis set were employed for investigating the structure optimization and energy minimization of Aluminum-Nitride nanotube, (Fig. 1) have been summarized in Table 1 and Figure 1.

After full optimization of Aluminum-Nitride nanotube [zigzag (6, 0)], we had calculated optimized structure of interaction between nanotube and phosgene at the level of B3LYP/6-31G*theory (Fig. 2) and then performed Natural Bond Orbital (NBO) calculations for giving NBO important parameters, Table 2.

To study absorption behavior of phosgene gas on AlN nanotube, we perform absorption for two sites in nanotube (outside (external) of carbon nanotube). After a full optimization (Table 1), we found two structures of this absorption (Figs. 2 and 3).

In the NBO analysis, in order to compute the span of the valence space, each valence bonding NBO (σAB), must in turn, be paired with a corresponding valence antibonding NBO (σ*AB): Namely, the Lewis σ-type (donor) NBO are complemented by the non-Lewis σ*-type (acceptor) NBO that are formally empty in an idealized Lewis structure picture. Readily, the general transformation to NBO leads to orbitals that are unoccupied in the formal Lewis structure. As a result, the filled NBO of the natural Lewis structure are well adapted to describe covalency effects in molecules. Since the non-covalent delocalization

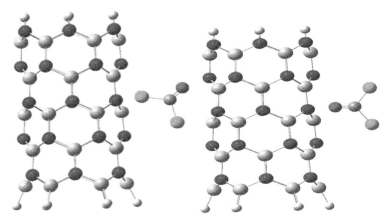

Figure 2. Absorption phosgene by nanotube in two position (0-down, Cl-down), optimized by B3LYP/6-31G*basis set.

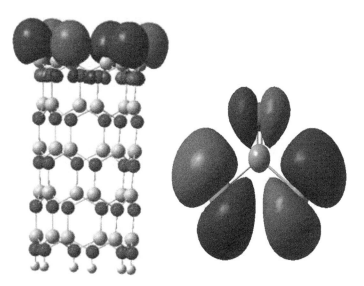

Figure 3. HOMO-LUMO phosgene (external) and nanotube after optimization.

Table 2. The second-order perturbation energies E(2) (kcal/mol) corresponding to the most important charge transfer interactions (donor → acceptor) in the compounds studied by using B3LYP/6-31G*method for Al29-N49.

Complex (Al-N, phosgene)	Donor → Acceptor		E(2), kcal/mol	$\varepsilon_j-\varepsilon_i$	F (i, j)
O-down	LP (Cl100)	σ* (N3-Al73)	0.11	0.67	0.008
		σ* (N3-Al77)	0.99	1.07	0.039
		σ* (N3-Al92)	0.19	1.25	0.014
	σ (C97-O98)	σ* (N37-Al92)	0.16	0.37	0.016
		σ* (N2-Al92)	0.29	0.37	0.022
	σ (N2-Al81)	σ* (C97-O98)	0.28	0.12	0.005
	σ (C97-O99)	σ* (C97-O98)	4.38	0.38	0.037

effects are associated with σ → σ* interactions between filled (donor) and unfilled (acceptor) orbitals, it is natural to describe them as being of donor–acceptor, charge transfer, or generalized "Lewis base-Lewis acid" type. The antibonds represent unused valence-shell capacity and spanning portions of the atomic valence space that are formally unsaturated by covalent bond formation. Weak occupancies of the valence antibonds signal irreducible departures from an idealized localized Lewis picture, i.e. true "delocalization effects". As a result, in the NBO analysis, the donor–acceptor (bond–antibond) interactions are taken into consideration by examining all possible interactions between 'filled' (donor) Lewis-type NBO and 'empty' (acceptor) non-Lewis NBO and then estimating their energies by second-order perturbation theory. These interactions (or energetic stabilizations) are referred to as 'delocalization' corrections to the zeroth-order natural Lewis structure. The most important interaction between "filled" (donor) Lewis-type NBO and "empty" (acceptor) non-Lewis is reported in Table 2 and in Figure 3, be observed between the bonding orbital (σ) C20-C50 and H49-C29 but don't see this bonding orbital for nano tube and internal phosgene.

4 CONCLUSION

1. We have modeled the nanotube with Nanotube Modeler program and then optimized at the B3LYP level of theory with 6-31G*basis set. After full optimization of nanotube, we have calculated adsorption of methane at the level of 6-31G*theory on outside (external) of Aluminum–Nitride nanotube.
2. The most important interaction between "filled" (donor) Lewis-type NBO and "empty" (acceptor) non-Lewis, be observed between the bonding orbital (σ) C20-C50 and H49-C29 but don't see this bonding orbital for nanotube and internal phosgene.

REFERENCES

Baughman, R.H., Cui, C., Zakhidov, A.A., Iqbal, Z., Barisci, J.N., Spinks, G.M., Wallace, G.G., Mazzoldi, A., DeRossi D.G. Rinzler A., Jaschinski, O, Roth, S, Kertesz, M. 1999. *Science* 284, 1340.
Ding, F., Bolton, K., Rosen, A., 2004 *J. Phys. Chem.* B 108, 17369.
Frisch, M.J., Trucks, G.W., Schlegel H.B., Scuseria, G.E., Robb, M.A., Cheeseman, J.R., ZZakrzewski, V.G., Montgomery, J.A., Stratmann, R.E., Burant J.C., Apprich, S., Millam, J.M., Daniels, A.D., Kudin, K.N., Strain, M.C., Farkas, O., Tomasi, J., Barone, V., Cossi, M., Cammi, R., Mennucci, B., Pomelli, C., Adamoc; Clifford, S., Ochterski, J., Petersson, G.A., Ayala, P.Y., Cui, Q., Morokuma, K., Malick, D.K., Rabuck, A.D., Raghavachari, K., Foresman, J.B., Cioslowski, J., Ortiz, J.V., Baboul, A.G., Stefanov, B.B., LIU, G., Liashenko, A., Piskorz, P., Komaromi, I., Gomperts, R., Martin, R.L., A., Fox, D.J., Keith, T., Al-laham, M. Peng, C.Y., Nanayakkara, A., Gonzalez, C., Challacombe, M., Gill, P.M.W., Johnson, B., Chen, W., Wong, M.W., Andres, J.L., Gonzalez, C., Head-Gordon, M., Replogle, E.S and Pople, J.A 1998. *Gaussian 98, Revision* A.7 Gaussian, Inc; Pi_sburgh PA.

Gao, H., Kong, Y, Cui, D., 2003. *NanoLe_* .3 471.

Hou. Shimin, Shen. Ziyong, Zhao. Xingyu, Xue. Zengquan. 2003. *Chemical Physics Le_ers* 373, 308–313

Iijima S, 1991. *Nature* 354,56.

Kong, J., Franklin, N.R., Zhou, C.W., Chapline, M.G., Peng, S., Cho, K.J, Dsai, H. 2000. J, *Science* 287, 622.

Liu, L.V., Tian, W.Q., Wang, Y.A., 2006. J. *Phys. Chem.* B 110, 13037.

Lu X., Chen Z.F., Schleyer, P.V., Am, 2005. J. *Chem. Soc.* 127, 20.

Mine, A, Atkinson, K., Roth, S., 2002 Carbon Nanotubes, in: F. Schüth, S.W. Sing, J. Weitkamp (Eds.), *Handbook of Porous Solids*, Wiley-VCH, Weinheim,.

Peng, S, Cho, K.J., 2000. *Nanotechnology* 11, 57.

Villalpando-Paez, F, Romero, A.H., Munoz-Sandoval, E., Martinez, L.M., Terrones, H., Terrones, M., 2004. *Chem. Phys. Le_.* 386, 137.

ValenEni, L. et al. 2004, *Chem. Phys. Le_.* 387, 356.

Yao, Z., Postma, H.W.C., Balents, L., Dekker, C., 1999. *Nature* 402, 273.

Zhao, J., MarEnez-Limia, A., Balbuena, P.B., 2005. *Nanotechnology* 16, S575.

Zhou, O, Shimoda, H, Gao, B.S., Oh, Fleming, L, Yue, G, 2002. *Acc. Chem. Res.* 35, 1045.

Manufacturing Engineering and Intelligent Materials – Lu & Abu Bakar (Eds)
© 2015 Taylor & Francis Group, London, ISBN 978-1-138-02832-6

Research on control strategy of single-phase photovoltaic grid-connected inverter with LCL filter

Z.X. Zhou & S. Shi

College of Electrical and Information Engineering, Beihua University, Jilin, China

ABSTRACT: In the same filtering effect, the single-phase photovoltaic grid-connected inverter with L-type filter needs large inductance, and the traditional PI controller of grid-connected inverter can't obtain zero steady-state error. A Quasi-PR control strategy by using LCL-type filter with current control outer loop based on grid current and current control inner loop based on capacitor current is proposed in this paper, which increases system damping, restrain oscillation and improves the stability of system. On this basis, the control system is modeled and analyzed, as well as the parameters are designed. The simulation model is designed to verify the theoretical analysis under MATLAB/Simulink. The results show that the grid current distortion rate is smaller, the grid current can be tracked with zero steady-state error, and the system output meets the requirements of grid.

1 INTRODUCTION

In recent years, due to the effects of the energy crisis, the distributed generation system with solar energy, wind energy and other clean renewable energy mainly has become the research focus. Grid-connected inverter is used as a core component of grid-connected PV system. Its main function is to convert the direct current of the PV array into alternating current with the same frequency and phase of the grid. Its control quality is directly related to the overall performance of the system, thus it has been recognized and studied by many researchers.

In order to obtain a low THD (Total Harmonic Distortion) of the grid current, the filter for voltage source inverter have L type, LC type and LCL type. Because the harmonic suppression of LCL type filter is better than the previous two kinds of filter under the condition of low switching frequency and small inductance, and it has been widely used (Xu et al. 2009), LCL, however, is the three-order system without damping and easy to produce resonance. The introduction of LCL to increase the order of the system, which requires higher requirements on the control strategy of control system. The grid-connected inverter with LCL filter that use the control strategy for direct closed-loop current is unstable. Zhang et al. (2007) proposed that connecting damping resistance to the capacitor branch to suppress the resonance and avoid the system instability, but it resulted in power losses and also decreased the degree of attenuation of high frequency components.

The traditional small power single-phase grid-connected inverter commonly use PI controller, but the PI controller in tracking the sinusoidal current instruction has large steady-state error and poor anti-interference ability. In order to solve the shortage of the PI controller, Peng et al.(2011) proposed control strategy for single-phase grid-connected inverter based on two-phase rotating reference frame. Although the system can achieve unity power factor, but it needed to go through complex coordinate transformation. Hu et al. (2014) proposed quasi proportional-resonant control in two-phase static frame, which achieved no steady error control of three-phase grid-connected inverter.

On the basis of these studies, a Quasi-PR control strategy by using LCL-type filter with current control outer loop based on grid current and current control inner loop based on capacitor current is proposed in this paper. The grid current loop is controlled by Quasi-PR controller because the traditional PI control of the sinusoidal reference quantity is difficult to eliminate the steady-state error. The use of high gain at the resonant frequency point provides to achieve static error-free tracking control of grid current. Finally, the Simulink simulation model is established and the simulation results verify the correctness of the proposed control strategy.

2   MODEL OF GRID-CONNECTED INVERTER WITH LCL FILTER

The photovoltaic grid-connected inverter control structure of Quasi-PR controller is shown in Figure 1 where U_{dc} is DC input voltage, i_1 is the inverter output current, i_C is capacitive current, i_2 is grid current, u_g is grid voltage. The resistance of the filter inductance and the parasitic resistance of capacitance are neglected in this paper.

The double closed loop control, that is current control outer loop based on grid current and current control inner loop based on capacitor current is used in this paper. The Quasi-PR controller is implemented in the outer loop and the conventional P controller is applied to the inner loop. The output of the inner loop after SPWM modulation drive switch to achieve the control of single-phase gird-connected inverter.

The system model of grid-connected inverter that is obtained from Figure 1 is shown in Figure 2 where $G(s)$ is the transfer function of Quasi-PR controller, I_{ref} is the grid-connected current reference. Since the switching frequency is much higher than the grid frequency and the impacts of switch motion on the system are neglected, the PWM inverter is approximated as a gain link.

$$K_{PWM} = \frac{U_{dc}}{U_{tri}} \tag{1}$$

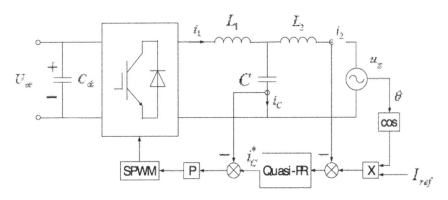

Figure 1.   The control structure of photovoltaic grid-connected inverter.

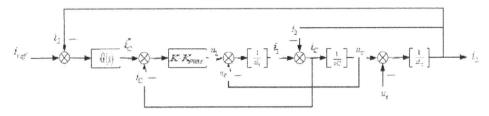

Figure 2.   The system model of grid-connected inverter.

26

where, U_{tri} is the peak of the triangular carrier.

The transfer function of grid current can be derived from this model:

$$i_2 = \frac{A(s)G(s)}{1+A(s)G(s)}i_{ref} - \frac{B(s)}{1+A(s)B(s)G(s)}u_g \qquad (2)$$

where,

$$A(s) = \frac{K \cdot K_{PWM}}{L_1 L_2 Cs^3 + K \cdot K_{PWM} L_2 Cs^2 + Ls}$$

$$B(s) = \frac{L_1 Cs^2 + K \cdot K_{PWM} Cs + 1}{L_1 L_2 Cs^3 + K \cdot K_{PWM} L_2 Cs^2 + Ls}$$

$$L = L_1 + L_2$$

3   THE SELECTION OF THE LCL FILTER PARAMETERS

The system parameters are designed as follows: grid voltage U_g = 220 V, grid frequency f = 50 Hz, dc bus voltage U_{dc} = 400 V, switching frequency f_s = 20 kHz, the system rated capacity P_n = 2 kVA.

The design method of inductance L_1 on the inverter side can refer to the traditional design of single-inductor, namely the current ripple is 10% to 25% of rated current. Because the LCL filter can achieve better filtering effect, it can be appropriate to relax the standards. This paper selects 20%, and the minimum value of the inductance can be calculated by formula (3).

$$L_1 \geq \frac{U_{dc}}{7 f_s I_n \times 20\%} \qquad (3)$$

The filter capacitor is chosen by reactive power that is generally no more than 5% of the rated power, and it can be calculated by formula (4).

$$C \leq 5\% \times \frac{P_n}{6\pi f U_g^2} \qquad (4)$$

Selection of inductance L_2 on the grid side needs comprehensive consideration. Qiu (2009) have demonstrated that L_1 determines the output current ripple, high-frequency current is shunted by L_2 and C, the capacitance provides a low impedance path to high frequency components. There must make X_C ☒ X_{L_2} to ensure that there is good effect of shunting, and generally take $X_C < X_{L2} \times 20\%$. In summary, this paper takes L_1 = 2 mH, L_2 = 0.2 mH and C = 2 μF [4].

The resonant frequency of the LCL filter is calculated by equation (5).

$$f_{res} = \frac{1}{2\pi}\sqrt{\frac{L_1 + L_2}{L_1 L_2 C}} \qquad (5)$$

According to calculation, f_{res} ≈ 8 kHz conforms to formula (6), namely the selected values of L_1, L_2 and C meet the requirements.

$$10f \leq f_{res} \leq 0.5 f_s \qquad (6)$$

4 THE QUASI-PR CONTROLLER

In order to solve the problem that the PI controller has large steady-state error and poor anti-interference ability in tracking the sinusoidal current instruction, the Quasi-PR controller is adopted to control the inverter in this paper and its transfer function is as follows:

$$G(s) = K_P + \frac{2K_R\omega_c s}{s^2 + 2\omega_c s + \omega_0^2} \tag{7}$$

where, K_P is the scaling parameter, K_P is the resonance parameter, ω_c is cut-off frequency, ω_0 is resonant frequency.

Equation (7) shows that the parameters of Quasi-PR controller are K_P, K_R and ω_c. For the analysis of these parameters on the influence of Quasi-PR controller, two parameters are fixed and another parameter is adjusted. By analysing the Bode diagram of controller under different parameters are known: K_R only has relation with system gain, which increase the system gain and improve the steady-state error; K_P also affect the system gain and resonant frequency of bandwidth, and the bandwidth of the resonant frequency becomes smaller and the system gain is increased with increasing K_P; ω_c affects only the bandwidth of the resonant frequency, which becomes larger with increasing ω_c. The parameters of the controller are $K_P = 1$, $K_R = 150$, $\omega_c = 5$ red/s and the bode diagram of Quasi-PR controller can be obtained by equation (7), as shown in Figure 3.

As shown in Figure 3, when the fundamental frequency $\omega_0 = 314$ rad/s, in which the gain at this frequency is 45 dB, while the non-based frequency signals have great attenuation. Thus, it can be achieved without steady-state error tracking on the fundamental frequency.

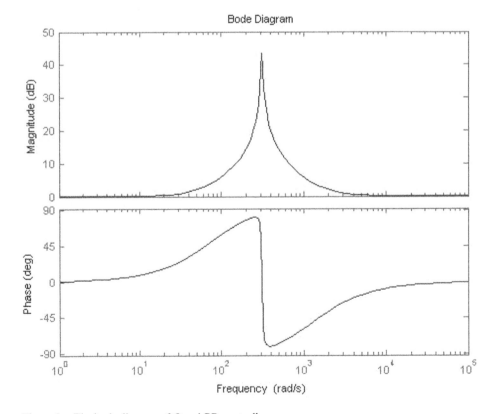

Figure 3. The bode diagram of Quasi-PR controller.

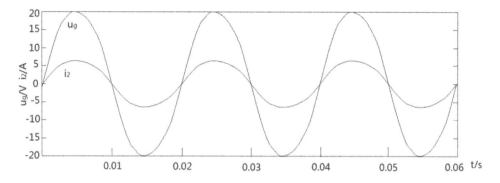

Figure 4. The simulation waveforms of the grid voltage and the output current.

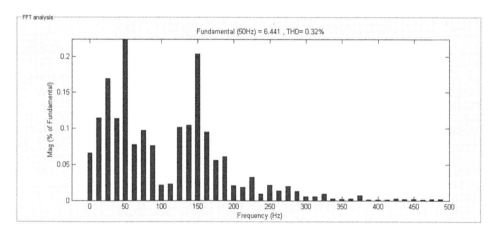

Figure 5. The frequency spectrogram of the output current.

5 THE ANALYSIS OF SIMULINK RESULTS

In order to validate the correctness of the control strategy and theoretical analysis, the simulation model of this system is established by MATLAB/Simulink.

The simulation waveforms of grid voltage and output current are shown in Figure 4. In order to express the phase relationship between current and voltage, the grid voltage $220\sqrt{2}\ V$ is normalized to 20 V.

Figure 5 is the frequency spectrogram and total harmonic distortion of the output current. The Total Harmonic Distortion (THD) of grid connected current is 0.32% (<5%), which is in compliance with the requirements of international standard IEEE 929-2000.

The simulation results show that the double closed loop control strategy of Quasi-PR with LCL filter can achieve that the output current of the grid-connected inverter in synchronization with grid voltage.

6 CONCLUSIONS

This paper introduces in detail the characteristics of LCL filter and gives out the criteria for selection of LCL filter parameters. It also introduces the characteristics of Quasi-PR controller and the parameter adjustment principles and methods of Quasi-PR controller. According to the characteristics of grid-connected inverter with LCL filter, the Quasi-PR control strategy by using LCL-type filter with current control outer loop based on grid current and current control inner loop based on capacitor current is proposed in this paper. The simulation results show that the grid-connected inverter with LCL filter has output current waveform with high quality and high power factor when using this control strategy, and that the output current of grid-connected inverter meets the requirement of grid power quality.

ACKNOWLEDGEMENTS

This work is financially supported by the National Natural Science Foundation of China (61179012), the Program of Jilin Provincial Department Education (2014167), the Scientific and Technological Innovation Team Program of Jilin Province (20150519023 JH). Corresponding author: Zhenxiong Zhou, professor, Ph.D., Master Instructor, his main research direction is the photovoltaic grid-connected control technology.

REFERENCES

Guoqiao Shen, Xuancai Zhu, Jun Zhang, et al. 2010. A new feedback method for PR current control of LCL-filter-based grid-connected inverter [J]. *IEEE Trans. on Industry Electronics*, 57(6): 2033–2041.

Hu Ju, Zhao Bin, Wang Jun, et al. 2014. Design for quasi proportional resonant controller of three-phase photovoltaic grid connected inverter [J]. *Renewable Energy Resources*, 32(2): 152–157. (In Chinese).

Loh P C, Holmes D G. 2005. Analysis of multiloop control strategies for LC/CL/LCL-filtered voltage-source and current-source inverters [J]. *IEEE Trans. on Industrial Electronics*, 41(2): 644–654.

Peng Shuangjian, Luo An, Rong Fei, et al. 2011. Single-phase Photovoltaic Grid-connected Control Strategy With LCL Filter [J]. *Proceedings of the CSEE*, 31(21): 17–24. (In Chinese).

Xu Zhi-ying, Xu Ai-guo, Xie Shaojun. 2009. Dual-loop Grid Current Control Technique for Grid-connected Inverter Using An LCL Filter [J]. *Proceedings of the CSEE*, 29(27): 36–41. (In Chinese).

Zhang Chenghui, Ye Ying, Chen Alian, et al. 2007. Research on Grid-Connected Photovoltaic Inverter Based on Output Current Control [J]. *Transactions of China Electrotechnical Society*, 22(8): 41–45. (In Chinese).

Zhang Xiaotian, Joseph W. Spencer, Josep M. Guerrero. 2013. Small-Signal Modeling of Digitally Controlled Grid-Connected Inverters With LCL Filters [J]. *IEEE Trans. on Industry Electronics*, 60(9): 3752–3765.

Zhang Xing, Cao Renxian. 2013. Solar Photovoltaic Power generation and Inverter Control [M]. Beijing:China Machine Press, (In Chinese).

Zhiling Qiu. 2009. Research on some key technologies of LCL-based Grid Connected Converter with three-phase three-wire [D]. Hangzhou, Zhejiang University, (In Chinese).

Manufacturing Engineering and Intelligent Materials – Lu & Abu Bakar (Eds)
© *2015 Taylor & Francis Group, London, ISBN 978-1-138-02832-6*

Research on outrigger site dynamic response during missile launching stage

X.H. Zhou, D.W. Ma, J. Ren & X.L. Hu
School of Mechanical Engineering, NUST, Nanjing, China

ABSTRACT: To obtain the outrigger site dynamic response during missile launching stage, the Hongnestad equation and improved Saenz equation were adopted to construct compressed concrete upward and downward stress-strain curve which suits asphalt concrete. Damage factor and Sidiroff energy equivalent principle were introduced to establish the plastic damage dynamic constitutive model of launch site surface layer, further establish the outrigger site numerical model, the outrigger site dynamic responses has been studied during missile launching stage. The results show that when the missile launched, the settlement of the site around the rear outrigger are far more serious than that around the front outrigger; the farther from the central interaction point of the rear outrigger, the less the displacement response of site is. The final damage on the rear outrigger is more serious than that on the front. The final damage responses value is: the site around the boundary of the acting surface of the chassis of the rear outrigger > the site around the acting surface of the chassis of the rear outrigger > the site within the acting surface of the chassis of the rear outrigger.

Keywords: missile launching stage; outrigger site; stress-strain relationships; plastic-damage; dynamic response

1 INTRODUCTION

The random emission happens on randomly selected field instead of on the prepared launching site. This has become an important development direction of the sub-grade strategic missile among domestic and foreign companies because of the high concealment and powerful mechanomotive characteristics. Regarding the highway as emission site not only satisfies the concealment and mechanomotive but also heightens the randomicity because of the construction's complexity.

Missile random emission requires safety on every grade pavement, but the existing pavement have uneven mechanical properties, and some low-grade pavement even cause great damage to the outrigger site during missile launching stage that would affect the missile emission precision and overall stability of the launch vehicle. Therefore, it is particularly important to study the coupling effect between site and launch vehicle outrigger during the missile emission. YAO Xiao-guang (Yao et al. 2008) has discussed the overall response and supported load condition of launch vehicle during missile standby phase, but missile emission mechanical analysis are not involved. Zhang Sheng-san (Zhang 2001) has studied the launch vehicle each states' outrigger reaction force and stability formulas through the theoretical calculation, but missile site supported load condition are not involved. Cheng Hong jie (Cheng et al. 2011) scaled the launch site into platform and has studied the missile launching each stages' site supported load formulas through the theoretical calculation, but the specific concrete constitutive relation are not involved and is also an incapable call for detailed research to the missile emission dynamic response.

In this paper the Hongnestad equation (Zhao & Nie 2009) and improved Saenz equation (Xu et al. 2012) will be used to construct compressed concrete upward and downward stress-strain curve which suits asphalt concrete, combined with damage factor and Sidiroff energy equivalent principle to establish the plastic damage dynamic constitutive model of missile launching site surface layer, further establish the outrigger site numerical model. Through numerical calculation obtain the dynamic responses during missile launching stage and mainly studies the settlement and damage evolution of the front and rear outriggers, interprets the coupling effect between launch platform and emission site. The research result can provide the theory evidence for the missile random emission.

2 THE ASPHALT CONCRETE PLASTIC DAMAGE MODEL

2.1 *The stress-strain relationship between asphalt concrete*

The stress-strain curve can be divided into three stages when the asphalt concrete is under compression (Jiang et al. 2012): the stress-strain curve is approximately straight when $\sigma \le 0.3$ σ_0, the stress is linear to the strain, the asphalt concrete is in stable crack growth stage when $0.3 \sigma_0 < \sigma \le \sigma_0$, the asphalt concrete is in stiffness degradation when $\sigma > \sigma_0$.

Firstly, the stable crack growth stage of the asphalt concrete is simulated by the Hongnestad equation. The equation simulated the stress-strain relationship approximately with parabola in the crack growth stage, and the equation is

$$\sigma = \sigma_0 \left[2\left(\frac{\varepsilon}{\varepsilon_0} \right) - \left(\frac{\varepsilon}{\varepsilon_0} \right)^2 \right] \tag{1}$$

where: σ_0 is the limit compression stress of the asphalt concrete, ε_0 is the relevant strain to the limit stress.

Secondly, the improved Saenz axial equation is used to simulate the decrease stage. The equation is

$$\sigma = \frac{\varepsilon}{A + B\varepsilon + C\varepsilon^2 + D\varepsilon^3} \tag{2}$$

where: the four parameters A, B, C, D can be controlled by the five controlling functions, the controlling functions are expressed as: $\varepsilon = 0$, $\sigma = 0$ is relevant to curve origin, $\varepsilon = 0$, $d\sigma/d\varepsilon = E_0$ is relevant to curve origin, $\varepsilon = \varepsilon_0$, $\sigma = \sigma_0$ is relevant to curve peak value, $\varepsilon = \varepsilon_0$, $d\sigma/d\varepsilon = 0$ is relevant to curve peak value, $\varepsilon = \varepsilon_u$, $\sigma = \sigma_u$ is relevant to curve peak value. where, ε_u is relevant strain of the failure limit point, σ_u is the relevant stress of the failure limit point. There exists linear stage when the asphalt concrete is under compression. The second equation of the controlling functions $d\sigma/d\varepsilon$ is equal to the initial elastic modulus E_0, the rest of the three conditions are substituted to the equation (2)

$$\sigma = E_0\varepsilon \Big/ \left[1 + \left(R + E_0/E_s - 2 \right)\left(\varepsilon/\varepsilon_0 \right) - \left(2R - 1 \right)\left(\varepsilon/\varepsilon_0 \right)^2 + R\left(\varepsilon/\varepsilon_0 \right)^3 \right] \tag{3}$$

where: E_0 is the initial elastic modulus, the equation of R is

$$R = \frac{\dfrac{E_0}{E_s}\left(\dfrac{\sigma_0}{\sigma_u} - 1 \right)}{\left(\dfrac{\varepsilon_u}{\varepsilon_0} - 1 \right)^2} - \frac{1}{\left(\dfrac{\varepsilon_u}{\varepsilon_0} \right)} \tag{4}$$

where: E_s is tangent modulus of the peak value of the curve.

As is known from equation (1), $E_0/E_s = 2$. Combine the (3) to (4) and σ_0, ε_0 are substituted, and the R is obtained. The value of R is substituted to (3) to obtain the stress-strain expression of the asphalt concrete in the decrease stage.

Combine equation (1), (3) to the equation of the linear stage, the dimensionless quantity $x = \varepsilon/\varepsilon_0$, $y = \sigma/\sigma_0$ are defined, the compression stress-strain curve of asphalt concrete is shown is Figure 1, the equation is

$$\begin{cases} y = x(E_0\varepsilon_0/f_c), & x \le 0.211; \\ y = 2x - x^2, & 0.211 < x \le 1; \\ y = 2x/[1 + (R + E_0/E_s - 2)x - (2R - 1)x^2 + Rx^3], & x > 1; \end{cases} \quad (5)$$

where: f_c is axial compression strength of asphalt concrete.

The stress-strain curve is hypothesis as straight line when the asphalt concrete is under tension, and then the stress decreases nonlinearly when the stress increases. The compression stress-strain curve of the asphalt concrete is shown in Figure 2. The curve equation is (GB50010-2002 Concrete structure design specification[S] 2002)

$$\begin{cases} y = 1.2x - 0.2x^6, & x \le 1; \\ y = \dfrac{x}{\alpha_t(x-1)^{1.7} + x}, & x > 1; \end{cases} \quad (6)$$

where: $\alpha_t = 0.312 f_t^2$, f_t is axial tension strength of asphalt concrete.

2.2 The deduction of damage factor

The energy equivalence hypothesis is deducted by the damage evolution equation. According to the energy equivalence law proposed by Sidiroff, the elastic complementary energy

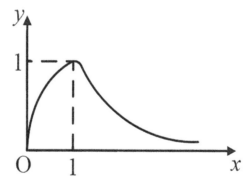

Figure 1.   Concrete compression stress-strain curve.

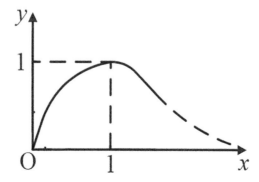

Figure 2.   Concrete tension stress-strain curve.

proposed by the stress on damaged material is in same form with that on the undamaged material.

The complementary energy of the undamaged material is

$$W_o^e = \frac{\sigma^2}{2E_0} \tag{7}$$

The equivalent complementary energy of the damaged material is

$$W_d^e = \frac{\bar{\sigma}^2}{2E_d} \tag{8}$$

The $E_d = E_o(1-d)^2$ is obtained

$$\sigma = E_0(1-d)^2 \varepsilon \tag{9}$$

where: d is damage factor.

Substitute (9) into (5) and (6), the damage factor calculation equation is deducted. The compression damage factor equation of asphalt concrete is:

$$\begin{cases} d = 0, & x \le 0.211; \\ d = 1 - \sqrt{k_c(2-x)}, & 0.211 < x \le 1; \\ d = 1 - \sqrt{2k_c/[1+(R+E_0/E_S-2)x-(2R-1)x^2+Rx^3]}, & x > 1. \end{cases} \tag{10}$$

The tension damage factor equation of asphalt concrete is

$$\begin{cases} d = 1 - \sqrt{k_t(1.2-0.2x^5)}, & x \le 1; \\ d = 1 - \sqrt{k_t/[\alpha_t(x-1)^{1.7}+x]}, & x > 1. \end{cases} \tag{11}$$

In (10) and (11): $k_c = f_c/(\varepsilon_c E_0)$, $k_t = f_t/(\varepsilon_t E_0)$.

3 OUTRIGGER SITE NUMERICAL MODEL

Outrigger site numerical model is shown in Figure 3. $H1$, $H2$, $H3$ and $H4$, respectively, represent the thickness of the asphalt concrete surface layer, base layer, sub-base layer and sub-grade layer from top to bottom of the site model. Numerical calculation uses the dynamic

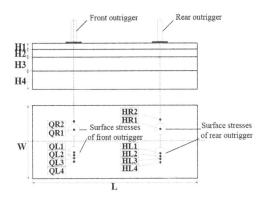

Figure 3. Launching platform structure.

Table 1. The material parameters and structure parameters of asphalt concrete layer.

$\rho/(t \cdot mm^{-3})$	E/Mpa	E_s/Mpa	ν	f_c/Mpa	f_t/Mpa	R	L/mm	W/mm	H_1/mm
2.4E-009	1200	600	0.2	0.66	0.113	0.21	25E+03	10E+03	20

Table 2. The material parameters and structure parameters of launching site.

	Base	Sub-base	Sub-grade
Height/mm	60	60	500
Modulus/Mpa	900	450	30
Poisson's ratio	0.30	0.30	0.35
Density/(t · mm⁻³)	2.2E-009	2.1E-009	1.85E-009

explicit algorithm and mm-tonne-s-Mpa units. In order to observe the outrigger site dynamic response more intuitively during missile launching stage, install dynamic response observation points to the different position of outrigger site as shown in Figure 3 (*HL* represent rear-left outrigger site, *HR* represent rear-right outrigger site, *QL* represent front-left outrigger site, *QR* represent front-right outrigger site). The defined observation points represent the near-certain range medium.

Plastic damage constitutive is adapted to numerically analyze the asphalt concrete surface layer, the material parameters and structure parameters are shown in Table 1 (Jiang et al. 2012). In order to mainly consider the dynamic responses of outrigger asphalt concrete surface layer during missile launching stage, base layer, sub-base layer and sub-grade layer are set to linear elastic material, the material parameters and structure parameters are shown in Table 2 (Wang et al. 2004, Liu et al. 2006). Surrounding of the each layer are free boundary, the bottom of sub-grade layer is fixed constraint.

In order to improve the accuracy of calculation, divide the process of dynamic responses numerical calculation into two steps.

The first step is missile standby stage. In this step, only loaded gravity and adopted static analysis technique, established balance of the initial stress field. The calculation results of boundary conditions and load equilibrium stress state as initial condition substitution to the dynamic analysis.

The second step is missile-launching stage. In this step, the first step calculation results as the initial stress field and applied the dynamic loads to the numerical model. By using the outrigger site observation points' settlements and damages, the working performance of the asphalt concrete pavement can be judged.

4 NUMERICAL SIMULATION OF THE OUTRIGGER SITE DYNAMIC RESPONSES

4.1 Settlement response

Vertical displacement of the outrigger site middle points during emission stage is shown in Table 3. Due to the self-weight of the launch platform, initial displacement at the outrigger site area existed, which is in accordance with the results in the simulation. The larger initial displacement of the rear outrigger makes the front-to-rear trend of the launch platform. The reason to for this situation is the rearward shift of the barycenter of the missile during the rising process, which results in the greater pressure on the rear outrigger at vertical launch stage.

Central point vertical displacement of the rear outrigger during the launching is shown in Figure 4. Substantial increase at t = 0.02 s appears and followed by large-scale degree surge. The front-to-rear status of the integral launching platform before launching will make the

Table 3. Vertical displacement of the outrigger site middle points.

	Initial displacement/mm	Maximum displacement/mm	Final displacement/mm
QL1	4.8	9.9	6.7
QR1	4.8	10.0	6.7
HL1	8.1	11.2	5.2
HR1	8.1	11.7	5.2

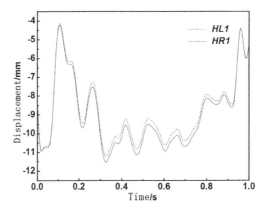

Figure 4. Vertical displacement curve of the back.

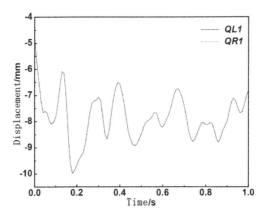

Figure 5. Vertical displacement curve of the front.

canister launcher inclined during launching. And the complex contact between the bottom of the canister launcher and the site leads to the subsiding and uplifting of the entire site. The different vertical displacement at the central point between front and rear outrigger is shown in Figure 4, which is caused by the torsional deformation of the beam of the launching platform during the launching.

Outrigger site middle points during emission stage outrigger site middle points during emission stage

Figure 5 shows the central point vertical displacement of the front outrigger during the launching. Vibration appears due to of the vibration of the launch truck caused by the contact between the bottom of the canister launcher and ground. However, in spite of misalignment, little difference of the vertical displacement at the left and right of the front outrigger is shown in Figure 5, illustrating the small torsional deformation of the beam of the launching platform at front outrigger.

The displacements of the left rear outrigger around a circle are investigated, considering the complex and large deformation of the rear outrigger during launching. Table 4 and Figure 6 show the vertical displacement of the left rear outrigger at different points of the entire period. As we can see, the displacements responses become less apparent as the distance between the interaction point and observation points increases.

4.2 Damage response

Damage and fracture of the site medium around the outrigger has an impact on the integral stability. The calculated damage results of the points around the rear outrigger located at *HL4*, *HR2*, *QL4* and *QL2* are analyzed and shown in Table 5 and Figure 7. It can be inferred that initial damage is caused by the self-weight of launch platform before being activated. Greater pressure on the site around the rear outrigger can cause serious damage as to the front. Obvious vibration is obtained as a result of the closing or cracking of the site under different stress conditions, caused by the complex contact between the bottom of the canister launcher and the site.

Due to the serious damage of the site at the rear outrigger during launching, the damage of the left rear outrigger under different radii is investigated. Table 6 and Figure 8 show calculated damage around the left rear outrigger at different observation points. As is shown, little damage and smooth damage curve appeared at *HL*1 and *HL*2 within the acting surface of the

Table 4. Vertical displacement responses of the left posterior outrigger site different observation points.

	Initial displacement/mm	Maximum displacement/mm	Final displacement/mm
*HL*1	8.1	11.2	5.2
*HL*2	6.7	9.2	4.4
*HL*3	5.5	7.6	3.7
*HL*4	4.6	6.4	3.2

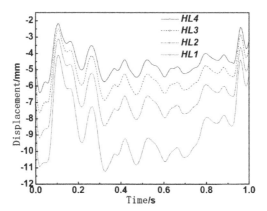

Figure 6. Vertical displacement of the left posterior outrigger site different observation points.

Table 5. Damage responses of the outrigger peripheral site different observation points.

	Initial damage	Maximum damage	Final damage
*HL*4	0.48	0.82	0.61
*HR*2	0.47	0.81	0.59
*QL*4	0.19	0.49	0.49
*QR*2	0.19	0.50	0.50

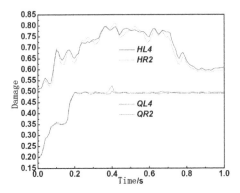

Figure 7.   Damage curve of the outrigger peripheral site different observation points.

Table 6.   Damage responses of the left posterior outrigger site different observation points.

	Initial damage	Maximum damage	Final damage
HL4	0.48	0.82	0.61
HL3	0.59	0.88	0.69
HL2	0.04	0.12	0.12
HL1	0.03	0.07	0.07

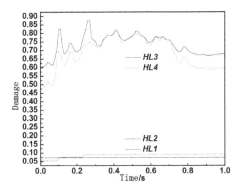

Figure 8.   Damage curve of the left posterior outrigger site different observation points.

chassis of the outrigger, as a result of the state of compression under the gravity of launch platform and the launching load. However, the site located at the boundary of the acting surface of the chassis of the outrigger (HL3) has serious damage, due to the self-weight of the launch platform and the shear stress applied by the chassis of the outrigger on the site.

5   CONCLUSIONS

Numerical analysis of the dynamic response of the site around the outrigger during launching is presented in this work.

1. The rearward shift of the barycenter of the missile during the rising process, resulting in the greater pressure on the rear outrigger at vertical launch stage, will lead to front-to-rear trend of the launch platform.

2. Subsiding of the site around the rear outrigger are far more serious than that around the front outrigger. The farther from the central interaction point of the rear outrigger, the less the displacement response of site is, the more the stability of the displacement curve.
3. The final damage on the rear outrigger is more serious than that on the front. The final damage responses value: the site around the boundary of the acting surface of the chassis of the rear outrigger > the site around the acting surface of the chassis of the rear outrigger > the site within the acting surface of the chassis of the rear outrigger.

REFERENCES

Cheng Hong-jie, Qian Zhi-bo, Zhao Yuan, et al. 2011. Study on the load to ground on missile erection process [J].*Ordnance Industry Automation*, 30(11):1–3, 19. (in Chinese).

GB50010-2002 Concrete structure design specification [S]. Beijing: China Architecture and Building Press, 2002. (in Chinese).

Jiang Jian-jing, Lu Xin-zheng, Ye Lie-ping. 2004. Finite element analysis of concrete structures [M]. Beijing: Tsinghua University Press, 2004. (in Chinese).

Liu Zhi-jun, Liu Chun-rong, Hu Peng, et al. 2006. Experiment study on reasonable evaluation of rebound modulis of subgrade [J]. *Journal of Chongqing Jiaotong College*, 25(3):62–64. (in Chinese).

Wang Jin-chang Zhu Xiang-rong. 2004. Dynamic analysis of asphalt concrete pavement on soft clay ground [J]. *Highway*, (3):6–11. (in Chinese).

Xu Bin, Chen Jun-ming, Xu Ning. 2012. Test on strain rate effects and its simulation with dynamic damaged plasticity model for rc shear walls [J]. *Engineering Mechanics*, 2012, 29(1):39–45,63. (in Chinese).

Yao Xiao-guang, Guo Xiao-song, Feng Yong-bao, et al. 2008. Load analysis on missile erection [J]. *Acta Armamentarii*, 29(6):718–722. (in Chinese).

Zhang Sheng-san. 2001. Calculation impulsive force of combustion-gas flow by measurement in test [J]. *Missiles and Space Vehicles*, (4):27–31. (in Chinese).

Zhao Jie, Nie Jian-guo. 2009. Nonlinear finite element analysis of steel plate-concrete composite beams[J]. *Engineering Mechanics*, 26(4):105–112. (in Chinese).

Manufacturing Engineering and Intelligent Materials – Lu & Abu Bakar (Eds)
© 2015 Taylor & Francis Group, London, ISBN 978-1-138-02832-6

A piezoelectric energy harvester based on magnetic coupling effect

J.N. Kan
Jilin Business and Technology College, Changchun, China

S.Y. Wang & D.L. Liu
Institute of Precision Machinery, Zhejiang Normal University, Jinhua, China

S.Z. Song
Jilin Business and Technology College, Changchun, China

ABSTRACT: In order to study the influence of Separated Distance between the Magnetic Dipoles (SDMD) and piezoelectric energy harvesting performance. A piezoelectric energy harvester based on vibration coupling magnetic force was fabricated and tested. The research results show that the energy generation performance of a Piezoelectric Vibration Energy Harvester (PVEH) can be tuned with changing SDMD and tip mass. When the upper magnet is placed above the magnet bonded on piezo-cantilever's free-end (PZT magnet) and produce attractive force (prototype I), with the reducing of SDMD from 40 mm to 15 mm, the optimal frequency decrease from 32.75 Hz to 30.5 Hz, the effective bandwidth for PVEH to generated a voltage of 15 V rises from 2.5 Hz to 7.5 Hz, and the generated voltage rise from 30.4 V to 44.4 V. When the lower magnet is placed beneath the PZT magnet to produce repulsive force (prototype II), with the SDMD increasing from 15mm to 40mm, the generated voltage rises from 24.4 V to 40 V, and the resonant frequency increases from 33.75 Hz to 38.75 Hz, the effective bandwidth for PVEH to generate a voltage of 10 V rises from 3.75 Hz to 7.75 Hz. Thus, with a reasonable SDMD, one can obtain more electrical energy and a desired resonant frequency namely to improve the piezoelectric energy harvesting performance achieve optimal.

1 INTRODUCTION

Micro-ElectroMechanical System (MEMS) and wireless sensor system are becoming extremely popular and used widely. At present, most of these systems are using batteries to provide electric energy. But batteries cannot satisfy the application in some special occasion due to their size, weight and limited lifetime. Thus, energy harvesting is becoming a very attractive technique for a wide variety of self-powered MEMS. Vibration energy exists widely in nature and it is not affected easily by temperature/pressure and other factors. The vibration energy can be converted to electric energy using three types of electromechanical transducers: electromagnetic, electrostatic, and piezoelectric (Saadon & Sidek 2011). Compared with these three types, piezoelectric materials can convert ambient vibration energy to electric energy efficiently without any extra power (Cook-Chennault et al. 2008) in addition to their large energy density, ease of applications. On the other hand, piezo-harvesters can be configured as a compact system and therefore it can be much suitable for MEMS (Piorno et al. 2010). Thus, piezoelectric has attracted the greatest attention (Qian et al. 2008).

The relative studies indicate the feasibility of using piezoelectric devices as power sources and a PZT harvester can be much effective if it is operated at resonance frequency (Uzun & Kurt 2013). A PVEH has the simplest structure, consisting of only a piezoelectric element and electrodes. But the energy conversion capability of the conventional PVEHs is still too low to be used widely. Previous researches show that the electrical energy generation of a piezoelectric generator based on vibration depends mainly on the structural parameters, external force (or acceleration), driving frequency and so on (Ferrari et al. 2011, Roundy 2005, Roundy et al. 2003, Kan et al. 2008). Thus, a tip mass bonded on piezo-cantilever's free-end should be utilized. However, too heavy tip mass will lead to an exceeded static displace and low reliability of the piezo-cantilever. For this reason, different kinds of PVEHs based on magnetic coupling have been proposed (Tang & Yang 2012, Challa et al. 2008).

It is well known that magnetic forces comprise the repulsive and attractive force. And magnetic force depends on the initial SDMD, exerts great influence on all of the generated voltage, optimal frequency, and effective bandwidth. Conclusions from some of the previous researches of PVEHs with magnetic coupling are incomplete, and even contradictory to each other. In this work, a novel PVEH based on vertical magnetic force was presented and its influence was investigated experimentally. The test results indicate that the SDMD exerts great influence on energy generation, optimal frequency, and even bandwidth. With a reasonable SDMD, one can obtain more electrical energy, wide bandwidth, and a desired frequency.

2 EXPERIMENTAL DESCRIPTION OF THE HARVESTER

The presented PVEH consists of a piezo-cantilever and a permanent magnet bonded on the piezo-cantilever's free-end, and another excitation permanent magnet fixed on vibration structure. The piezo-cantilever consists of substrate plate and a PZT plate bonded on it and the displacement of the beam tip is caused by vibrator and magnets. The two prototypes' (I & II) experimental structures are sketched in Figure 1.

In prototype I, the two magnets are placed with opposite magnetic poles facing each other to produce attractive force. And in prototype II, the two magnets are facing each other with the same magnetic poles to produce repulsive force. Previous researches show that the generated open-circuit voltage of a cantilever-type PVEH depends on its exciting force, which is given as Kan et al. (2008)

$$V_g = \frac{3\alpha(1-\alpha)\beta g_{31}L}{AhW}F \tag{1}$$

where $\alpha = h_m/h$, $\beta = E_M/E_P$, $A = \alpha^4(1-\beta)^2 - 2\alpha(2\alpha^2 - 3\alpha + 2)(1-\beta) + 1$, h_m and h are the thickness of the substrate plate and the piezo-cantilever, respectively. E_m and E_p are the Young's modulus of the substrate plate and the PZT plate, respectively. L and W is the length and width of the piezo-cantilever.

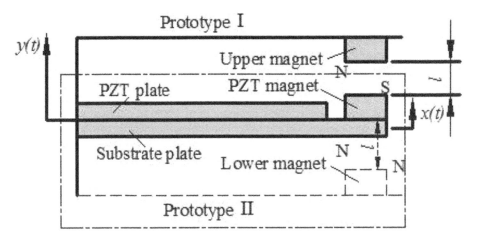

Figure 1. Structures of piezoelectric vibration energy harvester.

For a PVEH subjected to structure motion $y(t) = H\sin\omega t$, there will be Tang & Yang (2012).

$$M_e \ddot{x} + C_e \dot{x} + k_e x = M_e H \omega^2 \sin(\omega t) + F_m \qquad (2)$$

where m_e, C_e, and k_e the equivalent mass, damping coefficient, and stiffness respectively; ω is circular frequency, H is the amplitude of the vibration source, $F_m = -3\mu_0/2\pi \, (m_1 m_2/X^4)$ is magnetic force, m_1 and m_2 are the moments of the magnetic dipoles, μ_0 is the vacuum permeability, $X = l \pm x(t)$ is the distance between the magnetic dipoles, l is the initial Separated Distance between the Magnetic Dipoles (SDMD), $x(t)$ is the generated vibration displacement of the tip mass. For attractive force, $m_1 = m_2$, while for repulsive force, $m_1 = -m_2$. Substituting equation of F_m into Eq. (2) resulting in

$$m_e \ddot{x} + k_e x = m_e H \omega^2 \sin(\omega t) - \frac{3\mu_0}{2\pi} \frac{m_1 m_2}{(l \pm x)^4} \qquad (3)$$

According to Eq. (3), we can obtain the natural frequency of energy harvester with magnetic coupling, and there is [11]

$$\omega_{eff} = \sqrt{\frac{k_e}{m_e}} \qquad (4)$$

where $k_e = k_{beam} + k_{mag}$, $\overline{k_{mag}} = |6\mu_0 m_1 m_2/(\pi X^5)|$ (the sign of k_{mag} depends on the mode of the magnetic force), $k_{beam} = W(E_p h_p^3 + E_m h_m^3)/(4L^3)$, W is the width of the beam, E_p and h_p are the Young's modulus and height of PZT plate, E_m and h_m are Young's modulus and height of substrate plate; $m_e = m_t + m_m = m_t + 0.23WL(\rho_p h_p + \rho_m h_m)$, ρ_p and ρ_m are the density of the PZT and substrate materials and m_t is the tip mass.

Obviously, the magnetic force results in an additional stiffness and exerts influence on natural frequency of the harvester.

3 RESULTS AND DISCUSSION

The analytical equation derived above indicates that many factors have effect on the output performance of PVEH, such as frequency, geometrical parameters of piezo-cantilever and magnets, and magnetic force (denoted by SDMD). To find out the influencing regular of SDMD on energy generation, a PVEH was fabricated and tested. A vibrator (DC-1000-15) made in Austria was used to generate constant vibration amplitude of 0.5 mm. The photo of system is shown in Figure 2. The piezo-cantilever is in size of $70 \times 10 \times 0.5$ mm^3, with the thickness of PZT plate and substrate plate of 0.2 mm and 0.3 mm, respectively. The size and mass of a magnet are $\varnothing 10 \times 2$ mm^3 and 2.37 g, respectively.

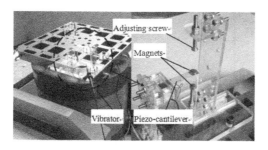

Figure 2.   Photo of the PVEH and test system.

Figure 3 shows the two prototypes' (I & II) relationships between generated peak-peak voltage and vibration frequency at different SDMDs. Symbol ∞ in the figures stands for no magnetic force exerting on the PZT magnet (regard as SDMD is infinite). From the prototype I's test curves we can find that with the decreasing of SDMD, the optimal frequencies decreased, the effective bandwidth to generate a certain voltage widens, and the voltage at the optimal is increased. With the reducing of SDMD from 40 mm to 15 mm, the optimal frequency decrease from 32.75 Hz to 30.5 Hz, the effective bandwidth for PVEH to generated a voltage of 15 V rises from 2.5 Hz to 7.5 Hz, and the generated voltage rise from 30.4 V to 44.4 V.

Test results of prototype II indicate that with the increasing of SDMD, the optimal frequency decreased, and the voltage increased except SDMD at II/∞. With the SDMD increasing from 15 mm to 40 mm, the generated voltage rises from 24.4 V to 40 V, the resonant frequency increases from 33.75 Hz to 38.75 Hz, and the effective bandwidth for PVEH to generate a voltage of 10 V rises from 3.75 Hz to 7.75 Hz.

The difference between prototype I and prototype II is the place of the exciting magnet. Prototype I puts upper magnet above the PZT magnet to produce attractive force. And prototype II puts lower magnet beneath PZT magnet to provide repulsive force. Figure 4 presents the influence trends of SDMD on the maximum peak-peak voltages as well as optimal frequencies. Here, I and II stand for prototype I and prototype II, respectively. Obviously, the voltage-distance curves are nonlinear. In prototype I, with the increasing of SDMD, voltage decreases first and then increases, finally tend to a certain value. And the frequency decreasing first, tend to a certain value at last. But the trends in prototype II are contrary.

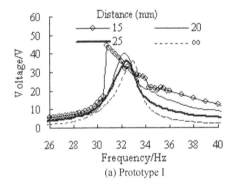 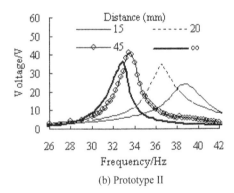

(a) Prototype I   (b) Prototype II

Figure 3.   Generated voltage vs frequency.

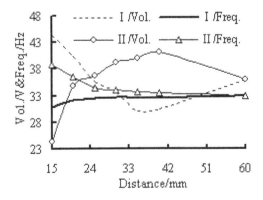

Figure 4.   Maximal voltage and optimal frequency vs SDMD.

4 CONCLUSIONS

A piezoelectric energy harvester based on vibration coupling magnetic force was designed and tested. The results were obtained by the magnets placed in different positions (above or beneath PZT magnet) to produce attractive/repulsive force and the constant vibration amplitude of the vibrator. The open-circuit output voltage was measured at different values of the distance (SDMD) between the magnet and the tip of the piezo-cantilever. The testing results show that the magnetic force, denoted by SDMD, exerts great influence on all of the generated voltage, optimal frequency, and effective bandwidth.

In prototype I, with the decreasing of SDMD, the optimal frequencies decrease, and the effective bandwidth widens. With the reducing of SDMD from 40 mm to 15 mm, the optimal frequency decrease from 32.75 Hz to 30.5 Hz, the effective bandwidth for PVEH to generate a voltage of 15 V rises from 2.5 Hz to 7.5 Hz, and the generated voltage rise from 30.4 V to 44.4 V. And in prototype II, with the SDMD increasing from 15 mm to 40 mm, the generated voltage rises from 24.4 V to 40 V, the resonant frequency increases from 33.75 Hz to 38.75 Hz, and effective bandwidth for PVEH to generate a voltage of 10 V rises from 3.75 Hz to 7.75 Hz.

The magnetic force coupling in the piezoelectric energy harvester based on vibration exerts great influence on all of the generated voltage, optimal frequency, and effective bandwidth. Thus, a desirable SDMD can be used to improve energy harvesting performance of a PVEH.

ACKNOWLEDGEMENTS

This work was supported by the National Natural Science foundation of China (Grant No. 51377147, and 51277166).

REFERENCES

Challa V.R., Prasad M.G., Shi Y. and Fisher F.T. 2008: *Smart Mater. Struct.*, Vol. 17, pp. 015035–01504.
Cook-Chennault K.A., Thambi N. and Sastry A.M. 2008: S*mart Mater. Struct.*, Vol. 17(2008), pp. 043001.
Ferrari M., Baù M., Guizzetti M. and Ferrari V. 2011: *Sensors and Actuators A: Physical*, 172, pp. 287–292.
Kan J.W., Tang K.H. and Wang S.Y. 2008: *Optics and Precision Engineering*, Vol. 16, pp. 71–75.
Piorno J.R., Bergonzini C., Atienza D., and Rosing T.S. 2010: *Journal of Intelligent Material Systems and Structures*, Vol. 21, pp. 1317–1335.
Qian L., Bera D. and Holloway P.H. 2008: *Smart Mater. Struct.*, Vol. 17, pp. 045009–045022.
Roundy S. 2005. On the effectiveness of vibration-based energy harvesting. *Journal of intelligent material systems and structures*, Vol.16, pp. 809–823.
Roundy S., Wright P.K. and Rabaey J. 2003: *Computer Communications*, Vol. 26, pp. 1131–1144.
Saadon S. and Sidek O. 2011: *Energy Conversion and Management*, Vol. 52. pp. 500–504.
Tang L.h. and Yang Y.W. 2012: *Applied Physics Letters*, Vol. 101, pp. 094102.
Uzun Y. and Kurt E. 2013: *Sensors and Actuators A: Physical*, Vol. 192, pp. 58–68.

Manufacturing Engineering and Intelligent Materials – Lu & Abu Bakar (Eds)
© 2015 Taylor & Francis Group, London, ISBN 978-1-138-02832-6

A magnetic-clamping Piezoelectric Energy Generator excited by rotating magnets

J.M. Kan
Jilin Business and Technology College, Changchun, China

S.Y. Wang
Zhejiang Normal University, Jinhua, China

F.S. Huang
University of Science and Technology of China, Hefei, Anhui, P.R. China

S.Z. Song
Jilin Business and Technology College, Changchun, China

ABSTRACT: To meet the demands of rotating machine monitoring system for self-power, a Piezoelectric Energy Generator (PEG) excited by rotating magnets under magnetic clamping is presented. Its structural design and test are processed, and the distribution of rotating magnets and the clamping distance between magnets for the influence of the energy generator are studied. The test results show the feasibility of the novel PEG and validity of theoretical results. The novel PEG excited by rotating magnets has wide effective frequency band and high reliability, and can work normally under high/constant speed. The change of the distribution of rotating magnets (mainly the change of the exciting frequency and exciting time) clamping distance between magnets (mainly the change of the pre-tightening force) have great influence on output voltage. With the increase of the clamping distance between magnets, the optimum frequency for maximal voltage decreases, and the maximal voltage increases.

1 INTRODUCTION

With development of the portable electronics and wireless sensors, energy harvesting is becoming a very attractive technique for a wide variety of self-powered micro-electro-mechanical systems. To satisfy the self-powering demands of micro-power electronics, the scholars have successfully developed various kinds of PEGs, such as the vibration generators (Saadon & Sidek, Harb 2011, Gua & Livermore 2010, Priya et al 2005, Qian & Holloway 2008) and rotary generators (Tien, Goo 2010, Janphuang, Isarakorn, Briand & De Rooij 2011, Cavallier, et al 2005). According to the exciting methods, the existing rotary PEGs can be divided into three categories: (1) inertia incentive type (Tien & Goo,2010), the bi-directional bending deformation of a piezoelectric cantilever being caused by the relative changing of gravity direction in the rotating process. The structure of this method is simple, but it is only suitable for low-speed applications; (2) striking type (Janphuang, ed al. 2011), using a rotation mechanism to strike the piezoelectric vibrator, these methods require relative rotation and causes large noise at high speed; (3) impact type (Cavallier, et al. 2005), using a rotary falling ball to impact the piezoelectric element, which is only suitable for lower-speed applications and is apt to cause impact noise as well as possible shock damages. Obviously, all of the above PEGs are excited along the rotating direction of piezoelectric elements, which has lots of disadvantages and will become the technology bottleneck restricting its practical applications.

Aiming at solving problems of the existing rotary piezoelectric generator, a PEG excited by the magnetic coupling force between rotating magnets fixed on a rotator and those fixed on piezo-beam under magnetic clamping was presented in this article. The influence of distribution of rotating magnets/clamping distance between magnets on energy generation was investigated experimentally.

2 STRUCTURE AND WORK PRINCIPLE OF THE PIEZOELECTRIC GENERATOR

The presented PEG (as shown in Fig. 1) consists of a piezo-beam, magnetic clamping device, rotator, rotating axis, fixed magnets on the cantilever end, and rotating magnets bonded on the rotator. The piezo-beam consists of substrate plate and a PZT plate bonded on it. A group or multi-group of the rotating magnets are fixed on the rotator. When the rotating magnets rotate and close to the fixed magnets on the piezo-beam, magnetic force between the magnets cause the piezo-beam to bend in axial direction. Repulsive force is generated when the same magnetic poles closed to each other while, attachment force happens between the different magnetic poles. With the rotating magnet rotating away from the fixed magnets, magnetic force disappears and the peizo-beam deforms reversely under its own elastic force (or continued free vibration). In the course of work, the clamping magnets provide effective limit to prevent the damage of the PZT and offer a soft fulcrum. In this way, the mechanical energy is converted into electrical energy.

For the non-contact magnetic force is used to excite the piezo-beam, the proposed piezoelectric generator has advantages of no impacts and noises in the work process. At the same time, it is easy for the exciting force to be adjusted with changing the strength of the magnetic field or distance between the rotating/fixed magnets. And the stiffness of piezo-beam can be adjusted by the distance between clamping magnets. Theoretically, the piezoelectric generator is able to generate effectively electrical energy in a variety of speeds. When the magnetic force between the two magnets is given, the open-circuit voltage and power generated by the piezo-beam can be given as (Kan, et al 2008, Kan, et al 2011).

$$V_g = -\frac{3\alpha(1-\alpha)\beta g_{31}L}{\lambda Wh}F \tag{1}$$

$$U_g = \frac{1}{2}C_f V_g^2 \tag{2}$$

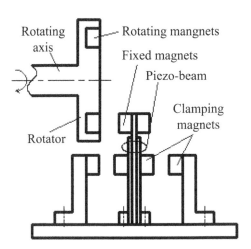

Figure 1. Structure of the PEG under magnetic clamping.

where $\lambda = \alpha^4(1 - \beta)^2 - 2\alpha(2\alpha^2 - 3\alpha + 2)(1 - \beta) + 1$, C_f is the free capacitance of piezo-beam, L/W/h are the length/width/thickness of the piezo-beam, respectively, $\alpha = h_m/h$ is the thickness ratio, h_m is the thickness of the metal substrate, g_{31} is the piezoelectric voltage constant, $\beta = E_m/E_p$ is the Young's modulus ratio, E_p and E_m are the Young's modulus of piezoelectric and substrate element, respectively, $k_{31}^2 = E_p g_{31}^2 / \beta_{33}^T$, $\beta_{33}^T = 1 / \varepsilon_{33}^T$ is the dielectric isolation rate, $\varepsilon_{33}^T = 1300\varepsilon_0$ is the dielectric constant of the piezoelectric ceramic in the thickness direction, F is the exciting magnetic force, and there is

$$F(t) = \begin{cases} F_m \lambda_s = F_m \sin\left(\dfrac{\pi}{T_m}t\right) & (0 \leq t \leq T_m) \\ 0 & (t > T_m) \end{cases} \tag{3}$$

where $T_m = 120/n\pi \arcsin(r/R)$ is the load time, r is the radius of the circular magnets, R is the radius of gyration of the magnet center, $T = 60/(nn_0)$ is the excitation cycle (the size of $x(t)$ depends on the structural parameters of piezo-beam, the magnetic force, and excitation frequency between two magnets), n is the rotary speed, n_0 is the number of rotating magnets fixed in circumferential direction of the rotator, $\lambda_s = \sin(\pi/T_m\ t)$ is the ratio of overlapping area of moving magnets and magnets to the total area of the magnets, F_m is the maximum impact load, and there is (Tang & Yang, 2012)

$$F_m = \frac{3\mu_0}{2\pi} \frac{m_1 m_2}{[l \pm x(t)]^4} \tag{4}$$

where μ_0 is the vacuum permeability, $X = l \pm x(t)$ is the distance between the magnetic dipoles, l is the initial separated distance between the magnetic dipoles, $x(t)$ is the generated vibration displacement of the fixed magnets (the size of $x(t)$ depends on the structural parameters of piezo-beam, the magnetic force, and excitation frequency between two magnets), m_1 and m_2 is the moment of the magnetic dipoles of the magnets fixed on the piezo-beam and rotator, respectively.

Above equations indicate that the required excitation force can be obtained utilizing a reasonable magnet size and magnet spacing. But the actual distance ($X = l \pm x(t)$) between the magnets will change when the flexible piezo-beam vibrated. Thus, it is difficult to obtain the actual distance and coupling force between two magnets in the analytical method. And there are few correlational researches for clamping magnets, so in this article, many experiments are processed to study the performance of the PEG.

3 EXPERIMENT AND ANALYSIS

In order to verify the feasibility of the presented PEG and obtain the effects of related factors on its power-generation performance, an experimental prototype is fabricated and tested (as shown in Fig. 2). The main test equipment include mainly an AC motor (maximum speed of 1390r/min), a frequency converter (the range of adjustment is 0–50 Hz, FM step 0.1 Hz), and an oscilloscope. The piezo-beam measures $90 \times 20 \times 0.6$ mm^3, the permanent magnet is ø12×2 mm^3 and 2.37 g, the diameter of the rotator is 80 mm and the poles of rotating magnets measure 70 mm form the center of the rotator. And there are 24 magnets fixed up uniformly on rotator at most. Because of the reciprocating bending deformation of the piezo-beam caused by magnetic force (repulsive force), the distance between the fixed magnet and the rotating magnet under free holding is determined to be 30 mm to avoid the fixed magnets and rotating magnets touch each other.

Under the different clamping distance between magnets, the generated voltage and rotating speed of the energy generator have great difference. Figure 3 shows the relationship between the generated voltage and rotating speed under different clamping distance such as 25/20/15/10 mm and two rotating magnets equispaced at the rotator. Figure 3 shows that

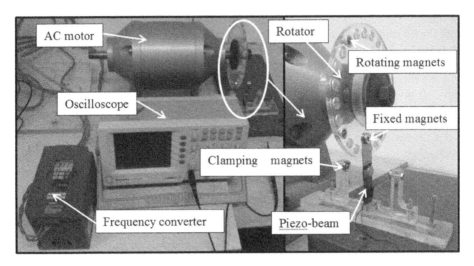

Figure 2. Photo of the PEG and test system under magnetic clamping.

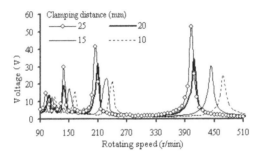

Figure 3. The relationship between the generated voltage and rotating speed.

there are several optimal rotating speeds for the PEG to achieve peak voltage at speed range of 0 to 1200 r/min under different clamping distance and the optimal rotating speed under corresponding optimal voltage is different from one another. The peak voltage under optimal rotating decreases with the decrease of the clamping distance. Preliminary experiments concluded that the decrease of peak voltage because of the restrain from the magnetic clamping and the deformation of the piezo-beam limited. For peak voltage under fourth stage, and the magnetic clamping distance are 25/20/15/10 mm, peek voltages from the PEG are 52.6/34.8/30.4/24.6 V at 402/408/444/468 r/min, respectively. When magnetic clamping distance is 25 mm, under the range of experiment speed the achieved peek voltages from the PEG are 14.8/29.8/41.6/52.6 V at102/138/204/402 r/min, respectively.

Figure 4 shows that both the clamping distance and the number of rotating magnets exert great influence on the performance of the PEG. For example, at the magnetic clamping distance of 25/20/15/10/5 mm, the optimal frequency and max voltage are 17.4/15.2/14.4/13.6/13.4 Hz and 20.6/32/35.2/34.8/52.6, respectively.

Figures 5 and 6 present the relationship between the optimal frequency/the max voltage and the magnetic clamping distance. Figure 5 shows that the optimal frequency decrease with the increase of the magnetic clamping distance. It may be explained by the change of the stiffness of the piezo-beam because of the interference of the magnetic force. Under the effect of magnetism-stress coupling, the piezo-beam get extra stiffness, and the increase of the magnetic clamping distance is equivalent to the decrease of the stiffness of piezo-beam. The curves in Figure 6 show that the max voltage increase with the increase of the magnetic clamping distance. It may be explained that the increase of the magnetic clamping distance

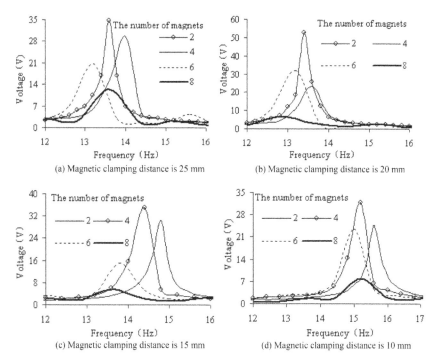

Figure 4. Voltage vs frequency under 2/4/6/8 rotating magnets.

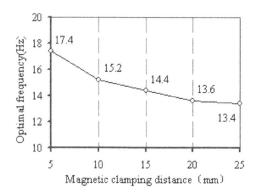

Figure 5. Optimal frequency vs clamping distance.

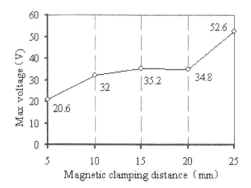

Figure 6. Max voltage vs clamping distance.

enlarge the space of the deformation, so the max voltage increase with the increase of the deformation of the piezo-beam.

4 SUMMARY

The paper presents a novel PEG excited by the magnetic coupling force between rotating magnet fixed on a rotator and one magnet fixed on piezo-beam under magnetic clamping. Its structural design and test are processed, and the distribution of rotating magnets and the clamping distance between magnets for the influence of the energy generator are studied. The test results show the feasibility of the novel piezo-cantilever generator and validity of theoretical results, and then there will be created a piezoelectric energy generator excited by rotating magnets which have wide effective frequency band, high reliability, can work normally under high/constant speed. The results show that the change of the distribution of rotating magnets ((mainly the change of the exciting frequency and exciting time) clamping distance between magnets (mainly the change of the pre-tightening force) have great influence on output voltage. For example, the optimum frequency of output voltage decreases with the increase of the clamping distance between magnets, but the optimum voltage increases with it.

ACKNOWLEDGEMENTS

This work was supported by the National Natural Science Foundation of China (Grant No. 51277166, 51377147).

REFERENCES

Cavallier B., Berthelot P., Nouira H., et al. 2005: *IEEE Ultrasonics Symposium*, Vol. 2(2005), pp. 943–945.
Gua L. and Livermore C. 2011: *Applied Physics Letters*, Vol. 97, pp. 081904.
Harb A.: *Renewable Energy*, Vol. 36, pp. 2641–2654.
Janphuang P., Isarakorn D., Briand D. and De Rooij N.F. 2011: *16th International Solid-State Sensors, Actuators and Microsystems Conference*, Beijing, China, June 5–9, p. 735–738.
Kan J.W., Tang K.H., Wang S.Y., et al. 2008: *Optics and Precision Engineering*, Vol. 16, pp. 71–75.
Kan J.W., Wang S.Y., Peng S.F., et al. 2011: *Optics and Precision Engineering*, Vol. 19, pp. 2108–2116.
Priya S., Chen C.T., Fye D., et al. 2005: *Japanese Journal of Applied Physics*, Vol. 44, pp. L104–L107.
Qian L., Bera D. and Holloway P.H. 2008: *Smart Mater. Struct.*, Vol. 17, pp. 045009–045022.
Saadon S. and Side k O. 2011: *Energy Conversion and Management*, Vol. 52, pp. 500–504.
Tien C.M.T. and Goo N.S. 2010: *Proc. of SPIE*, Vol. 7643(2010), pp. 371–379.
Tang L.H. and Yang Y.W. 2012: *Appl. Phys. Lett*, Vol. 101, pp. 094102.

Manufacturing Engineering and Intelligent Materials – Lu & Abu Bakar (Eds)
© 2015 Taylor & Francis Group, London, ISBN 978-1-138-02832-6

Modeling and analysis on vice-frame structure of fuel-cell-vehicle

L. Shan
Zhejiang Tongji Vocational College of Science and Technology, Zhejiang, China
Zhejiang Sci-Tech University, Hangzhou, China

L. Jun
Zhejiang Tongji Vocational College of Science and Technology, Zhejiang, China

ABSTRACT: In order to lighten fuel-cell-vehicle's weight, using optimizational software, vice-frame's model are built and vice-frame's stiffness and strength analyzed in different work conditions. According to the analysis results, the vice-frame model was modified. The modified-vice-frame's stiffness and strength are tested and analyzed. The analysis results show that the modified vice-frame has the advantages of light weight, good stiffness and good strength characteristics.

Keywords: lightweight; structural optimization; sub-frame

1 INTRODUCTION

As the global energy crisis and environmental pollution become more and more serious in the world, people have put forward strict request for new energy automobile research and development. In new energy technology, one of the most promising solutions is fuel-cell due to its dual advantages of energy saving and clean energy. Nowadays, materials of vice-frame structure are used in steel. But the structure is not reasonable and cannot meet actual running requirement of fuel-cell vehicle in certain aspects. Quality of fuel-cell vehicle must have more redundancy, which has a greater optimization space to lighten. Vice-frame structure of fuel-cell vehicle is chosen as analysis object to study common methods of optimization of lightweight.

2 CONSTRUCTION OF FINITE ELEMENT MODEL

Vice-frame structure of fuel-cell vehicle is the part above the front axle and under axle shaft, which is different from the one of a traditional vehicle. Vice-frame structure carries electrical motor cooling pump, fuel-cell stack cooling pump, air-conditioner compressor, gear box and electrical motor frame, which is connected with controlling arm of front axle by lifting lug in the front and behind as shown in Figure 1.

Figure 2 shows the Finite model which is constructed in Hypermesh condition by UG software with technical dimension parameters.

The number of elements of finite model of frame are 11624. In the major parts, four node shell elements are used, and three node elements are used in others. The detail is shown in Table 1.

Inspection standard of Element Mass is shown in Table 2.

Element mass inspection results of vice-frame are shown in Table 3.

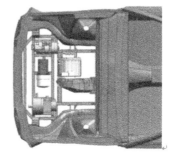

Figure 1. The location of sub-frame.

Figure 2. Finite element model of sub-frame.

Table 1. Scale of sub-frame finite element model.

Element numbers	11624
Node numbers	11643
Triangle element numbers (%)	220
Triangle element rate (%)	1.9

Table 2. Inspection standard of element mass.

Quadrilateral elements warping value	>5.000
Minimum length	<7.500
Quadrilateral element minimum interior angle	>45°
Quadrilateral element maximum interior angle	<135°
Triangle element minimum interior angle	>20°
Triangle element maximum interior angle	<120°
Jacques ratio	<0.700
Length to width ratio	>5.000

Table 3. Inspection results of element mass.

Quadrilateral elements warping value	0% failed
Minimum length	0% failed
Quadrilateral element minimum interior angle	0% failed
Quadrilateral element maximum interior angle	0% failed
Triangle element minimum interior angle	0% failed
Triangle element maximum interior angle	0% failed
Jacques ratio	0% failed
Length to width ratio	0% failed

3   LOADING DETERMINATION FOR VICE-FRAME CALCULATION

Figure 3 shows that vice-frame not only carries a lot of accessories, but also hinges with controlling arm. During driving time, vice-frame loads are caused not only by all accessories mounted on itself, but also by force and torque in all directions transmitted from front wheels to junction of controlling arm connected from point A to point B.

Vice-frame is assembled with electrical motor cooling pump, fuel-cell stack cooling pump, air-conditioner compressor, electrical motor gear box, electrical motor support and so on. All of these are fixed in axles of vice-frame fastened by screw. Detail positions of all accessories are shown in Figure 4 and parameters are shown in Table 4.

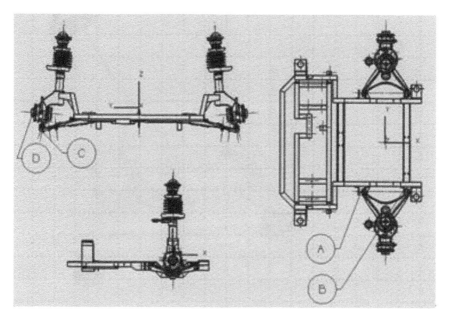

Figure 3.   The connections between sub-frame and controlling arm.

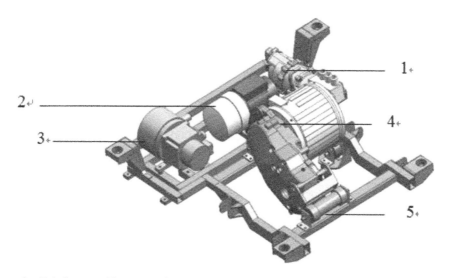

Figure 4.   Sub-frame and its accessories.

Table 4. Loading conditions of sub-frame in fuel cell vehicles.

Serial number	Part name	Quality (kg)
1	Electrical motorcooling pump	6.5
2	Fuel-cell stack cooling pump	9.0
3	Air-conditional compressor	12.0
4	Electrical motorgear box	71.5
5	Electrical motorsupport	10.0

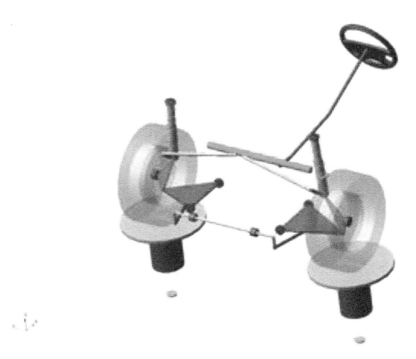

Figure 5. Multi-body model of front suspension.

The centroid position of each accessory is obtained by software. Forces in work condition are separately carried on centroid position and centroids are connected with assembling holes of accessories. Force carried on contact of controlling arm and vice-frame are analysed by MSC. Adams software is used in three typical work conditions: braking condition, acceleration condition (0.8 g) and turning condition (0.6 g) (as shown in Fig. 5).

4 CALCULATION CONSTRAINT LOADING IN FOUR WORKING CONDITIONS

1. Bending conditions
   In bending conditions, loading coefficient value is 1.5.
   In this condition, vertical load is caused by 1.5 times gravity of accessories mounted on vice-frame. Detail load conditions are shown in Figure 6(a).
2. Braking conditions
   Simulation is carried out in braking conditions when brakes in flat road at 0.8 g. Not only gravity and braking force are carried by accessories, but also braking forces and torques are carried by lifting lug. Detail load conditions are shown in Figure 6(b).

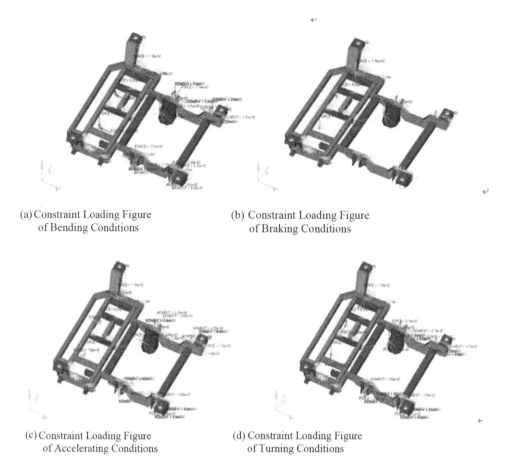

(a) Constraint Loading Figure
   of Bending Conditions

(b) Constraint Loading Figure
   of Braking Conditions

(c) Constraint Loading Figure
   of Accelerating Conditions

(d) Constraint Loading Figure
   of Turning Conditions

Figure 6.   Sub-frame constraint loading figure.

3. Accelerating conditions
   The same as braking conditions, in accelerating conditions, simulation is carried out when accelerate in flat at 0.8 g. Accelerating forces and torques are carried by accessories. Details are shown in Figure 6(c).
4. Turning conditions
   Just like the conditions, in turning conditions, simulation is carried out when turn in flat road is at 0.6 g, force is carried by vice-frame and shown in Figure 6(d).

5   STIFFNESS AND STRENGTH ANALYSIS OF VICE-FRAME

Before stiffness and strength of vice-frame is analysed, material property parameters and constraint conditions should be determined. Detail results are listed below.

At present, vice-frame are made by steel; material parameters are shown in Table 5.

Constraint conditions of finite model of vice-frame is that four assembling holes are limited all degrees of freedom. Stiffness and strength analysis results are shown below:

1. Bending conditions
   Figure 7 shows that the largest deformation in bending conditions occurs in the middle of front axle where vice-frame supports electrical motor, and the value of deformation is 0.06 mm. The largest stress occurs in the rear axle where vice-frame supports gear box.

Table 5.   Material parameters.

Elastic modulus (Mpa)	2.068×105
Poisson's ratio	0.29
Allowable stress (Mpa)	2.1×102
Density (g/cm³)	7.8

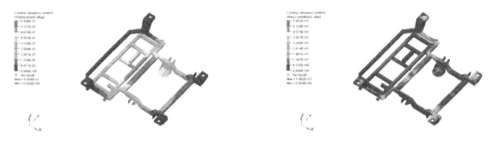

Figure 7.   Analysis of stiffness and strength in bending conditions.

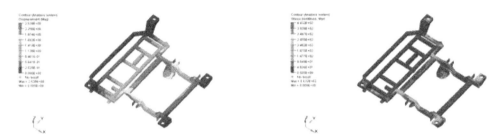

Figure 8.   Analysis of stiffness and strength in braking conditions.

Value of stress is 58.8 Mpa, which is great less than yield strength of 210 Mpa. There is no danger of fracture.

2. Braking conditions

   Figure 8 shows that the largest deformation in bending conditions occurs in right side of front axle where vice-frame supports electrical motor, and value of deformation is 2.45 mm. The largest stress occurs in assembling holes of vice-frame which value attaches 443 Mpa while the material yield strength is 210 Mpa, which means the largest stress is out of yield strength. In these conditions, the vice-frame cannot meet requirements of strength. It would fracture.

3. Accelerating conditions

   Figure 9 shows that the largest deformation in accelerating conditions occurs in contact of controlling arm and lifting lug of vice-frame, and value of deformation attaches 2.28 mm. The largest stress occurs in assembling holes of vice-frame which value attaches 397 Mpa while the material yield strength is 210 Mpa, which means the largest stress is out of yield strength. In these conditions, the vice-frame cannot meet requirements of strength. It would fracture.

4. Turning conditions

   Figure 10 shows that the largest deformation in accelerating conditions occurs in contact of controlling arm and lifting lug of vice-frame, and value of deformation attaches 0.48 mm. The largest stress is occurred in left side of rear axle of vice-frame, value of stress is 156 Mpa, which is less than yield strength 210 Mpa. There is no danger of fracture.

   Above all, in braking and accelerating conditions, the largest deformation values of vice-frame exceed 2 mm. The stiffness problems should be modified and optimized.

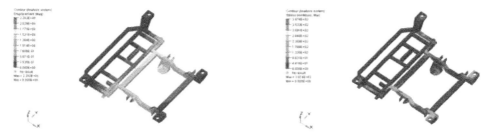

Figure 9.   Analysis of stiffness and strength in accelerating conditions.

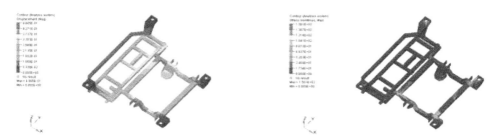

Figure 10.   Analysis of stiffness and strength in turning conditions.

6   CONCLUSION

It can be seen from analysis of stiffness and strength that vice-frame in braking and accelerating conditions cannot meet requirements of working strength. It means that vice-frame would fracture and can be dangerous. However, in other conditions, great stress are redundant. The value is less than allowable stress. Stress distribution is extremely unaverage, even in braking and accelerating conditions multiple structure stresses are small. Optimization of structure is necessary. Many lightweight potentialities can be discovered in the target to meet the performance requirements.

REFERENCES

Gong Zheng, Wang kuo-rong. 1998. The status and trend of Automobile's lightweight [J] *automobile's repair*. 2:45–47.
Li Minghui, Lu Xiaochun. 2002. CADCAECAM integration technology in Lightweight of automobile application [J] *highways & Automobile transportation*. 8:121–124.
Qi Guoqiang. 2001. Automobile Engineering Handbook [M] (Manufacture of articles) 5. Beijing: China Communications press.
Yu Zhisheng. 1989. Automobile theory [M] Beijing: mechanical industry press. 4.

Manufacturing Engineering and Intelligent Materials – Lu & Abu Bakar (Eds)
© 2015 Taylor & Francis Group, London, ISBN 978-1-138-02832-6

Research on equal-intensity design of gun shrink barrels

Y.Q. Zhang, G.G. Le, D.W. Ma, J.L. Zhong & G.T. Feng
School of Mechanical Engineering, Nanjing University of Science and Technology, Nanjing, China

ABSTRACT: Gun shrink barrels' design scheme and corresponding equal-intensity design scheme are presented. Firstly, the solution of stress function for plane axisymmetry problem is obtained and combined with equilibrant equation, geometrical equation and physical equation. Secondly, the numerical formulas of stress, strain and displacement for monocular barrel in elastic stage and the elastic design scheme are provided. Thirdly, the equal-intensity design scheme of shrink barrels is derived considering sleeve pressure. Finally, for certain navy gun barrels, the design approaches of monocular barrel and shrink barrels are obtained, and the finite element models are simulated and analyzed to verify the correctness and validity of the design schemes. The results show that: compared to the ordinary elastic design scheme of monocular barrel, the equal-intensity design scheme of shrink barrels can save materials by 11.5%, and the equal-intensity design scheme guides the design of gun barrels better.

1 INTRODUCTION

The tubular member giving muzzle velocity and firing direction to the projectile at launch is called Barrel. Common barrels (Pan & Guo, 2007) include monocular barrel, enhanced barrel and decomposable barrel, and the enhanced barrel can be divided into shrink barrels, autofrettaged barrels, etc. Research on barrel often simplifies it as a thick-walled cylinder. (Li, Liu & Ni 2007, Su et al. 2012, Lin & Hu 2006, Theocaris 1986) have studied the limit pressure of a single-thick-walled cylinder, combining with the unified strength criterion and elastic-brittle-plastic deformable models. (Wu 2000, Li, Wang & Chen 2001, Yin 2000) have studied assembly pressure and dynamic response of combined thick-walled cylinder, applying finite element method and linear elastic theory. Wang Ying-ze and his colleagues Wang et al. (2008) have established launch dynamic model of a gun and studied the key problems of modeling and analyzed the nonlinear launching process. Di et al. (2002) have established the reliability model of strength design of gun barrels based on analysis about distributions of material property, loads and dimensions, and presented the strength design scheme based on Consistent Reliability. Nevertheless, there are few research reports on design scheme and intensity design method of gun shrink barrels nowadays.

This paper simplifies the gun barrel under chamber pressure as plane-stress axisymmetry problem combining with equilibrant equation, geometrical equation and physical equation, and researched on it with stress function. The numerical formulas of stress, strain and displacement for monocular barrel in elastic stage are derived, and the elastic design scheme is provided. The Calculation Method of equal-intensity design for a gun shrink barrel is derived considering the scheme of combined gun shrink barrels and sleeve pressure. This paper presented elastic design and equal-intensity design formonocular barrel and gun shrink barrels based on a certain type of gun barrel. The finite element model of monocular barrel and the dynamics model of gun launcher are established. The finite element models are simulated and analyzed to verify the correctness and validity of the equal-intensity design methods. This research method offers a new reference for designs of gun barrels.

2 STRESS FUNCTION SOLUTION FOR PLANE AXISYMMETRY PROBLEM

2.1 *Basic equation for plane axisymmetry problem*

Little strength influenced on gun barrels for the little axial stress, the problem can be solved as plane-stress axisymmetry problem. The plane-stress axisymmetry problem satisfies the following equation:

1. Geometrical equation

$$\xi_r = \frac{du}{dr}, \xi_\theta = \frac{u}{r} \tag{1}$$

2. Equilibrant equation

$$\frac{d\sigma_r}{dr} + \frac{\sigma_r - \sigma_\theta}{r} = 0 \tag{2}$$

3. Physical equation

$$\xi_r = \frac{1}{E}(\sigma_r - \mu\sigma_\theta), \ \xi_\theta = \frac{1}{E}(\sigma_\theta - \mu\sigma_r) \tag{3}$$

where u, ξ, σ, E and μ radial displacement, strain, stress, elastic modulus and poisson ratio, respectively. The subscript 'r' stands for radial direction, and the subscript 'θ' indicates ring direction.

2.2 *Stress function solution for plane-stress axisymmetry problem*

To solve elasticity problems, geometrical equation, equilibrant equation and physical equation should be calculated simultaneously, and find solutions to meet boundary condition.

From Eq. (2), we can extrapolate

$$\frac{d}{dr}(r\sigma_r) = \sigma_\theta \tag{4}$$

Introduce stress function $\varphi(r)$, we have

$$r\sigma_r = \varphi(r), \frac{d\varphi}{dr} = \sigma_\theta \tag{5}$$

On the simultaneous Eqs. (1), (3) and (5), we can obtain compatible equation expressed by stress function, i.e.

$$r\frac{d^2\varphi}{dr^2} + \frac{d\varphi}{dr} - \frac{\varphi}{r} = 0 \tag{6}$$

According to Eq. (6), we have

$$r\frac{d}{dr}\left(\frac{1}{r}\frac{d}{dr}(r\varphi)\right) = 0 \tag{7}$$

Through integral solution of Eq. (7), we obtain

$$\varphi = Ar + \frac{B}{r} \tag{8}$$

where A and B are integral constants. Substituting φ into Eq. (5), and combining with Eq. (1) and Eq. (3), we have relationship between stress, strain, displacement and radius, i.e.

$$\sigma_r = A + \frac{B}{r^2}, \ \sigma_\theta = A - \frac{B}{r^2} \tag{9}$$

$$\xi_r = \frac{1}{E}\left[(1-\mu)A + (1+\mu)\frac{B}{r^2}\right], \ \xi_\theta = \frac{1}{E}\left[(1-\mu)A - (1+\mu)\frac{B}{r^2}\right] \tag{10}$$

$$u = \frac{r}{E}\left[(1-\mu)A - (1+\mu)\frac{B}{r^2}\right] \tag{11}$$

where integral constants can be gained according to boundary conditions. As stated above, we obtained stress, strain and displacement of plane-stress axisymmetry problem.

3   ELASTIC ANALYSIS AND DESIGN OF MONOCULAR BARREL

3.1   *Elastic analysis of monocular barrel*

Simplify monocular barrel of a gun as a thick-walled cylinder, as shown in Figure 1. The radius of inner and outer thick-walled cylinder are a and b. Internal pressure on inner surface is p_1, at the same time external pressure on outer surface is p_2. For monocular barrel, p_2 is zero. Corresponding boundary conditions as follows:

$$\sigma_r\big|_{r=a} = -p_1, \ \sigma_r\big|_{r=b} = -p_2 \tag{12}$$

Substituting boundary conditions above into Eq. (9), we obtain two constants, i.e.

$$A = \frac{a^2 p_1 - b^2 p_2}{b^2 - a^2}, \ B = \frac{a^2 b^2 (p_2 - p_1)}{b^2 - a^2} \tag{13}$$

Substituting Eqs. (9), (10) and (11) into Eq. (13), we have stress component (Lame Formula), strain component and displacement component as follows

$$\sigma_r = \frac{a^2 p_1 - b^2 p_2}{b^2 - a^2} + \frac{a^2 b^2 (p_2 - p_1)}{r^2(b^2 - a^2)}, \ \sigma_\theta = \frac{a^2 p_1 - b^2 p_2}{b^2 - a^2} - \frac{a^2 b^2 (p_2 - p_1)}{r^2(b^2 - a^2)} \tag{14}$$

$$\xi_r = \frac{1}{E}\left[(1-\mu)\frac{a^2 p_1 - b^2 p_2}{b^2 - a^2} + (1+\mu)\frac{a^2 b^2 (p_2 - p_1)}{r^2(b^2 - a^2)}\right]$$

$$\xi_0 = \frac{1}{E}\left[(1-\mu)\frac{a^2 p_1 - b^2 p_2}{b^2 - a^2} - (1+\mu)\frac{a^2 b^2 (p_2 - p_1)}{r^2(b^2 - a^2)}\right] \tag{15}$$

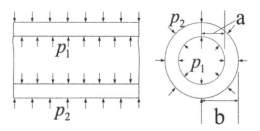

Figure 1.   Thick-walled cylinder with balanced pressure.

$$\mu = (1-\mu)\frac{a^2 p_1 - b^2 p_2}{b^2 - a^2} + (1+\mu)\frac{a^2 b^2 (p_2 - p_1)}{r^2 (b^2 - a^2)} \tag{16}$$

3.2 *Elastic design scheme of monocular barrel*

Elastically design a gun barrel with the second strength theory based on the elastic failure theory for materials. Take the Tresca yield criterion as failure criterion, i.e. the barrel reaches its elastic strength limit when meet $\sigma_r - \sigma_\theta = \sigma_s$.

For monocular barrel, balanced internal pressure on inner surface is p_1 and external pressure on outer surface is zero. Stress components as follows

$$\sigma_r = \frac{a^2 p_1}{b^2 - a^2}\left(1 - \frac{b^2}{r^2}\right), \ \sigma_\theta = \frac{a^2 p_1}{b^2 - a^2}\left(1 + \frac{b^2}{r^2}\right) \tag{17}$$

There is maximum of $(\sigma_\theta - \sigma_r)$ in the thick-walled cylinder's inner space where will be the first to enter the plastic state. Substituting $(r = a)$ and Tresca yield criterion into Eq. (17), we have

$$p_e = \frac{\sigma_s}{2}\left(1 - \frac{a^2}{b^2}\right) \tag{18}$$

where a, σ_s, p_e monocular barrel's inner radial, yield strength and limit chamber pressure, respectively. a, σ_s and p_e are known constants. Complete the elastic strength design of monocular barrel with solving its outer b Eq. (18).

4 ELASTIC ANALYSIS AND DESIGN OF SHRINK BARRELS

4.1 *Sleeve pressure between inner cylinder and outer cylinder*

Gun shrink barrels are combined with two thick-walled cylinders. As shown in Figure 2, the inner radial and outer radial are a and c. The inner radial and outer radial of inner cylinder are a and $(b + \delta_1)$, while outer cylinder's are $(b + \delta_2)$ and c. Magnitude of interference at the assembly position is $\delta(\delta = \delta_1 + \delta_2)$. Based on the assumption of same materials for two barrels, heat the outer cylinder to expand it and set on the inner cylinder. Uniformly distributed pressure will be generated at contact surface with the temperature decreasing.

Sleeve pressure is external pressure for inner cylinder, while internal pressure for outer cylinder. We can obtain the displacement of outside radius for inner cylinder and inside radius for outer cylinder from Eq. (16), i.e.

$$\mu_{r=b}^1 = -\frac{pb}{E}\left(\frac{b^2 + a^2}{b^2 - a^2} - \mu\right), \ \mu_{r=b}^2 = -\frac{pb}{E}\left(\frac{b^2 + c^2}{c^2 - b^2} + \mu\right) \tag{19}$$

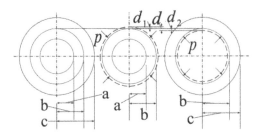

Figure 2. Combined thick-walled cylinder.

Sleeve pressure can be obtained, that is:

$$p = \frac{E\delta(b^2 - a^2)(c^2 - b^2)}{2b^3(c^2 - a^2)} \tag{20}$$

where

$$\delta = |u^1_{r=b}| + u^2_{r=b} \tag{21}$$

4.2 *Equal-intensity design for gun shrinking barrels*

Gun shrink barrels simultaneously bear internal pressure and sleeve pressure. Combinatorial stress $(\sigma_\theta - \sigma_r)$ will reach critical values at the location of $r = a$ and $r = b$, related to layered radial b and magnitude of interference. Equal-intensity design calls for inner surface and delamination region meet elastic limit simultaneously, so it's an ideal design.

For inner cylinder, P_1 and P_2 represent internal pressure on inner surface $r = a$ and external pressure on outer surface $r = b$. And satisfy:

$$p_2 = p - (\sigma_r)_{r=b} \tag{22}$$

where $(\sigma_r)_{r=b}$ is radial stress subjected to internal pressure at complete cylinder's $r = b$. a and c stand for inner radial and outer radial of complete cylinder. Substituting radial stress into Eq. (13), we have

$$(\sigma_r)_{r=b} = \frac{a^2(b^2 - c^2)}{b^2(c^2 - a^2)} p_1 \tag{23}$$

For outer cylinder, p_2 represents internal pressure. And outer surface at $r = c$ is free boundary, that is

$$\sigma_r|_{r=c} = 0 \tag{24}$$

Yield of thick-walled cylinder depends on magnitude of $(\sigma_r - \sigma_\theta)$. According to Eq. (14), $(\sigma_r - \sigma_\theta)$ at $r = a$ of inner cylinder and $r = b$ of outer cylinder are:

$$(\sigma_r - \sigma_\theta)_{r=a} = \frac{2b^2(p_1 - p_2)}{b^2 - a^2}, (\sigma_r - \sigma_\theta)_{r=b} = \frac{2c^2 p_2}{c^2 - b^2} \tag{25}$$

As $(\sigma_r - \sigma_\theta)_{r=a} = (\sigma_r - \sigma_\theta)_{r=b}$, we can extrapolate:

$$(\sigma_r - \sigma_\theta)_{r=a} = (\sigma_r - \sigma_\theta)_{r=b} = \frac{2b^2 c^2}{b^2(c^2 - b^2) + c^2(b^2 - a^2)} p_1 = f(b)p_1 \tag{26}$$

$f(b)$ should be minimum value to make shrink barrels with maximum bearing capacity. We can obtain condition to get minimum value of f based on the assumptions that a and c are fixed value. That is:

$$b^2 = ac \tag{27}$$

Substituting Eq. (27) into Eq. (26), we have

$$p_2 = \frac{p_1}{2} \tag{28}$$

65

If $(\sigma_r - \sigma_\theta)_{r=a} = (\sigma_r - \sigma_\theta)_{r=b} = \sigma_s$, we can extrapolate elastic limit pressure of shrink barrels, i.e.

$$p_e = \frac{c-a}{c}\sigma_s \tag{29}$$

Taking into account Eqs. (27) and (29), we get outer radial and layered radial of a barrel, if barrel's inner radial, limit yield strength and chamber pressure are known. As stated above, equal-intensity design of shrink barrels is completed.

5 ACTUAL EXAMPLE

5.1 Elastic design and equal-intensity design schemes for monocular barrel and shrink barrels

Design a gun barrel with average pressure curve. Preliminary design scheme of gun barrel can be obtained from p_m, p–l, l_g, l_{ys}, r_a, d and theoretical strength curve of gun barrel calculated from classical interior ballistics. Where p_m, p–l, l_g, l_{ys}, r_a and d are maximum chamber pressure obtained from interior ballistics, interior ballistics curve, projectile trip, length of powder chamber, inner radial of each section and caliber of barrel. Figure 3 gives detailed sketch map for preliminary design scheme of gun barrel.

Assumptions are that fixed outer radial of powder chamber as shown in Figure 3, unchanged inner radial of shrink design domain and elastic design for outer cylinder in this design domain. Figure 4 shows design shapes of the gun barrel in elastic design of monocular and equal-intensity of shrink barrels. Parameters of gun barrel in different parts are shown in Table 1. The weight of gun barrel in elastic design scheme of monocular is 478.8 kg, while 423.9 kg in equal-intensity design scheme of shrink barrels. Comparing the elastic design scheme of monocular barrel, the equal-intensity design scheme of shrink barrels can save materials by 11.5%;

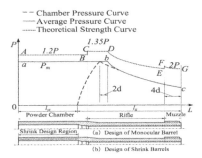

Figure 3. Sketch map for preliminary design scheme of gun barrel.

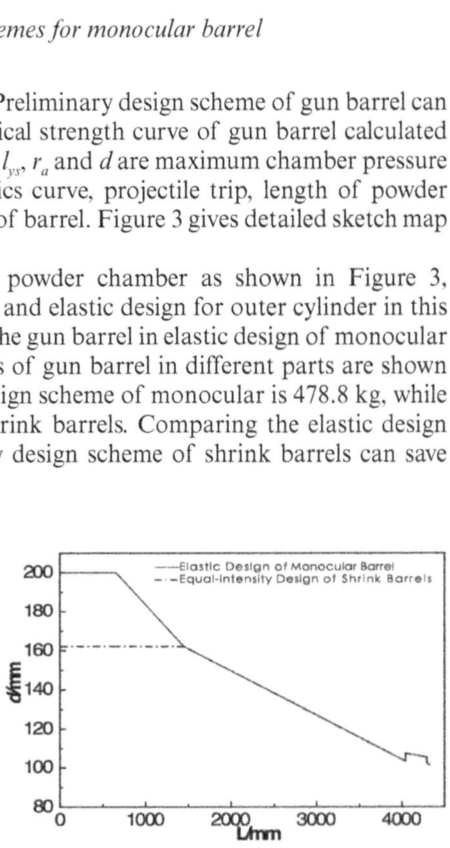

Figure 4. Design shape of the gun barrel.

Table 1. Parameters of the gun barrel in different parts.

Distance from bottom of powder chamber L/mm	Powder chamber 650	Rifle 1450	4040	Muzzle 4040	4290	4325	Mass/kg
Inner radial d_1/mm	102.3	66.7	66.7	66.7	66.7	66.7	–
Theoretical strength p_1/Mpa	241.6	271.8	191.2	201.3	195.7	192.3	–
Out radial d_2/mm							
Elastic	201.2	162.1	103.5	107.6	102.5	101.5	478.8
Equal-intensity	162.2	162.1	103.5	107.6	102.5	101.5	423.9

66

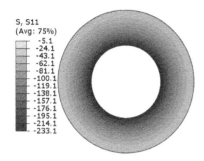

Figure 5.   Stress nephogram in radial direction.

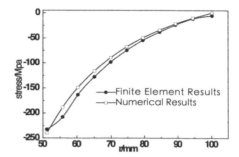

Figure 6.   Comparison between finite element calculation and numerical calculation.

Figure 7.   Launch dynamics model.

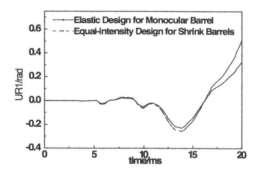

Figure 8.   Angular displacement of the muzzle.

5.2   The finite element model of monocular barrel

Establish finite element model of monocular barrel based on elastic design scheme, and analyse the model to obtain stress distribution in radial direction of powder chamber. Figure 5 shows stress nephogram in radial direction of powder chamber. Compare with the numerical results of Eq. (13); finite element results agree well with numerical results, as Figure 6 shows.

5.3   Launch dynamics models of whole gun system

Establish dynamics models of gun launchers based on two different design schemes as Figure 7 show. Figure 8 shows angular displacement curves of the muzzle with the time variation that is gained from simulation. The muzzle vibration of two different design schemes having little visible difference. And both of them can ensure firing accuracy.

6   CONCLUSION

1. Finite element results agree well with numerical results in radial stress of monocular barrel, which verified the validity of the design schemes and the feasibility of numerical method for combined shrink barrels' design scheme based on it.
2. Comparing the ordinary elastic design scheme of monocular barrel, the equal-intensity design scheme of shrink barrels can save materials efficiently and improve the mobility of guns.
3. The muzzle vibration of two different design schemes having little difference can be obtained from simulation of guns' launch dynamics model. Both of them meet the demands of intensity. As stated above, the validity of design schemes are indicated.

REFERENCES

Di Chang-chun, Wang Xing-gui, Qin Jun-qi.2002. Strength Design of Gun Barrel Based on Consistent Reliability [J]. *Gun Launch &Control Journal*, (2): 58–60.

Li Bao-feng, Liu Xie-quan, Ni Xin-hua.2007. Application of twin shear strength theory in gun barrels' intensity calculation [J]. *Science Technology and Engineering*, 7(13): 3235–3237.

Li Zhi-kang, Wang Wan-peng, Chen Jian-jie. 2001. Linear Elastic Dynamic Response Similarity of Multi-layer Anti-blast Structure [J]. *Journal of Liaoning Technical University*, 20(4): 396–398.

Lin Tai-qing, Hu Xiao-rong. 2006. Limit pressures for thick wall cylinders based on the triple shear unified failure criterion [J]. *Journal of Fuzhou University*, 34(5): 727–731.

Pan Yu-tian, Guo Bao-quan. 2007. Design of Gun Barrels [M]. Beijing: The Publishing House of Ordnance Industry.

Su Zhong-ting, Xu Da, Yang Ming-hua, Xue Jing. 2012. Finite-element model updating for a gun barrel based on modal test [J]. *Journal of vibration and shock*, 31(24): 54–59.

Theocaris P S.1986. A general yield criterion for engineering materials [J]. *Depending on Void Growth Mechanical*, (21): 91–105.

Wu Ying. 2000. The Assembly Pressure of Circular Tubes with Different Lengths [J]. *Journal of Taiyuan University of Technology*, 31(2): 215–217.

Wang Ying-ze, Zhang Xiao-bing, Yuan Ya-xiong. 2008. Flexible Dynamics Model Analysis of Gun Tube [J]. *Gun Launch & Control Journal*, (4): 49–52.

Yin Xiao-chun. 2000. Research on Collisions of Multilayered Hollow Cylinders [D]. Nanjing University of Science and Technology.

Manufacturing Engineering and Intelligent Materials – Lu & Abu Bakar (Eds)
© 2015 Taylor & Francis Group, London, ISBN 978-1-138-02832-6

Hysteresis characteristics modeling of the piezoelectric actuators

C. Qiang
State Key Laboratory of Advanced Design and Manufacturing for Vehicle Body, Hunan University, Changsha, China
Hunan University College of Mechanical and Vehicle Engineering, Changsha, China

J.Q. E, Y.W. Deng & H. Zhu
State Key Laboratory of Advanced Design and Manufacturing for Vehicle Body, Hunan University, Changsha, China
Hunan University College of Mechanical and Vehicle Engineering, Changsha, China
Institute of New Energy and Energy-Saving & Emission-Reduction Technology, Hunan University, Changsha, China

ABSTRACT: To solve the problem of the hysteretic character of the piezoelectric material in real application, the initial weighted factors of the hysteretic units was calculated based on the theory of Preisach and the FORCs test data, then the combined method of RBF NN and FLS-SVM was introduced to calculate more weighted factors of the hysteretic units. Thus, the final model was established. The results of comparison between RBF result, FLS-SVM result, combined result and test result show that the combined model was better than any methods of RBF NN or FLS-SVM. So, the combined model is an effective method.

1 INTRODUCTION

The advantages of piezoelectric actuator are as follows: high resolution, strong driving force, highly efficient electromechanical coupling, swift response and noiseless (Janaideh 2009), but the piezoelectric materials have the inherent characters of hysteresis, nonlinearity. Until now, lots of research papers and researchers had proposed some of the hysteretic models such as Preisach model (Wolf 2012), Maxwell model (Choi 1999), Dahl model (Xu 2010), etc.

In order to solve the problem of the hysteresis, a new method was proposed and certified in this paper. Firstly, relatively few weighted values were obtained based on the FORCs (First order reversal curves); secondly, the combined method of RBF neural network and FLS-SVM was introduced in the paper to compute more weighted values than that in the first step; at last, the final model was established and validated.

2 THEORY OF THE GENERALIZED PREISACH HYSTERETIC MODEL

The mathematical generalized Preisach hysteretic model can be expressed as equation (1).

$$f(t) = R[u(t)] = \iint_{\alpha \leq \beta} \mu(\alpha, \beta) \gamma_{\alpha, \beta}[u(t)] d\alpha d\beta \qquad (1)$$

where $R[u(t)]$ is the computational operator; t is the non-dimensional time; α and β are the lower and upper threshold values, respectively; $\mu(\alpha, \beta)$ is the weighted function; $u(t)$ is the input; $\gamma_{\alpha, \beta}[u(t)]$ is the hysteretic unit. The principle of the hysteretic unit is shown in Fig. 1(a).

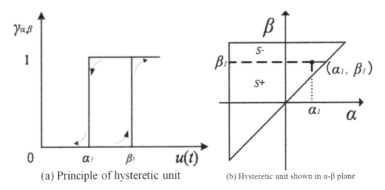

(a) Principle of hysteretic unit   (b) Hysteretic unit shown in α-β plane

Figure 1.   Hysteretic unit.

The hysteretic unit (α_1,β_1) which is shown in Figure 1(a) can also be expressed in $\alpha - \beta$ plane, the schematic diagram is shown in Figure 1(b).

The discrete hysteretic model can be expressed by equation (2).

$$f = \sum_{i,j}^{n} \boldsymbol{\mu}_n(i,j) \cdot \gamma_{i,j}(t) \tag{2}$$

where, $\boldsymbol{\mu}_n$ is a $n \times n$ matrix, the elements of the matrix are the weighted factor of each hysteretic unit; $\gamma_{i,j}(t)$ is the hysteretic computational unit matrix.

3   ESTABLISHMENT OF THE LINEAR COMBINATION MODEL

3.1   *Modeling of the RBF neural network*

The RBF neural network has three layers: input layer, hidden layer and the output layer. The Gaussian function was used in this paper as the nonlinear activation function, so the jth neuron equation of the hidden layer can be expressed as equation (3).

$$z_j = \phi\left(\|\mathbf{x} - \boldsymbol{\mu}_j\|\right) = \exp\left(-\|\mathbf{x} - \boldsymbol{\mu}_j\|^2 \big/ 2\sigma_j^2\right) \tag{3}$$

where, $x = (x_1, x_2, ..., x_n)$ is the input sample data of the RBF neural network; the $\boldsymbol{\mu}_j$ is the jth center of the basis function; the σ_j is the jth width of the basis function.

The linear mapping from the hidden layer to the output layer can be expressed as equation (4).

$$y = \sum_{j=1}^{m} w_j \cdot z_j = \mathbf{w}^T \cdot \mathbf{z} \tag{4}$$

where w_j represents the weighted value from the hidden layer to the output layer.

3.2   *Modeling of the FLS-SVM*

The basic principle of the Fuzzy Least Square Support Vector Machine (FLS-SVM) is as follows (Wang 2005): the fuzzy logic theory has been introduced into the SVM to contrast the corresponding objective function. Thus, when contrasted to the objective function, the contribution of different samples has different factors.

The objective function of the FLS-SVM is modified after introducing the fuzzy subjection degree.

$$\min J(w, \xi) = \frac{1}{2} \| w \|^2 + \frac{C}{2} \sum_{i=1}^{n} \lambda(x_i) \cdot \varepsilon_i^2 \tag{5}$$

where, w is the weighted vector of the hyperplane, $\lambda(x)$ is the fuzzy membership, ε_i is the slack variable, c is the penalty factor.

In order to solve the optimal problem, the Lagrangian factor a_i ($i = 1, 2, ..., n$) is introduced, and the corresponding Lagrangian function is as equation (6).

$$L(w, \varepsilon_i b, a) = \frac{1}{2} \| w \|^2 + \frac{C}{2} \sum_{i=1}^{n} \lambda(x_i) \cdot \varepsilon_i^2 - \sum_{i=1}^{n} a_i (w_i \cdot \varphi(x_i) + b + \varepsilon_i - y_i) \tag{6}$$

Making the partial derivative of the parameters w, ε_i, b, a in the equation (6) equal to 0, then the optimal problem is transferred into the problem of solving the linear system of equation (7)

$$\begin{bmatrix} 0 & E^T \\ E & \Omega + C \cdot \lambda(x_i)^{-1} \end{bmatrix} \begin{bmatrix} b \\ a \end{bmatrix} = \begin{bmatrix} 0 \\ y \end{bmatrix} \tag{7}$$

Defining $K(x_i, x_j) = \varphi(x_i)\varphi(x_j)$ and $K(x_i, x_j)$ can be called the kernel function, thus the equation (6) can be expressed as equation (8) which is the FLS-SVM model.

$$f(x) = \sum_{i=1}^{n} a_i K(x, x_i) + b \tag{8}$$

Kernel function is as equation (9).

$$K(x, x_i) = \exp(-|x - x_i|^2 / \sigma^2) \tag{9}$$

where, σ is the kernel parameter.

3.3 Modeling of the combined method

The combined method was used to establish a combined model by using a combination of FLS-SVM and the RBF neural network, the model can compromise between structural risk minimization and empirical risk minimization. The model can be expressed as equation (10).

$$f(x) = w_1 \cdot f_1(x) + w_2 \cdot f_2(x) \tag{10}$$

where, the $f_1(x)$ and w_1 represents the FLS-SVM model and error, $f_2(x)$ and w_2 represents the RBF neural network and error;

Assuming Z_i were the combination forecast value and the actual observation vector were $(X_1, X_2, ..., X_n)$; the forecast values of the FLS-SVM were $Z_{1i}(i = 1, 2, ..., n)$ and errors were $e_{1i}(i = 1, 2, ..., n)$; the forecast values of the RBF NN were $Z_{2i}(i = 1, 2, ..., n)$ and errors were $e_{2i}(i = 1, 2, ..., n)$. After combination:

$$Z_i = w_1 \cdot Z_{1i} + w_2 \cdot Z_{2i} \tag{11}$$

$$e_i = w_1 \cdot e_{1i} + w_2 \cdot e_{2i} \tag{12}$$

where, $w_1 + w_2 = 1$, $e_{1i} = X_i - Z_{1i}$, $e_{2i} = X_i - Z_{2i}$.

Thus, the minimum value of the combination forecast error variance was:

$$\min\left(\sum e_i^2\right) = \min\left[w_1^2 \sum e_{1i}^2 + 2w_1 w_2 \sum e_{1i} e_{2i} + w_2^2 \sum e_{2i}^2 \right] \tag{13}$$

71

The following contrast can be proved through the equations above: min $(\sum e_i^2) \leq$ min $(\sum e_{1i}^2)$, min $(\sum e_i^2) \leq$ min $(\sum e_{2i}^2)$. Thus, it also shows that the combined forecasting method is better than single prediction methods.

4 CALCULATION OF THE WEIGHTED FUNCTIONS

The FORCs method is introduced to obtain the test data and computing the weighted coefficients with these data. According to equation (1), the integrand is the weighted factor. So, we can find a suitable weighted factor value in the domain of integration as a mean value. Multiplying mean weighted factor value with the corresponding area as the total value of output and the mean weighted factor value is as the weighted matrix elements in equation (2). Thus, equation (2) can be expressed as equation (14).

$$f(t) = R[u(t)] = \iint_{\alpha \leq \beta} \mu(\alpha,\beta)\gamma_{\alpha,\beta}[u(t)d\alpha d\beta] = \bar{\mu}(\alpha,\beta) \cdot (\alpha,\beta) \tag{14}$$

where, $\bar{\mu}(\alpha,\beta)$ represents the mean value matrix of each integral unit and $A(\alpha, \beta)$ represents the area matrix of the integral unit.

After applying the combined model, the result of the calculation is shown in Figure 2.

For obtaining the weighted value which can be used in the discrete hysteretic model, we should multiply the mean weighted value with the corresponding area. Thus, the initial piezoelectric hysteretic model was established.

5 VALIDATION OF THE HYSTERETIC MODEL

In order to prove the feasibility and accuracy of the hysteretic model established in this paper, numerical simulation and test platform were established. Figures 3 to 5 show the outer loop error of RBF NN, FLS-SVM and the combined model.

From the error figures, the error of the outer loop based on combined model was minimum, so the combined model was better.

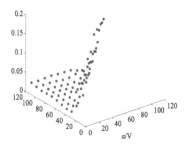

Figure 2. The calculated mean weighted values.

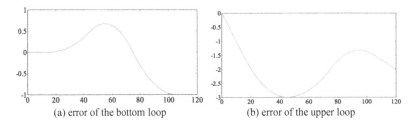

(a) error of the bottom loop   (b) error of the upper loop

Figure 3. Error of the outer loop based on RBF NN.

72

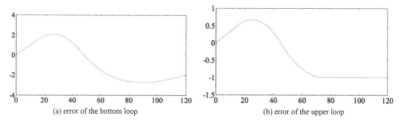

(a) error of the bottom loop   (b) error of the upper loop

Figure 4.   Error of the outer loop based on FLS-SVM.

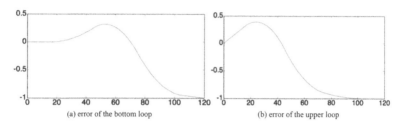

(a) error of the bottom loop   (b) error of the upper loop

Figure 5.   Error of the outer loop based on combined model.

6   CONCLUSIONS

For the problem of the hysteresis in the piezoelectric actuator, the combined method of RBF NN and FLS-SVM was introduced to calculate the weighted functions. The comparison between the test data and simulation data show that the combined model was better than any method of RBF NN or FLS-SVM. The combined model is an effective method.

REFERENCES

Choi G.H., Oh J.H., Choi G.S., 1999. Repetitive tracking control of a coarse-fine actuator. *Proceedings of the 1999 IEEE/ASME International Conference on Advanced Intelligent Mechatronics*, 335–340.
Janaideh M.A., Rakheja S., Su C.Y., 2009. Experimental characterization and modeling of rate-dependent hysteresis of a piezoceramic actuator. *Mechatronics*, 19(5): 656–670.
Wang Y.Q., Wang S.Y., Laik K., 2005. A new fuzzy support vector machine to evaluate credit risk. *IEEE Transactions on Fuzzy Systems*, 13(6): 820–831.
Wolf Felix, Alexander Sutor, Stefan J. Rupitsch, et al. 2012. A generalized Preisach approach for piezoceramic materials incorporating uniaxial compressive stress. *Sensors and Actuators A: Physical*, 186(10): 223–229.
Xu Qingsong, Li Yangmin. 2010. Dahl Model-Based Hysteresis Compensation and Precise Positioning Control of an XY Parallel Micromanipulator With Piezoelectric Actuation. *Journal of Dynamic Systems, Measurement, and Control*, 132(4): 1–12.

Manufacturing Engineering and Intelligent Materials – Lu & Abu Bakar (Eds)
© *2015 Taylor & Francis Group, London, ISBN 978-1-138-02832-6*

Static pressure pipe field tests and simulation analysis

C.W. Yan
Shaying River Valley Authority of Henan Province, China

Z.Q. Huang, X.C. Huang & H. Li
School of Resources and Environment, North China University of Water Resources and Electric Power, Zhengzhou, China

ABSTRACT: We conducted a field static load test on the static pressure pile, and got the relationship curves and rebound curves between the pile settlement and the grading curve pressure. We used FLAC 3D program on the pile numerical simulation analysis, the distribution of the pile in the settlement laws under loading conditions, as well as changes in ground plastic zone. The numerical simulation results and field static load test results were compared and found both largely coinciding with each other. Therefore, FLAC 3D simulation analysis and predictive analysis can be used on pile settlement research.

Keywords: PHC pipe pile; field test; numerical modeling; finite difference

1 INTRODUCTION

With the growing awareness of environmental protection and urban construction of more stringent noise control, pile foundation design and construction methods have become an issue of primary concern (Huang & Wang 2003, Huang et al. 2009). To solve the strong noise and air pollution generated by pile sinking, as well as water pollution problems generated by drilled pile and digging pile, static pressure prestressed pipe technology was invented. The static pressure prestressed pipe has a lot of advantages such as high pile capacity, fast construction progress, low noise, less pollution, low pile supplies, low cost and so on. In many large cities, hydrostatic prestressed pipe pile has been more widely used. Prestressed high strength concrete pile (PHC pile) is produced by professional manufacturers.

Advanced technology such as pre-tensioned prestressed and adding finely ground material are used to build this kind of pile. This kind of pile is precast concrete which is made of slender, hollow, equal section and have some bending, compression performance of precast concrete (Liu & Han 2005).

2 PROJECT OVERVIEW AND GEOLOGICAL CONDITIONS

2.1 *Project overview*

The subject of this paper is the residential project with a total construction area of 13538 m², located in Guangzhou city. PHC tubular pile foundation is used. Total number of piles is 799 and concrete strength of the pile is c80. Pile bearing stratum of coarse sand layer, pile length is 7 meters from the ground. Hydraulic press is used to penetrate piles; the pressure of the final pressure is 900 kn. Ultimate static-bearing capacity of single pile is determined according to the site load test. Geological conditions of the site are so complicated that there are a small amount of underground caverns in local area. To test the quality of the design and construction of the piles, eight of the PHC pipes are proceeded on-site static load test.

Table 1. List of physical and mechanical properties of the layers of soil.

| Soil layer | Index | | | | | | |
	Specific weight (kN/m³)	Moisture content (%)	Void radio	Compression modulus (MPa)	Compression coefficient (MPa⁻¹)	Cohesion C	Internal frictional angel
Mucky soil	25.87	1.252	1.51	2.55	1.06	7.3	3.4
Silty clay①	26.17	0.949	1.32	3.25	0.71	14.9	7.4
Silty clay②	26.17	0.621	1.10	4.23	0.51	22.2	12.5
Silty clay③	26.36	0.366	0.78	5.18	0.34	30.3	16.2

Table 2. Results of PHC piles vertical compressive static load test results.

Pile number	Pile diameter (mm)	Depths of penetration (m)	Design value of bearing capacity (KN)	Ultimate bearing capacity (KN)	Largest settlement (mm)	Residual settlement (mm)	Settlement (mm)	Resiliency (%)
1#	Φ400	7	400	≥800	8.19	1.98	1.28	75.82
2#	Φ400	8	400	≥800	8.58	1.14	2.27	86.71
3#	Φ400	7	400	≥800	7.44	1.17	1.90	84.27
4#	Φ400	7	400	≥800	8.50	1.38	2.22	84.14
5#	Φ400	7	400	≥800	11.35	2.49	2.13	78.06
6#	Φ400	7	400	≥800	7.03	0.87	1.49	87.62
7#	Φ400	7	400	≥800	8.79	2.48	2.30	71.79
8#	Φ400	7	400	≥800	9.72	3.12	2.19	67.9

2.2 Engineering geological conditions

According to the geological survey data, geological conditions of the project are shown in Table 1.

3 STATIC LOAD TEST OF SITE

3.1 Static load test and results

Using a slow load compression method to test the bearing capacity of single pile, the manual pump is used to test pile progressively loaded. The load is divided into ten loads and unload five per class load for 800 kn.

Static pressure pile machine is used to load values measured by the gauge, then converted calibration curve is given by the jack, test pile settlement through symmetrically arranged on the pile head forward dial readout. Results of the tests are shown in Table 2. Q ~ s curve of eight test piles are shown in Figure 2.

3.2 Static load analysis results

From the Q ~ s curve (Fig. 2) of test piles we can see that4 # test pile under load settled levels can stabilize and the shape of the curves are similar, showed slow variant. When loaded to 800 kn, there was no significant downward turning segment; the pile did not reach the limit state.

Rebound after unloading is 84.14%, settling in the elastic range. When the ultimate bearing capacity is greater than 800 kn, designed to meet the load value claim. We see 4# pile static load test reached 8.5 mm, which is the largest settlement. The analysis may have the following reasons:

1. According to the geological survey map, the upper part of the soil from top to bottom of the pile are in the order of artificial fill, about 2.1 m, silty soil, about 2.5 m and grit, thickness of about 2.1 m. Pile penetrates into the softer side silty clay layer.

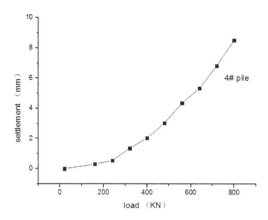

Figure 1. Comparison between the load and the settlement.

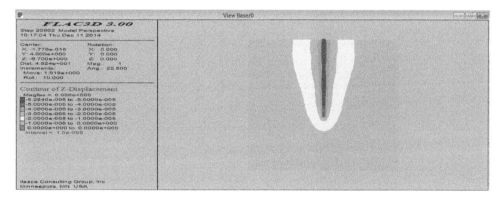

Figure 2. The generalized entirety deformation nephogram under load 160 KN.

2. At the lower partial pile, there is a smaller range of underground cave. The soil is loose, and has large porosity. When the load is applied on the pile, the pile side of the lower part of the soil has squeezed the role of the caves smaller. Lower part of the pile soil is greatly compressed, so that is a greater pile settlement.

3. The pile is at the edge of the building. Soil compaction effect produced is not hard, the pile of soil compaction is not obvious. When the load on the pile to produce, compared with other experimental pile, soil around the pile sinking restraints small, so the amount of the settlement pile produced is larger than the others.

4 STATIC LOAD TEST OF SITE

Model clamped bottom, sides simply supported on piles. Vicinity encrypted processing. Soil and pile Moerkulun model adopted in the pile. Add interface unit between the soil, in order to better monitor the impact on the settlement of friction. z-axis of the model is set to 13.5 m, x-axis is 16 m, y-axis is 8 m, because of symmetry, take half of the model to simulate. Model clamped bottom, sides simply supported on piles near the area were encrypted. Moerkulun model are adopted for the soil and pile. In the soil between piles, interface unit are added to monitor the impact of the friction to settlement.

By using FLAC 3D numerical calculation to the 2nd pile, vertical displacement cloud were obtained in the load 160 kn, 240 kn, 320 kn, 400 kn, 480 kn, 560 kn, 640 kn, 720 kn, 800 kn. Figures 2 to 5 are vertical displacement contours of 160 kn, 320 kn, 640 kn and 800 kn.

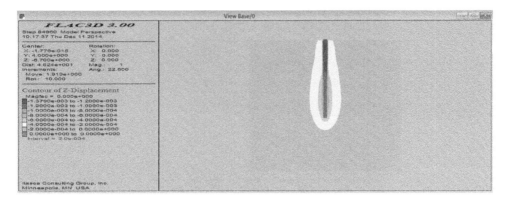

Figure 3. The generalized entirety deformation nephogram under load 320 KN.

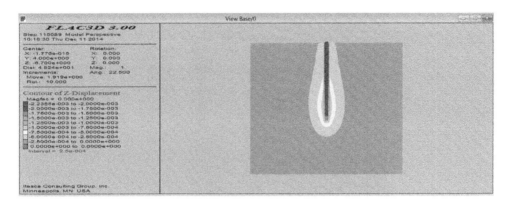

Figure 4. The generalized entirety deformation nephogram under load 640 KN.

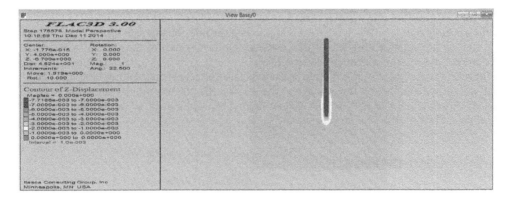

Figure 5. The generalized entirety deformation nephogram under load 800 KN.

From Figure 2 to Figure 5 it can be seen that when the load is gradually increased, the soil around the pile near the affected range increases no obvious plastic zone after calculating equilibrium. From the vertical displacement of the case, the load increases, the maximum vertical displacement is gradually increased, when loaded to 800 kn, maximum vertical displacement is 8.93 mm. The FLAC 3D numerical results with the results of static load test for comparison can be seen, the maximum absolute error is 0.52 mm. The maximum relative error is 27.3%. So, we can say that FLAC 3D conduct pile settlement analysis is an effective method.

5 CONCLUSIONS

The static pile settlement analysis is carried out. The results of static load test compared with the calculated results by using FLAC 3D agree well under a pile loaded settlement laws. Loaded to 800 kn test pile the settlement value of the measured values of static load test pilot is closer, indicating the pile settlement conducted using FLAC 3D analysis is an effective method.

It should be noted that, due to a number of factors, were simplified during modeling, without considering the impact of groundwater and underground cave that may exist on the pile settlement. Resulting in the results with the experimental results of the above method to calculate, there are still some errors, but errors within an acceptable range, these results are only in conjunction with other related analyses together in order to be close to the actual situation.

ACKNOWLEDGEMENTS

This work was financially supported by the Plan For Scientific Innovation Talent of Henan Province, Natural Science Foundation of China (41202246) and Key technology research project of Henan province.

Corresponding author: Zhiquan Huang, (1970-), Professor, Mainly engaged in geotechnical engineering work. e-mail: huangzhiquan@ncwu.edu.cn.

REFERENCES

Experimental study of strength characteristics of unsaturated soil of landslide 1# in Xiaolangdi Reservoir, *Rock and Soil Mechanics*, 2009 Vol. 30 No. 3: 640–644.
Huang Zhiquan, K. Tim Law, Liu Handong & Jiang Tong. The chaotic characteristics of landslide evolution: a case study of Xintan landslide. *Environmental Geology*: (2009), Volume 56, Issue 8:1585–1591.
Huang Zhiquan, Wang Sijing. 2003. Synergetic-bifurcated prediction model of slope occurrence and its application, *Science in China*, Ser. E, 2003, Vol. 46: 69–77.
Liu Bo, Han Yanhui. 2005. FLAC Theory, examples and application notes [M]. Beijing: China Communications Press.
Stability analysis of expansive soil slope based on in-situ shear test, *Rock and Soil Mechanics*, 2008 Vol. 29 No. 7: 1764–1768.

Manufacturing Engineering and Intelligent Materials – Lu & Abu Bakar (Eds)
© *2015 Taylor & Francis Group, London, ISBN 978-1-138-02832-6*

Energy losses and temperature distribution of giant magnetostrictive high power ultrasonic transducer

H.Q. Zeng

School of Mechanical and Automotive Engineering, Xiamen University of Technology, Xiamen, China

ABSTRACT: Giant magnetostrictive high power transducers usually work under a high intensity and frequency magnetic field. Hysteresis losses and eddy current losses are very heavy. At the same time, giant magnetostrictive materials (GGM) are sensitive to temperature. Therefore thermal analysis is very important for the transducer's design. In this paper a high power ultrasonic transducer was designed. The magnetization behavior of GGM was characterized by the modified Jiles-Atherton hysteresis model which considers eddy current losses and anomalous losses. The model parameters were determined from the least square fit between experimental values and model values. Energy losses of the transducer were calculated. Temperature distribution of the transducer was calculated with finite element method. Calculation results and experimental results show a good consistency.

1 INTRODUCTION

Ultrasonic transducers are widely used in industrial processes such as industrial cleaning, chemical processing, and ultrasonic welding, etc. However, the largest limitation in many of these applications is the inability of traditional drive materials to provide a single transducer with sufficient power to make important laboratory processes commercially succesful.

GGM technology addresses the existing limitations and enables next-generation high power ultrasonic transducers to fill the technology gap between laboratory process demonstration and successful commercial viability. The unique attibutes of the material, such as high energy density and thermal handing capabilities make it possible to build large-scale high power ultrasonic transducers. An ultrasonic transducer rated at 6 kW continnuous duty had been developed by ETREMA Products, Inc. And a predicted 25 kW ultrasonic transducer is under development with funding from the National Institute of Standards and Technology (NIST) Advanced Technology Program Julie (2000). However, giant magnetostrictive high power ultrasonic transducers usually work under a magnetic field of high intensity and high frequency. The heat from energy loss is huge. GGM is sensitive to temperature. Therefore thermal analysis is an important part for design of the transducer.

In this paper a high power ultrasonic transducer was designed. The magnetization behavior of GGM in the transducer was characterized by the modified Jiles-Atherton hysteresis model which includes eddy current losses and anomalous losses. Model parameters were determined from the least square fit between experimental values and model values. Energy losses of the transducer were calculated. Temperature distribution schemes of the transducer were calculated with finite element method. Calculation results and experimental results show a good agreement.

2 TRANSDUCER STRUCTURE AND WORKING MECHANISM

The sketch of magnetostrictive high power ultrasonic transducer is illustrated in Figure 1. A Giant Magnetostrictive Rod (GMR) was placed within an stainless steel house and

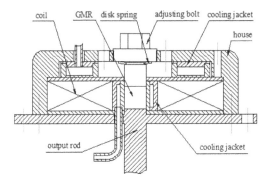

Figure 1. Sketch of the magnetostrictive high power ultrasonic transducer.

surrounded by a coil through which alternating current passed. The GMR is repressed with a disk spring which applies a constant compressive preload to protect the moderately brittle GMR from dynamic tensile stresses and get a maximum output. The GMR elongates or shortens and the output rod outputs ultrasonic vibration as the intensity of magnetic fields produced by the solenoid changes. To prevent the double-frequency phenomena, a bias current was input at the same time. A pickup coil wound around the GMR was used to measure the magnetic induction B. Thermal couples were put at some key positions to monitor the temperature. There are cooling jackets surrounding the GMR and on end-face of the solenoid to remove the heat. The GMR was laminated to reduce eddy current losses.

3 HYSTERESIS LOSSES

3.1 *Hysteresis model*

The transducer power losses consist of electrically resistive, eddy current, hysteresis, anomalous losses and mechanically resistive losses Fredrik (2000). Hysteresis losses are most important part. The calculation of hysteresis losses bases on hysteresis model. There are two extensively used hysteresis models. They are Preisach model and Jiles-Atherton model. Preisach model provides an empirical relationship between variables and are commonly employed to describe many hysteretic processes. Models specifically targeting giant magnetostrictive materials are included in the references Smith (1998). The model has the advantage that it is easier to develop and implement. However, the intent of the Preisach model is to empirically fit the input and output of experimental data, independent of the nature of the physical process. Jiles-Atherton model is a physically based model (Jiles 1986, Jiles 1995). It has the advantage of relying on intuitive development, which helps with design and facilitates extension of the model to new types of transducers. Jiles-Atherton model bases on domain wall translation and domain wall bowing. The model can disclose the underlying mechanism and provide insights into the physical process. In order to use the model at high frequencies, Jiles proposed modified Jiles-Atherton model which include eddy current losses and anomalous losses Jiles (1994). The model is presented as Eq. 1

$$\left(\frac{\mu_0 d^2}{2\rho\beta}\frac{dH}{dt}\right)\left(\frac{dM}{dH}\right)^2 + \left(\frac{\mu_0 GdwH_0}{\rho}\right)^{1/2}\left(\frac{dH}{dt}\right)^{1/2}\left(\frac{dM}{dH}\right)^{3/2}$$

$$+ \left(k\delta - \alpha\left(M_{an} - M + k\delta c\frac{dM_{an}}{dH_e}\right)\right)\left(\frac{dM}{dH}\right) - \left(M_{an} - M + k\delta c\frac{dM_{an}}{dH_e}\right) = 0 \qquad (1)$$

The first term in Eq. 1 represents classical eddy current losses, anomalous losses are represented by the second term, and the remaining terms constitute the Jiles-Atherton magnetization model.

In this equation M is magnetization, H is applied magnetic field, μ_0 is the permeability of free space, d is a dimension of the material (diameter for cylinders, thickness for laminations, here $d = 0.002$ m), ρ is the resistivity of the material, here $\rho = 6.0 \times 10^{-7}$ Ωm, β is a geometry factor (6 for laminations, 16 for cylinders, for this paper $\beta = 6$), G is a dimensionless constant of value 0.1356; w is the width of laminations in mm, here $w = 0.006$ m, H_0 is a parameter representing the internal potential experienced by domain walls, H_0 has dimensions of A/m, and so is equivalent to a magnetic field, $\delta = $ sgn (dH/dt), k is average pinning energy loss factor, A/m, determined directly from experimental data, α is a mean field parameter representing interdomain coupling, dimensionless, which has to be determined experimentally, c is domain wall bending coefficient, dimensionless, determined directly from experimental data, M_{an} is the anhysteretic magnetization. A Langevin function is used to model the anhysteretic magnetization,

$$M_{an} = M_s \left[\coth\left(\frac{H_e}{a}\right) - \frac{a}{H_e} \right] \tag{2}$$

where M_s is the saturation magnetization here $M_s = 7.65 \times 105$ A/m, a is effective domain density coefficient, controls the slope of anhysteretic magnetization, A/m, determined directly from experimental data, H_e is the effective magnetic field. The effective magnetic field includes the applied magnetic field H, the Weiss mean field interactions αM, and magnetoelastic interactions H_σ as follows,

$$H_e = H + \alpha M + H_\sigma \tag{3}$$

Magnetoelastic interactions depend on the stress state of the material and magnetostriction λ as follows

$$H_\sigma(t) = \frac{3\sigma}{2\mu_0}\left[\frac{\partial \lambda}{\partial M}\right] \tag{4}$$

According to quadratic law Eq. 5,

$$\lambda = \frac{3\lambda_s}{2M_s{}^2}M^2 \tag{5}$$

where λ_s is the saturation magnetostriction, here $\lambda_s = 996$ με. Substituting Eq. 5 into Eq. 4 yields

$$H_\sigma(t) = \frac{9}{2}\frac{\sigma}{\mu_0}\frac{\lambda_s}{M_s^2}M \tag{6}$$

3.2   *Parameters identification and losses calculation*

The tests were run at fixed-frequency sinusoidal input current to the transducer. Data collected included current, voltage, flux density. A linear amplifier operating in current-controlled mode was used to drive the transducer. A 25-turn pickup coil wound around the GMR was used to obtain the magnetic induction B. The voltage signal from this coil was fed into an integrating voltmeter calibrated so that

$$B(t) = -\frac{1}{NA}\int V(t)dt \tag{7}$$

with A the cross sectional area of the rod. N the turns of the pickup coil. $V(t)$ induced voltage. Magnetic field and flux density were used to calculate the magnetization of the GMR,

$$M = \frac{B}{\mu_0} - H \tag{8}$$

Applied magnetic field, H, was computed from the input current. So $M-H$ hysteresis loop was obtained.

Preload the GMR to $\sigma_0 = 6.1$ MPa. Input an alternating current at frequency of 20 kHz (the transducer usually works under this frequency). Hysteresis loop was obtained. Then parameters were determined from the least square fit between experimental values and the model values as follows, $a = 10000$ A/m, $k = 6000$, $c = 0.21$, $\alpha = 0.75$, $H_0 = 0.009$ A/m.

Suppose magnetic field in the GMR is uniform, input sinusoidal current of 10 A, 20 kHz to the transducer. A period of sine current is divided into 180 discrete time steps. For each time-step, the magnetic field intensity H and magnetic field density B was calculated. After finishing a period's calculation, $B-H$ hysteresis loop was obtained. The energy losses unit volume per period is the integral of the magnetic field intensity and magnetic field density. The total losses (include eddy current loss and anomalous loss) is the product of the losses unit volume per period with volume of the GMR and frequency of driving current. Here the total losses are equal to 75.71 W.

3.3 Electrically resistive losses

Electrically resistive losses can be given by

$$Q = I^2 R \tag{9}$$

where I is driving current passing through the coil, R is electrical resistance of the coil. Substituting the length of the wire, resistivity of the wire, etc. into equation (9) yields $Q = 8.23$ W.

3.4 Temperature distribution calculation

Finite element model of the transducer was built. Neglecting mechanically resistive losses, export the losses calculated above to the finite element model, temperature distribution of the transducer can be obtained as Figures 2 and 3.

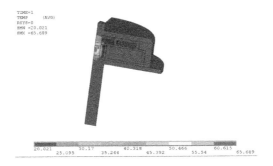

Figure 2. Temperature distribution map of the transducer.

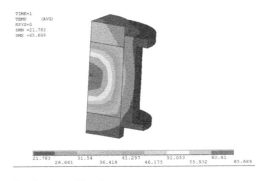

Figure 3. Temperature distribution of local area.

From temperature distribution map of the transducer, it was found that the maximum temperature point of the transducer was situated at the center of the GMR. The temperature value was approximately 65.7°C, far below the Curie temperature of the GMR (380°C). So the cooling system's design is correct. To get more accurate temperature distribution surrounding GMR, a temperature distribution of local high-temperature area was obtained as Figure 3.

4 TEMPERATURE CONTROL AND TEST

4.1 *Temperature control system design*

To measure accurately temperature of the key points on the transducer. A microcomputer-based control system for temperature with 4 Channels was designed. The system consist of a computer, a microcomputer, a AD converter, a relay and 4 thermocouples. The 4 thermocouples were put respectively on the side of the GMR, two end of coil, outside of the coil. Temperature signal after AD conversion enter the microcomputer. There converted temperature was compared with the set value. If the temperature is below the set value, the transducer continues to work. If the temperature is higher than the set value, the computer alarms, the relay switches off. So the transducer is protected. At the same time, the computer also records the temperature.

4.2 *Experimental verification*

Sinusoidal current of 10 A, 20 kHz was input into the transducer. Cooling water of 0.2 Mpa, 18°C was pumped into the cooling jackets of the transducer. At the beginning, the temperature of the transducer rise slowly. Some time later, the temperatures stop rising. Then the temperature beside GMR was taken. The value is 34°C, almost same as the value shown in Figure 3. The discrepancy between calculation value and experiment value is small.

5 CONCLUTIONS

1. The modified Jiles-Atherton hysteresis model which includes eddy current losses and anomalous losses and considers the preload exerted on the GMR is suitable for characterization of the magnetization behavior of GGM in the transducer. The model can be taken as one of the theoretical basis of Giant magnetostrictive high power transducers design.
2. When the high power ultrasonic transducer designed in this paper works under expecting working condition, the highest temperature of the GMR was approximately 65.7°C, far below the Curie temperature of the GMR (380°C). This indicates that the transducer cooling system's design is correct.

REFERENCES

Fredrik S., et al. 2000. Dynamic simulation and performance study of magnetostrictive transducers for ultrasonic applications. *Proceedings of SPIE*, vol. 3992: 594–602.
Jiles, D.C. & Atherton, D.L. 1986. Theory of ferromagnetic hysteresis. *Journal of Magnetism and Magnetic Materials*, 61: 48–60.
Jiles, D.C. 1994. Modelling the effects of eddy current losses on frequency dependent hysteresis in electrically conducting media. *IEEE Transactions on Magnetics*, 30(6): 4326–4328.
Jiles, D.C. 1995. Theory of the magnetomechanical effect. *Journal of Physics D: Applied Physics*, 28(8): 1537–1546.
Julie C.S., et al. 2000. Modeling of a TERFENOL-D ultrasonic transducer. *Proceedings of SPIE*, vol. 3985: 366–377.
Smith R.C. 1998. Hysteresis modeling in magnetostrictive materials via Preisach operators. *Journal of Mathematical Systems*, Estimation, and Control, 8(2): 249–252.

Manufacturing Engineering and Intelligent Materials – Lu & Abu Bakar (Eds)
© 2015 Taylor & Francis Group, London, ISBN 978-1-138-02832-6

Research on the characteristics and preparation of multi-component cementitious materials

L.B. Bian

School of Civil and Environmental Engineering, University of Science and Technology, Beijing, China
School of Civil Engineering and Transportation Engineering, Beijing University of Civil Engineering and Architecture, Beijing, China

J.H. Liu

School of Civil and Environmental Engineering, University of Science and Technology, Beijing, China

D. Guo

School of Civil and Environmental Engineering, University of Science and Technology, Beijing, China
School of Civil Engineering and Transportation Engineering, Beijing University of Civil Engineering and Architecture, Beijing, China

ABSTRACT: In this paper, we tested the cement properties, the hydration heat properties and the shrinkage properties of composite cementitious material with different rations of the clinker, fly ash, slag, and limestone power. The results showed that it can formulate the 42.5 cement of the ratio 60% used fly ash and slag with the specific surface area 310 m^2/kg of clinker, 410 m^2/kg of fly ash, 460 m^2/kg of slag and 430 m^2/kg of limestone power. It still can formulate the 32.5 cement of the ration 60% used fly ash, slag and limestone power. The curve of hydration heat of multi-component cementitious materials owns two peaks while the traditional owns one. The multi-component cementitious materials greatly reduce the heat and the hydration temperature peak. And the performance of shrinkage is also greatly improved.

Keywords: multi-component cementitious materials; admixture; grinding; heat of the hydration; shrinkage

1 INTRODUCTION

The production and use of modern cementitious materials have many problems. The traditional user does not know the proportion of clinker or mineral admixtures in cement. According to their experience or experimental data, they add all kinds of mineral admixtures again. Widely used in mineral admixtures of cementitious material system, to meet the requirements of strength and the pursuit of the maximization of economic benefit, as cement clinker is finer, specific surface area is becoming more and more big. The national standard for cement specific surface area of the requirement is no less than 300 m^2/kg. Actually, high-tech materials after grinding machine, the specific surface area is often up to 450 m^2/kg. Although it can make up for the early strength, the late strength growth is limited, even reduced in recent years after carefully mixing of clinker and mineral admixture. The question is also more and more due to the cement clinker grinding interloper in recent years. Early strength is too high and growth rate of the intensity of the long-term reduces, even shrinks (Liang 2006). Tong-sheng zhang (Zhang 2015) designed the matching proportion of cement clinker with auxiliary cementitious material optimization. They got high performance composite cement in the laboratory which has low dosage of clinker and good volume stability and durability. They prepare the composite cement strength grade 42.5, in which cement clinker accounted for 25%, slag accounted for

25% and low activity auxiliary cementitious material accounted for 39%. Barbara Lothenbach has offered effect of composite cementitious with different mineral admixtures (Barbara & Karen 2011) in the superimposed material system and studied its performance.

This subject adopts the technical route "coarse clinker which has low specific surface area and fine admixture which has high specific surface area". After pure clinker matched with gypsum powder grinded, using the many kinds of mineral admixture, fly ash, limestone powder, slag powder, and mixing to acquire a new multivariate composite cementitious material, then tests the standard consistency water consumption and its strength. At the same time, test and analysis of hydration heat and shrinkage performance of multiple cementitious material which has different proportion of admixture.

2 EXPERIMENT

2.1 Raw materials

2.1.1 Clinker
The clinker used in this test is clinker of Portland cement, with a planetary ball mill grinding to powder for 30 minutes. Its specific surface area is 310 m^2/kg. The chemical composition of clinker is as shown in Table 1.

2.1.2 Fly ash
Specific surface area of fly ash is 410 m^2/kg. It belongs to type II and its performance indicators are as shown in Table 2.

2.1.3 Slag
Specific surface area of slag is 460 m^2/kg and it belongs to S95 grade.

2.1.4 Limestone powder
See Table 3.

2.1.5 Gypsum
Ordinary dihydrate gypsum

2.1.6 Particle size distribution
The analysis of grading use melvin dry grain size and grain shape analyzer. Four kinds of particle size distribution of raw materials value are shown in Table 4.

Table 1. The chemical composition of Portland cement clinker.

Component name	SiO_2	Al_2O_3	Fe_2O_3	CaO	MgO	K_2O	Na_2O	SO_3	P_2O_5	Cl	TiO_2	MnO	SrO
Content (%)	19.46	4.26	3.42	67.68	2.13	1.15	0.23	0.93	0.19	0.02	0.39	0.07	0.08

Table 2. The performance indexes of fly ash.

Sieving residue of 45 um (%)	Density (g/cm³)	Specific surface area (m²/kg)	Water demand ratio (%)	Loss on ignition (%)	Content of SO_3 (%)
1.2	2.2	419	96	3.2	0.68

Table 3. The performance indexes of limestone powder.

Density (g/cm³)	Specific surface area (m²/kg)	Fluidity of mortar ratio (%)
2.81	430	105

Table 4. The value of the particle size distribution.

Grain size/μm	10	20	30	40	50	60	70	80	90	100
Clinker	29.55	49.00	61.77	70.87	79.52	83.41	86.88	89.87	92.34	94.28
Fly ash	44.73	69.12	79.11	84.98	89.05	90.91	92.55	94.00	95.27	96.37
Slag	44.45	74.56	86.70	91.73	94.39	95.06	95.45	95.66	95.76	95.82
Limestone power	31.59	57.60	74.17	84.11	91.56	94.16	96.03	97.27	98.00	98.36

Table 5. The mixture ratio of composite cementitious materials.

Number	Clinker + gypsum	Fly ash	Slag	Limestone powder	Number	Clinker + gypsum	Fly ash	Slag	Limestone powder
N	100%	0%	0%	0%	M2	50%	30%	20%	0%
L1	60%	20%	10%	10%	M3	50%	20%	10%	20%
L2	60%	20%	20%	0%	S1	40%	20%	40%	0%
L3	60%	10%	10%	20%	S2	40%	40%	20%	0%
M1	50%	20%	30%	0%	S3	40%	20%	20%	20%

Note: N group is benchmark group which is grinding pure clinker and the rest of the group are dosage of mineral admixture, respectively, 40%, 50%, 60%, account for the proportion of different groups in the branch office is different, specific mixture ratio is shown in Table 5.

It can be seen from Table 4 that particle size distribution of clinker is different with fly ash, slag and limestone powder. Particle size distribution of fly ash, slag and limestone flour is far lower than the clinker. Mixing mineral admixtures which are finer than cement particles can improve cementitious material's grain size distribution. Particle size distribution of cementitious materials powder can be improved by mixing four kinds of powder materials in proportion, so superimposed effect of different powder materials can give full play. After different fineness of mineral admixtures composited, the particles can complement each other, so as to make the particle size distribution of the cementitious material closer to the request of accumulation state (Yang 2003).

2.2 Test method and mixture ratio

Ingredients used in the experiment as shown in Table 5.

3 THE EXPERIMENTAL RESULTS AND DATA ANALYSIS

3.1 The cement performance results

Water requirement of normal consistency's test method refers to GB/T1346-2011 "water requirement of normal consistency, setting time, stability test method", the strength test refers to GB/T17671-1999 "the cement mortar strength testing method", the results are as shown in Table 6.

Pure clinker can be seen from Table 6, the normal consistency of cement water consumption of pure clinker is the highest, with the increase of proportion of mineral admixtures, standard consistency water gradually decreased, pure clinkers' is 29.9%, while added 60% (fly ash, slag, limestone powder ratio is, respectively, 20%) mineral admixtures is 21.9%. From this perspective, the mineral admixtures added played an excellent effect on reducing water; another reason is that specific surface area is lesser after the clinker coarse grinding. Cement clinker as water requirement of most components in the cement, the decrease of its specific surface area is bound to lead to reduce the unit volume of particles, thus resulting in a loss of hydration and the determination of water consumption standard consistency.

Table 6. Water requirement of normal consistency of different mixture ratio.

Number	N	L1	L2	L3	M1	M2	M3	S1	S2	S3
Water requirement of normal consistency/%	29.9	25.3	25.4	25.0	23.3	23.6	23.1	22.3	22.6	21.9

Table 7. The fluidity of cement mortar and strength of different ages.

Number	The fluidity of cement mortar (mm)	Flexural strength (MPa)		Compressive strength (MPa)		Number	The fluidity of cement mortar (mm)	Flexural strength (MPa)		Compressive strength (MPa)	
		3 d	28 d	3 d	28 d			3 d	28 d	3 d	28 d
N	180	4.5	8.6	23.2	53	M2	200	3.3	7.6	13.6	38.4
L1	170	4.3	7.9	19.6	44.8	M3	205	2.9	6.7	12.9	29.0
L2	160	4.8	8.1	21.5	51.6	S1	215	2.9	8.8	13.4	48.2
L3	185	4.2	8.1	18.6	41.5	S2	215	2.5	6.9	11.1	34.4
M1	195	3.6	8.4	15.3	46.9	S3	225	2.8	7.2	11.3	32.9

It can be seen from the chart that the pure clinker cement strength of 28 days is 53 MPa, strength reduces after the addition of mineral admixtures. 28 days' strength of cement materials which added 40% of the proportion of mineral admixtures is up to 51.6 MPa. When the proportion of fly ash and slag is 20% each, the strength is highest. 28 days' strength of cement materials which added 50% of the proportion of mineral admixtures is up to 46.9 MPa. When the proportion of fly ash is 20% and slag is 30%, the strength is highest. 28 days' strength of cement materials which added 60% of the proportion of mineral admixtures is up to 48.2 MPa. When the proportion of fly ash is 20% and slag is 40%, the strength is highest.

In analysis, increasing the proportion of slag in all the use of mineral admixtures is good for the improvement of strength; it is concerned with nature of slag whose activity is higher than fly ash and lime powder. The whole of the strength of the cementitious system is reduced because of application of limestone powder. This is mainly due to the composition of limestone powder and its chemical composition. The main components of the limestone powder is inert $CaCO_3$ whose active effect is much lower than other two kinds of mineral admixtures (active Al_2O_3 and SiO_2).

The cap on the limestone powder add also makes a breakthrough of traditional cement admixtures. The addition of limestone powder is feasible, the proportion of limestone powder added to the 20% will still be able to meet the requirements of 32.5 cement though intensity decreases.

In short, the integration of mineral admixture and clinker makes the admixture ratio to reach 60% during traditional cement production process; in the case of admixture use 60%, strength index will still be able to meet the requirements of 42.5 cement. The addition of limestone powder makes the new mineral admixtures to use, but strength decreases.

3.2 Hydration heat test

In general, the hydration process of cement can be divided into three stages from heat evolution rate: the first stage is called the dormant period. At the initial stage, cement particles contact and react with water and the heat releases quickly. While because of the existence of gypsum, a layer of passivation mode will form on the surface of cement particles, which lowers the heat release rate; the second stage can be called a phase-boundary reaction stage, in which cement hydration heat release rate is fastest and cement particles also grow quickly;

the third stage can be called diffusion control stage. With the hydration products of cement on the surface of cement particles gradually accumulating, cement hydration exothermic rate gradually decreased, so the reaction is controlled by diffusion.

Put 10 g samples in the test tube and shake. Take needle filled with 5 ml of pure water and insert it in composite cement, then put the three water pipes in the cement hydration heat meter. After all is ready, inject the water in sample. We can get the result of the hydration heat after three days.

The results are as shown in Table 8.

It can be seen from the chart and graph (Figs. 1 and 2), the hydration heat peak and eventually total calories of multiple mineral admixtures are lower compared with pure clinkers. At the same time, the hydration heat peak of pure clinker appears relatively abrupt and unimodal exothermic, while adding mineral admixtures of multivariate cementitious material appears bimodal exothermic. The reason of the first exothermic peak is that the C_3A content in cement clinker is more and the hydration heat quantity is big, the secondary peak can be considered that the addition of mineral admixtures decreases C_3A. At the same time, the mineral admixtures of Al_2O_3 and SiO_2 also need hydration in alkali environment. After the first peak, it is embodied in the heat quantity of C_3S, the adding of mineral admixtures makes dispersion and time delay of the hydration heat release.

On the other hand, the incorporation of mineral admixtures can significantly reduce the hydration heat value, as can be seen from the above, accumulative hydration of the benchmark set N put calorific value reached 72 J/GH within 72 h, and the cumulative values of the hydration heat obviously decrease after adding mineral admixtures. According to the different proportion of mineral admixtures which adding 50% of mineral admixtures multivariate cementitious material, hydration heat accumulated value, respectively, only: 342 J/GH, 197 J/GH, 151 J/GH, significantly reduced. According to the data, the effect of reducing the hydration heat can be the ranked as follows: limestone powder > fly ash > slag. This is because the inertia of limestone powder and stable structure in the multivariate composite cementitious material filling effect and ball effect which is greater than its hydration effect, compared with fly ash and mineral powder, hydration performance and active of slag is higher, so that heat effect is slightly lower. The multivariate cementitious material is a good choice for reducing heat of hydration.

3.3 Autogenous shrinkage experiment

Select number N, S1, S2 and S3 and use test method of embedded probe nail. Firstly, put clean probe pin in the center hole of model. Secondly, molded according to the mortar, especially to prevent the probe pin falling off in process. Test the shrinkage rate after forming specimen and set to zero. The experimental results are shown in Table 8 and Table 9.

Use par to correct zero before the test, and check 1 to 2 times at least in the process of determination. If you discover the original value more +/–0.01 mm than zero, deviation should be zero and determination again. The location of the specimen should be consistent at a time on the shrinkage instrument placed direction. So it should indicate the corresponding symbols on the tested pieces, such as sequence and direction etc. The test piece should

Table 8. The results of hydration heat test.

Number	Hydration heat J/GH			Time to peak/h	Number of peak	Number	Hydration heat J/GH			Time to peak/h	Number of peak
	24 h	48 h	72 h				24 h	48 h	72 h		
N	287.67	492.73	567.14	11	1	S2	57.21	107.46	133.63	12/29	2
M1	179.74	291.03	342.68	11/24	2	S3	42.6	90.07	113.06	15/35	2
M2	98.74	166.06	197.57	11/25	2	M1	179.74	291.03	342.68	12/24	2
M3	84.4	134.48	151.33	11/25	2	M2	98.74	166.06	197.57	11/25	2
S1	137.26	233.76	282.79	11/27	2	M3	84.4	134.48	151.33	12/25	2

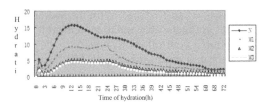

Figure 1. The incremental hydration heat values changing with time.

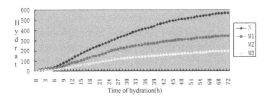

Figure 2. The cumulative hydration heat values changing with time.

Table 9. The results of shrinkage experiment (%).

| Number | Date | | | | | |
	1 d	4 d	7 d	14 d	28 d	56 d
N	−0.0020	−0.0033	−0.0045	−0.0053	−0.0057	−0.0059
S1	−0.0019	−0.0024	−0.0032	−0.0047	−0.0048	−0.0050
S2	−0.0014	−0.0028	−0.0033	−0.0049	−0.0052	−0.0053
S3	−0.0015	−0.0018	−0.0021	−0.0028	−0.0034	−0.0037

be carefully placed and taken out. It can't collision table and table bar, otherwise should be recalibrated to zero.

It can be seen from Table 9, sample's shrinkage rate of multivariate composite cementitious material which adding mineral admixtures is decreased obviously than pure clinker. This is because that hydration speed of admixture is slower than pure clinker's relatively strong hydration speed. At the same time, after mixing with limestone powder, lime powder in cementitious material plays the role of more the filling effect and compensation contract, while hydration participation is limited.

4 CONCLUSIONS AND PROBLEMS

1. Limestone powder, fly ash, slag, clinker are four types of gelling systems, with technology routes which clinker coarse grinding and admixture fine grinding to meet the performance qualified cementitious material. At the same time, it can make up for a series of problems which is brought by the cement technical route. 28 days' strength of cement which added 60% of the proportion of mineral admixtures is up to 48.2 MPa. In the case of mixed with 20% limestone powder, 28 days' compressive strength of cement can reach 32.9 MPa.
2. The hydration heat peak of pure clinker appears unimodal exothermic, while adding mineral admixtures of multivariate cementitious material appears the bimodal exothermic and calorific value is lower than the unimodal's. Exothermic peak time is also delayed. At the same time, multiple cementitious material is a good solution for traditional cement hydration's high heat. Hydration heat quantity significantly reduced has great reference value and significance for the control of mass concrete construction and hydration temperature rise.

3. Mineral admixtures, especially the use of limestone powder can compensate shrinkage and properly reduce the shrinkage of cement base material.
4. Hydration degree and the mechanism of the composite cementitious material system need further research.

REFERENCES

Barbara Lothenbach, Karen Scriveder. 2011. Supplementary cementitious materials. *Cement and concrete research.* 2011:1244–1256.
Liang Huizhen. 2006. What kind of cement modern concrete need. *Cement*, 2006.9:8–13.
Yang Jing. 2003. The grain size of mineral admixtures impacts on mechanical properties of cement paste and high-performance concrete material. *Industrial Architecture.* 33:6–8.
Zhang Tongsheng. 2015. Optimization matching of cement clinker and auxiliary cementitious material. *South China University of Technology. PhD dissertation*, 5:15–19.

Manufacturing Engineering and Intelligent Materials – Lu & Abu Bakar (Eds)
© 2015 Taylor & Francis Group, London, ISBN 978-1-138-02832-6

Rolling defects diagnosis method of Fuzzy Petri Nets

Y.Y. Zhao & Y. Wang
University of Jinan, Jinan, Shandong, China

ABSTRACT: Fuzzy Petri net method is applied to the rolling defect diagnosis. A defect knowledge base and the corresponding fuzzy Petri net model is established through the study of a H-beam steel rolling quality defects and then a combination of positive and negative reasoning is used to achieve the quality defects of steel rolling expert system reasoning. Finally, the experiment verifies the correctness of this method.

1 INTRODUCTION

Products of hot rolling strip steel may not be completely in accord with the demand of customers in the whole production process due to the influence from various aspects like high temperature, high pressure, high speed and hardware equipment and so on. Several different kinds of defects will be produced. So, it is of great significance for the quality improvement of steel and the study of steel quality defects type and finding the specific causes of defects in time according to the actual production situation. The current research of steel quality defect is mainly divided into the surface quality defect recognition and the flatness defect recognition and major progress has been made in introducing intelligent methods into the research. With the development of the world steel production technology, H-beam becomes widely used in all walks of life as a kind of energy-saving steel. The steel structure made by the hot rolled H section steel is light weight, quick construction and good seismic performance as well as the advantages of saving energy and reducing consumption. Researches on H-beam product quality defects caught the attention of relevant scholars and technical experts, such as the research of H-beam web folding defects, researches on the surface quality defects, etc.

Petri net is a kind of the system model which can be represented by the mesh graphics. It uses the graphical language to describe the system characteristics. In the real life, especially in the knowledge system, many production rule is fuzzy and difficult to be described accurately. The Fuzzy Petri Net (FPN) is a good tool to express such IF-THEN structure production rules. Because of the strong ability of modeling and analysis, FPN can be applied to different fields, such as knowledge representation, reasoning and dynamic description of a causal relationship. FPN obtained a widespread application in knowledge representation and reasoning, using the fuzzy Petri nets as knowledge expression and reasoning tools to combine with expert system is an important aspect in the research of fuzzy Petri nets. Fuzzy Petri net can be used for the knowledge representation and reasoning of expert system mainly has several reasons:

1. The fuzzy Petri net can construct knowledge based on the known rules and describe the relationship between rules so that fuzzy reasoning can be realized;
2. Graphical nature of the Petri net provides a visual dynamic behavior based on the rule reasoning;
3. Fuzzy Petri net can make the reasoning algorithm more efficient;
4. The reasoning ability of the Petri net provides basis for the development of knowledge verification technology;

5. Fuzzy Petri net can be used in establishing models which have rules of excitation potential concurrent relation. This is of vital importance in real-time.

This article establishes a defects knowledge base and a fuzzy Petri net model through a detailed analysis for the situation of the scrap steel of H-beam rolling quality defects and a combination of positive and negative reasoning method is used to achieve the surface defects of steel rolling expert system reasoning.

2   FUZZY PETRI NET

2.1   *The definition of fuzzy Petri net*

Fuzzy Petri net is a combination of fuzzy theory and Petri net theory. It is a fuzzification for ordinary Petri nets. Fuzzy Petri net generally is made up by places, transitions and directed arcs elements and so on and have very good description for the uncertainty. FPN has unique characteristics compared with ordinary Petri net: 1) Each place of FPN has a degree of confidence and represented by a real number between 0 and 1; 2) Each transition is given a transition factor. It is used to show the the probability of the occurrence of transition. The definition of fuzzy Petri net is Definition 2.1.

Definition 2.1: $FPN = (P, T, D, A, S, cf, \alpha, \beta, \theta)$. $P = (P_1, P_2, \ldots P_n)$ is the finite set of places. $T = (T_1, T_2, \ldots T_n)$ is the finite set of transitions. $D = (K, G, U)$ is the thesis set. A is the set of directed arcs. $S = (S_p, S_t)$ is the state set, S_p is the state set of transitions, cf is the confidence level of the result. α is the one-one mapping from the place to truth value. β is the one-one mapping from place to proposition. θ is threshold value of transitions.

2.2   *Fuzzy production rules*

Fuzzy production rules are often used to describe uncertainties and are expressed by the IF-THEN statement. The basic expression of fuzzy production rule is:

$$R_k: \text{IF } A_k \text{ THEN } B_k \ (cf = \sigma_k)$$

A_k represents symptoms and its scope is [0, 1]. B_k represents the causes. σ_k is the degree of confidence. The larger the value of σ_k, the more trusted of the $k th$ rule. What's more, fuzzy production rule have "AND rule" and "OR rule". Their representations are as follows:

$$R: \text{IF } A_1 \text{ AND } A_2 \ldots \text{ AND } A_n \text{ THEN } B \ (cf = \sigma_k)$$
a. AND rule
$$R: \text{IF } A_1 \text{ OR } A_2 \ldots \text{ OR } A_n \text{ THEN } B \ (cf = \sigma_k)$$
b. OR rule

The graphical representation is shown in Figure 1.

In Figure 1(a), suppose the validity of A_k is $M(A_k)$, the degree of confidence of the rule is σ_k, the threshold value is θ. When $M(A_k) * \sigma_k > B_k$, the transition is active. In Figure 1(b),

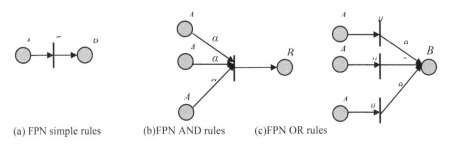

(a) FPN simple rules   (b)FPN AND rules   (c)FPN OR rules

Figure 1.   FPN production rules.

$\alpha = (\alpha_1, \alpha_1, \ldots \alpha_n)$ is the weight vector, the threshold value is θ, when $\min(M(A_i)^* \alpha_i > \theta$, $i = 1, 2, \ldots n$, the transition is active. In Figure 1(c), $\theta = (\theta_1, \theta_2, \ldots \theta_n)$ is the weight vector, when $\{M(A_1)^* \alpha_1 > \theta_1 \cup M(A_1) \alpha_1 > \theta_1 \cup M(A_n)^* \alpha_n > \theta_{n,}\} \neq 0$, the transition is active.

2.3 Establishment of the fuzzy Petri net

The following is a general algorithm which is used to transform a set of fuzzy production rule into fuzzy Petri nets

1. Each production rule corresponding to a transition in Petri net T.
2. Each variable in production rules corresponding to a place P.
3. If the variable V belongs to the production rules R (t, V), it is put into the figure drawing as a flow relationship.
4. The end.

According to this algorithm, the fuzzy Petri net of rolling steel surface defect recognition is established.

2.4 Fuzzy Petri net reasoning

There are three kinds of fuzzy reasoning algorithm of fuzzy Petri net.

1. Forward reasoning. Its general form is as follows:

A_{11} AND $A_{12} \ldots$ AND $A_{1n} \rightarrow B_1$

......

A_{n1} AND $A_{12} \ldots$ AND $A_{1n} \rightarrow B_1$
and A_1^* AND $A_2^* \ldots$ AND A_m^* is known
Solve B^*.

The forward reasoning start with the known phenomenon and find a before rule match with the fact in the rule base. When the rule conditions part match the known facts, the rules listed in the candidate rules queue. Select a rule to reasoning through conflict reduction strategy, put the reasoning conclusion as the reasoning result and continue until no rule can be matched. Then the reasoning is over. This reasoning algorithm can provide users with explanation, reasoning is more intuitive, but efficiency is not high.

2. Backward reasoning. The general form is:

A_{11} AND $A_{12} \ldots$ AND $A_{1n} \rightarrow B_1$

......

A_{n1} AND $A_{12} \ldots$ AND $A_{1n} \rightarrow B_1$
B^* is known
Solve A_1^* AND $A_2^* \ldots$ AND A_m^*.

Table 1. Fuzzy Petri net and the corresponding relation of fuzzy production rule.

Fuzzy Petri Net	Fuzzy production rule
Transition	Rule
Net	Rule base
Place	Proposition
Lable set	Proposition set
Transition active	Rule application
Credibility of transition	Credibility of rule

Backward reasoning first assumes a reason, then start from the assumption to find evidence that is in favor of this hypothesis. When the evidence is found, the reasoning ends. This kind of reasoning can provide users with good explanation but reasoning process is blind, efficiency is not high.

3. Hybrid reasoning. Such reasoning method combine the forward reasoning and backward reasoning together, and have advantages of the two. The efficiency of reasoning is improved significantly. First, begin with the phenomenon, then backward reason from the reasons and find the sustaining evidence. This reasoning method is in accord with thinking habit of the person, process is easier to understand. This article uses the hybrid reasoning method.

3 THE APPLICATION OF FUZZY PETRI NET IN THE DIAGNOSIS OF ROLLING QUALITY DEFECT

The following is the formation of surface quality defects of steel rolling "scrap" diagnosis knowledge rules, the credibility cf of the fuzzy production rule is given by fuzzy production rule, the credibility of the cf by domain experts according to the experience is given by domain experts according to the experience:

R_1: IF problems in ingredients quality THEN Perforation $cf = 0.7$
R_2: IF stretch steel seriously THEN Perforation $cf = 0.9$
R_3: IF leg and waist stretch not balance THEN Perforation $cf = 0.88$
R_4: IF problems in ingredients quality THEN Cracked edge $cf = 0.65$
R_5: IF over burning THEN Cracked edge $cf = 0.83$
R_6: IF low finishing temperature then Cracked edge $cf = 0.92$
R_7: IF large temperature difference in billet heating THEN Leg wave $cf = 0.89$
R_8: IF the leg bigger than the waist THEN Leg wave $cf = 0.9$
R_9: IF low temperature of the tail THEN Leg wave $cf = 0.92$
R_{10}: IF straightening temperature low THEN Web waves $cf = 0.85$
R_{11}: IF leg and waist stretch not balance THEN Web waves $cf = 0.8$
R_{12}: IF web waves THEN the inclined flange $cf = 0.9$
R_{13}: IF straightening adjustment error THEN The inclined flange $cf = 0.89$
R_{14}: IF the straightening roller crack THEN the inclined flange $cf = 0.7$
R_{15}: IF perforation THEN Scrap $cf = 0.93$
R_{16}: IF cracked edge THEN Scrap $cf = 0.95$
R_{17}: IF leg wave THEN Scrap $cf = 0.90$
R_{18}: IF the inclined flange THEN Scrap $cf = 0.95$

The rules of FPN is shown in Figure 2.

P1 ingredients quality P2 stretch steel seriously P3 not balance in leg and waist stretching P4 perforation P5 over burning P6 low finishing temperature P7 cracked edge P8 large temperature difference in billet heating P9 the leg bigger than the waist P10 low temperature of the tail P11 leg wave P12 straightening temperature low P13 not balance in leg and waist stretching P14 web waves P15 straightening adjustment error P16 the straightening roller crack P17 the inclined flange P18 scrap.

Table 2. h-beam rolling scrap defect name and the reason.

Defect name	Reason
Perforation	Ingredients quality, stretch steel seriously, not balance in leg and waist stretching
Cracked edge	Ingredients quality, over burning, low finishing temperature
Leg wave	Large temperature difference in billet heating, the leg bigger than the waist, low temperature of the tail
The inclined flange	Web waves, straightening adjustment error, the straightening roller crack
Web waves	Straightening temperature low, not balance in leg and waist stretching

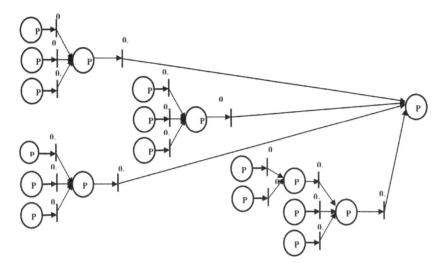

Figure 2.　FPN model of quality defects of steel rolling scrap.

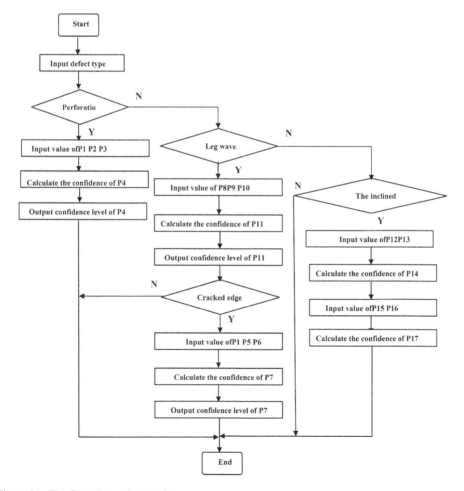

Figure 3.　The flow chart of reasoning.

In the reasoning process, start reasoning from the start place and object place at the same time using forward and backward reasoning algorithm. Find the evidence for the intermediate output. The reasoning process is shown in Figure 3. Specific reasoning process is as follows:

Make judgment to find the actual defect type belongs to which defect type according to the actual situation and find corresponding places.

1. Find the various causes and corresponding places for the defect types.
2. Blur the actual data which is the causes of defects to get the input values for the corresponding places.
3. Using the activation conditions for the transitions $M(A_i) * \alpha_i > \theta$ to calculate the credibility of each reason which can cause the defect. Choose 0.5 as the threshold value.
4. Choose the maximum value of credibility as the credibility of next place.
5. The rest can be done in the same manner, until one gets the credibility value of the corresponding defects.

4 THE EXPERIMENTAL RESULTS

Using VC software to write FPN reasoning process program. Input four different defect types and the value of corresponding conditions which is fuzzily processed. Calculate the causes of the defect and credibility through reasoning programming. Then validate and compare with the actual situation.

In order to validate the rationality of the model, get the input fuzzy processing value using four set data in the actual production conditions in a steel plant. Four cases obtained from reasoning results are in line with the actual situation, and proved the correctness of the reasoning process.

Table 3. The experimental results.

Defct type	Input	Output: *cf*	Cause analysis	The actual situation
Perforation	P1 = 0.9, P2 = 0.88, P3 = 0.92	P14 = 0.63, P24 = 0.792, P34 = 0.8096, P4 = MAX = 0.8096	Not balance in leg and waist stretching cf = 0.8096	Not balance in leg and waist stretching
Cracked edge	P1 = 0.88, P5 = 0.4, P6 = 0.67	P17 = 0.572, P57 = 0, P67 = 0.6164, P7 = MAX = 0.6164	Low finishing temperature cf = 0.6164	Low finishing temperature; over burning
Leg wave	P8 = 0.7, P9 = 0.8, P10 = 0.65	P811 = 0.623, P911 = 0.72, P1011 = 0.598 P11 = MAX = 0.72	The leg bigger than the waist cf = 0.72	The leg bigger than the waist
The inclined flange	P12 = 0.9, P13 = 0.56, P15 = 0.9, P16 = 0.75	P1214 = 0.765 P1314 = 0.448 P1417 = 0.6885 P1517 = 0.801 P1617 = 0.525 P17 = MAX = 0.801	Straightening adjustment error cf = 0.801	Straightening adjustment error

4 CONCLUSION

The fuzzy Petri nets is used in steel rolling defect recognition and establish fuzzy Petri net model for defects of steel rolling. The experimental simulation results verify the accuracy of this method and provide a new way for the defect of H-beam diagnosis. The limitation of this method lies in only one reason which can reason the largest value of confidence be given in the result. The following research will focus on how to put the variety of causes that have larger probability in diagnosis.

REFERENCES

Chongyi Yuan. 2005. *Modeling mechanism and applications of Petri nets*. Beijing: Publishing House of Electronics Industry.
Gao Haijian, Xi Tie, Sun Feifei. 2002. Hot-rolled h-beam rolling and its engineering applications. *Construction steel structure development* 4(1):32–35.
He Xingui. 1994. Fuzzy Petri nets. *Chinese J. Computers.* 17(12):946–950.
Hua Jianxin. 1998. The application of artificial neural network in flatness defect recognition. *New technology digestion and the application* 3:45–47.
Jiang Haitao, Tong Lizhen, Wang Junsheng, Zhang Tao. 2012. Research on large h-beam web fold defects of the rolling mill of Laigang. *Laigang Technology* 5:40–43.
Jiao Hong, Fu Huili, Luan Zhaoliang, Zhang Yu. 2009. Research and control of the surface quality defects of hot-rolled h-beam. *China academic journal electronic publishing house*. China: Beijing.
Li Hongji. 2005. *Fuzzy mathematical basis and practical algorithm*. Beijing: Beijing Science Press.
Liang Wenhao. 2010. Research on Recognition of Steel Strip Surface Defects Based On Machine Vision. Wuhan: Wuhan university of science and technology.
Xiang Yang, Chen Li, Zhang Xiaolong. 2012. Research on Recognition of Strip Steel Surface Defect Based on Support Vector Machine. *Industrial control computer* 25(8):99–101.
Xu Hui. 2012. *Blast Furnace Fault Diagnosis Based on Fuzzy Neural Network*. Wuhan: Wuhan university of science and technology.

Manufacturing Engineering and Intelligent Materials – Lu & Abu Bakar (Eds)
© 2015 Taylor & Francis Group, London, ISBN 978-1-138-02832-6

UV laser coaxial vision system based on wavelet transform

L.H. Chen & Y.Q. Zhou
College of Digital Media, Shenzhen Institute of Information Technology, Shenzhen, China

S.Y. Zhao, H. Zhong & S.L. Shang
Shenzhen Hymson Laser Technology Co. Ltd., Shenzhen, China

ABSTRACT: UV laser micron machining is one of the development trends of laser machining. Through the introduction of air cooling system of UV laser generator and its laser cavity, the UV laser coaxial vision system is studied and developed, to make laser structure in compact and machining in stable. Based on wavelet transform, the filter of noise CCD image is developed and the algorithm to calculate outline checkpoints of real CCD image of workpiece is studied. The machining case shows that the coaxial vision system can meet the need of intelligent micron machining such as laser marking and laser cutting etc., and improve accuracy of UV laser micron machining.

1 INTRODUCTION

UV laser machining is a cold-etching process instead of the process of hot melting, to which infrared laser machining relates, and makes machining size smaller and more accurate (Andrew et al. 2014, Wu et al. 2012). UV laser is thus widely used in the field of accurate and fine machining, such as accurate laser marking, cutting, drilling, scribing, welding, electrical component enclosing etc. (Liu et al. 2010).

In order for UV laser to machine parts in high accuracy, vision system is often equipped to inspect machining size on line for the high intelligent machining effect in auto-optimization. However, most of UV laser generators use water-cooling system. It determines its vision system and often uses outerhanging structure in huge size, and makes high machining cost. The coaxial vision system has been started to be used in infrared laser cutting machine (Zhang et al. 2007), but it is not found in UV laser machining device yet, since the water-cooling system occupies too much space, the coaxial vision system could not be stalled inside any more.

In this paper, an air-cooling UV laser system with plug-in coaxial vision is developed based on wavelet transform, to achieve the performance of high machining accuracy with low machining cost.

2 AIR COOLING UV LASER GENERATOR

Current end-pump air-cooling laser generator is only realized in infrared stable output. The stability of power output gets down if retrofitting as frequency doubling in same structure. The first root cause is that the frequency doubling generator requires optical oscillation cavity higher mechanical accuracy than infrared, but the air-cooling heat sink often makes the cavity deformed. Second, although high accurate cooling sheet is used to control the temperature, the doubling efficiency of frequency doubling generator is still closely related with temperature, the heat source makes total cavity too hot, the generator changes the temperature of doubling crystals and affects the exchange efficiency of doubling frequency, then makes

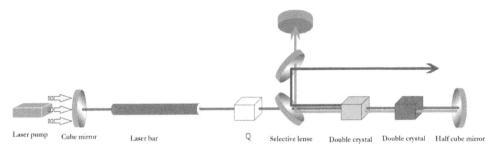

Laser pump   Cube mirror   Laser bar   Q   Selective lense   Double crystal Double crystal Half cube mirror

Figure 1. UV laser cavity covered by air cooling system.

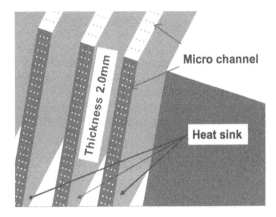

Figure 2. Heat sink with micro channels.

power output of doubling unstable. Such instability is more obvious in frequency tripling (UV laser) than in frequency doubling (green laser).

Figure 1 shows an UV laser generator with 12 W power, it is covered by a air-cooling heat sink with a series of micro channels (Fig. 2). Also, a strengthened rib which is vertical to heat sink, exists to enhance cavity strength. In another aspect, the main heat source can be spread out in the rear of strengthened rib, the micro-channel can make heat radiating area more efficiently so that the heat of source can be spread over near the cavity, rather than transferring to doubling modules. Therefore, the power output of air-cooling doubling generator keeps stable since the temperature of doubling crystal varies a little.

3 UV LASER COAXIAL VISION SYSTEM

A CCD vision device, which is installed inside UV laser generator, can catch visible image of workpiece to be machined by UV laser. Both visible paths share same lens and galvanometer scanning system with UV laser, but the selective lens identifies the two lights of different wavelengths, the UV laser beam pass through the selective lens (beam splitter plate), the visible light is reflected by the lens to be caught by CCD as an image, which could be post-processed for laser to accurately mark or cut workpiece.

Figure 3 shows the air-cooling UV laser coaxial vision system. The workpiece is put in the position of focus of lens (6), the light source of coaxial vision CCD provides measuring light, through both the refraction offocusing lens and the total refection of X/Y (5) galvanometer scanning system, the light of workpiece becomes parallel and incident to the splitter (4), then totally reflects to coaxial version CCD (3) to be a clear image which is captured and studied by the CCD monitor system.

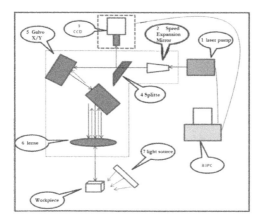

Figure 3. UV laser co-axial vision system.

The developed machining software based on wavelet transform of noise filtering and outline inspecting compares the captured workpiece image to the standard model, in order to verify whether the captured workpiece is the just one which should be machined at the current work station. After the verification, it acquires position data of workpiece, save all related coordinates, and trigger UV laser generator (1) to emit UV laser. The laser beam passes through the speed expanding mirror (2) and splitter (4), then diverges rapidly until it enters one or two focusing lenses. The beam, now converging, passes through and is directed by a set of X and Y mirrors moved by X/Y (5) galvanometer scanners. The orthogonal arrangement of the X and Y mirrors direct the beam down towards and over the length and width of the working field to meet the needs of accuracy machining according to the coordinates provided by coaxial vision system.

4 COAXIAL VERSION SYSTEM BASED ON WAVELET TRANSFORM

Wavelet transform has been studied from 1980s (Canny 1986), and widely developed in the field of image and signal process recently (Donoho 1995, Filipe et al. 2012). CCD image signal is a 2D function, if $f(x, y)$ is an image in wavelet analysis, $L^2(R^2)$ presents square integrable function, that is

$$f(x,y) \in L^2(R^2) \Leftrightarrow \int_{-\infty}^{+\infty} \int_{-\infty}^{+\infty} |f(x,y)|^2 \, dxdy < \infty \qquad (1)$$

The key functions of intelligence control of coaxial version system are pertaining to both noise image $f_i(x, y)$ filtering and outline inspection of real image which is used to check whether the sizes of part machined by laser meet the requirements.

4.1 *Image noise filtering based on Wavelet transform*

No matter what coaxial version CCD or outerhanging CCD, their image signals contain noise regarding to photon, trap, reset and KTC etc. The sampling noise of image contains gauss noise; band limited white noise, quantization noise and red noise etc. There is no laser interference noise since UV laser is power off when coaxial version system is working. The noise image shall be eliminated, otherwise the laser machining software mistakes noise image for the area to be machined.

According to Mallat algorithm of biorthogonal multi-resolution analysis, $\{c^k, d_k^1, d_k^2, d_k^3,\}$ is the decomposition threshold sequence of first grade of 2D Wavelet transform, the decomposition algorithm is below,

$$\begin{cases} c_{k;n,m} = \sum_{l,j} h_{l-2n} h_{j-2m} c_{k+1;\,l,j} \\ d^1_{k;n,m} = \sum_{l,j} h_{l-2n} g_{j-2m} c_{k+1;\,l,j} \\ d^2_{k;n,m} = \sum_{l,j} g_{l-2n} h_{j-2m} c_{k+1;\,l,j} \\ d^3_{k;n,m} = \sum_{l,j} g_{l-2n} g_{j-2m} c_{k+1;\,l,j} \end{cases} \tag{2}$$

The filter group of removing noise corresponding to 2D Mallat algorithm is shown as in Figure 4.

4.2 *Outline checkpoint algorithm of vision image based on wavelet transform*

The real image is a 2D function after removing noise:

$$f_0(x,y) \in L^2 \, (R^2) \Leftrightarrow \int_{-\infty}^{+\infty}\int_{-\infty}^{+\infty} |f_0(x,y)|^2 \, dxdy < \infty \tag{3}$$

There is a Wavelet, respectively, along X and Y direction. The solution of partial derivative against $f_0(x, y)$ function along X and Y direction is below,

$$\psi^1(x,y) = \frac{\partial \theta(x,y)}{\partial x} \text{ and } \psi^2(x,y) = \frac{\partial \theta(x,y)}{\partial y} \tag{4}$$

The Wavelet transform could be exported from $\psi^1 (x,y)$ and $\psi^2 (x,y)$:

$$\begin{pmatrix} W^1_{2^j} f_0(x,y) \\ W^2_{2^j} f_0(x,y) \end{pmatrix} = 2^j \begin{pmatrix} \frac{\partial}{\partial x}(f_0 * \theta_{2^j})(x,y) \\ \frac{\partial}{\partial y}(f_0 * \theta_{2^j})(x,y) \end{pmatrix} = 2^j \vec{\nabla}(f_0 * \theta_{2^j})(x,y) \tag{5}$$

Equation (5) indicates that the partial derivative along X and Y direction is the Wavelet transform along row and column after $f_0(x,y)$ image becomes smooth. The two components of the vector of Wavelet transform are related to the two components of gradient vector $2^j \vec{\nabla}(f_0 * \theta_{2^j})(x,y)$. Regarding the vector scale, the module of the gradient vector is below,

$$M_{2^j} f_0(x,y) = \sqrt{|W^1_{2^j} f_0(x,y)|^2 + |W^2_{2^j} f_0(x,y)|^2} \tag{6}$$

The phase angle of gradient vector in horizontal direction is,

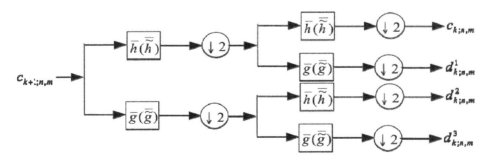

Figure 4.   Filter of removing noise image.

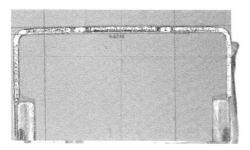

Figure 5. Original CCD image of workpiece.

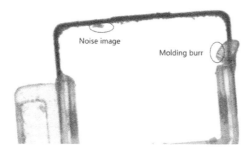

Figure 6. Workpiece by UV laser cut.

$$A_{2^j} f_0(x,y) = \arg(W^1_{2^j} f_0(x,y) + i W^2_{2^j} f_0(x,y)) \tag{7}$$

The break point of $f_0 * \theta_{2^j}(x,y)$ is the point (x, y) when $M_{2^j} f_0(x, y)$ is maximum along gradient direction, it is the outline checkpoint coordinates. Drawing each checkponit (x_i, y_i), the part outline to be machined is figured out. Making a comparison to the standard model, the unnecessary outline of part could be removed by UV laser.

4.3 Case study

Figure 5 is a part designed by a famous communication company; it is a plastic molded part with an inserted metal frame. Around metal enclosure, there is micro burr which impact assembly accuracy. The burr has to be removed by manual in low efficiency and poor quality before. By using UV laser cutting with coaxial vision system, the cutting efficiency (1.5 s/piece) is much higher than manual, the yield is over 99% without obvious heat effect (Fig. 6).

5 CONCLUSION

The CCD imaging system shares same light components and light path with UV laser to become coaxial vision system. Based on Wavelet transform, UV laser cutting system can intelligently machine workpiece in high efficiency and control the machining tolerance under 0.01 mm. Further study will focus on updating Wavelet transform algorithm for laser cutting system to distinguish much noise and recognize smaller burr by vision.

ACKNOWLEDGEMENT

This research was financially supported by Shenzhen New Pattern Industry Development Foundation in China (CYZZ20130402111433136), Shenzhen Science and Technology

Research foundation in China (CXZZ20130321145942564), and Shenzhen Science Plan Foundation in China (CXZZ20140416104115582) and Guangdong Natural Science Foundation in China (S2013010015726). The authors are greatly indebted to many field engineers for helping conducting lots of in-situ test.

REFERENCES

Andrew Dunn, Jesper V. Carstensen, KrystianL. Wlodarczyk. 2014. Nanosecond laser texturing for high friction applications. *Optics and Lasers in Engineering*. 62, 9–16.

Canny J. 1986. A computational approach to edge detection. *IEEE Trans. Pattern Anal. Machine Intell*, 8(6), 679–698.

Donoho D.L. 1995. De-noising by soft-thresholding. *IEEE Transactions on Information Theory*. 41(3), 613–627.

Filipe A. Apolonio, Daniel H.T, Franco. 2012. A note on directional wavelet transform: Distribution boundary values and analytic wavefront sets. *International Journal of Mathematics and Mathematical Science*. Article ID 758694.

Liu Qiang, Yan Xingpeng, Chen Hailong. 2010. New Progress In High-Power All-Solid_State Ultraviolet Laser. *Chinese Journal of Lasers*. 37(9). 2289–2298.

Wu Yuwen, Guo Liang, Zhang Qingmao. 2012. Study on process of marking two-dimentional codes on air circuit breakers by UV laser. *High Power Laser And Particle Beams*. 24(6), 1329–1334.

Zhang Yongqiang, Chen Wuzhu, Zhang Xudong. 2007. Self-optimizing control for laser cutting quality based on coaxial vision. *Transactions of the china Welding Institution*. 28(4), 58–60.

Manufacturing Engineering and Intelligent Materials – Lu & Abu Bakar (Eds)
© *2015 Taylor & Francis Group, London, ISBN 978-1-138-02832-6*

Research on paths of the in-situ urbanization and the evaluation of the sustainable development of You Yang County in Chongqing

Y.J. Cai
School of Management, Chongqing Jiaotong University, Chongqing, China

ABSTRACT: Now, in-situ urbanization has become one of the main paths in the new type of urbanization in China. This paper mainly studies the in-situ urbanization paths and its sustainable development. This paper establishes the evaluation index system of the sustainable development after doing research on a large number of related literatures and the investigation of You Yang County. This paper selected 6 main factors as the contestant factors and chose 3 factors as input indicators and another as output indicators. Then set up DEA model and use MATLAB to solve it. Finally, analyzing the results and putting forward the related suggestions on the in-situ urbanization path and its sustainable development in the southeast of Chongqing.

Keywords: in-situ urbanization; path; sustainable development; DEA model; Chongqing

1   INTRODUCTION

According to the development statistical bulletin in 2013, China's urbanization rate reached 53.7%, which has surpassed the average urbanization rate of the world. However the "Big city disease" is more and more serious. Jin-ping Xi stressed that the core of the new type of urbanization is people-oriented and improving the quality of development is the key. The new type of urbanization requires the medium cities, the small cities and towns and the new rural community to develop coordinated and promote mutually. It is subnational of the traditional urbanization and attaches great importance to the balance of the economic society, urban and rural. In December 2013, it emphasized that the development of the new urbanization should be based on the original village, aim to improve people's life, and finally realize the in-situ urbanization on the central working conference of urbanization. The in-situ urbanization is the supplement to the concept of traditional urbanization, and it is also one of the new urbanization paths. At present, the research on in-situ urbanization is insufficient, it is worth studying to realize better and faster development of in-situ urbanization. In this paper, we take You Yang autonomous county as an example to discuss the path of in-situ urbanization and do some evaluation for its sustainable development.

2   THE LITERATURE REVIEW

Zhang Yanming inspected the village urbanization in Jiangsu and Zhejiang economic developed area, and on this basis concluded that the development pattern of city urbanization in village edge area has "Industrial concentration type", "Trade market-oriented", "Tourist attractions"—the three typical development models (Zhang 2009). Yang Yiqing pointed out that to promote economic growth by increasing the production efficiency, based on regional advantages and characteristics, to achieve quality and efficient operation of agricultural industry (Yang 2013). Yu Hongsheng put forward three paths in the process of urbanization in China that is through the township and village enterprises, foreign investment and urban agglomerations to promote the development of in-situ urbanization, then analyzed the obstacles of the three paths, and finally pointed out that the new urbanization development process should pay attention to the coordinated development and the use of the city characteristics to promote the development of in-situ urbanisation (Yu 2013). Xuan Chao studied in-situ urbanization pattern in Henan province as a case, put forward the market-oriented, policy leading, people leading these three urbanization modes (Chao 2013).

3 THE CURRENT STATUS OF THE STUDIED AREA

Since 2003, with the rapid advance of new urbanization in Chongqing, the large, medium and small cities and towns system gradually formed. On September 13 to 14, 2013, the concept of the five functional areas of Chongqing was proposed on the basis of "lap wings". It can divide the functional regions scientifically and give a clear functional positioning of the counties. In this paper, the research object, unitary Yang autonomous county is a county in southeast of Chongqing.

By the above SWOT analysis and field studies of unitary Yang, the maybe paths for in-situ urbanization are as follows: Developing tourism; establishing agricultural sightseeing garden to carry out the "rural tourism"; tourism improves the added value of agricultural products so as to increase the proportion of the tertiary industry.

Table 1. The SWOT analysis on the present situation of You Yang County.

S Strengths	W Weakness
Regional strengths: In the suburban of Changsha and Chongqing, so the transportation is convenient; Tourism resources strengths: Several geographical humanities landscape, the Peach Garden is a 5 A class tourist scenic spot; Resources strengths: Water, many varieties of mineral resources;	The natural environment: too much mountain, less ground, the contradiction between human and land, serious soil and water loss, frequent natural disasters; infrastructure: Roads, power grids, water conservancy projects are not perfect; Talent weakness: Low quality of the labor force, talent shortage
O Opportunity	T Threats
Economic development opportunities: Be the road thoroughfare to the southeast of Chongqing, easy to further expand the market; Policy opportunities: "314" the overall deployment	Competitive threats: Zhang Jiajie in the east, the phoenix ancient town in the south. Make its threat has been marginalized

Table 2. The selection index of in-situ urbanization sustainable development.

Target	The first indicators	The second level indicators	The third level indicators
The sustainable development of urbanization	Input	Land resources	Urban construction land area (Square kilometres)
		Capital investment	Investment in fixed assets (hundred million RMB)
		Human capital	Non-agricultural population (ten thousand people)
	Output		Secondary and tertiary industry GDP (hundred million RMB)
			Fiscal revenue, (hundred million RMB)
			Social retail sales of consumer goods (hundred million RMB)

Table 3. The input index of in-situ urbanization sustainable development of unitary Yang County from 2008 to 2013.

Input indicators	Urban construction land area (square kilometres)	Investment in fixed assets (hundred million RMB)	Non-agricultural population (ten thousand people)
2008	6.62	57.71	11.42
2009	7.70	77.67	14.36
2010	10.24	95.27	15.87
2011	10.11	116.32	22.00
2012	9.63	104.64	23.11
2013	10.2	114.92	23.57

Data source: http://youy.cq.gov.cn/ Unitary Yang public information network.

Table 4. The output index of in-situ urbanization sustainable development of unitary Yang county from 2008 to 2013.

Output indicators	Secondary and tertiary industry GDP (hundred million RMB)	Fiscal revenue, (hundred million RMB)	Social retail sales of consumer goods (hundred million RMB)
2008	21.49	2.54	16.89
2009	30.30	3.61	21.21
2010	44.22	7.66	25.15
2011	59.58	12.25	29.71
2012	70.01	20.29	33.96
2013	79.50	23.57	38.39

The data source: http://youy.cq.gov.cn/ Unitary Yang public information network.

4 THE SELECTION OF INDEX AND DATA COLLECTION

About the selection of indicators, this paper reviewed a large number of relevant literatures, and then selected the six indicators about sustainable development of urbanization.

According to the input and output indicators in Table 2, this article query related data on unitary Yang public information network. Statistics are shown in Table 3 and Table 4.

5 RESEARCH METHODS

5.1 Introduction of the DEA method

Data Envelopment Analysis (DEA) was proposed by Charnes, Coopor and Rhodes in 1978. The method project every Decision Making Units whose input and output remain unchanged to the production frontier which is determined by means of mathematics programming and statistical data, then compare the extent of deviation of DMU and the production frontier to evaluate the relative effectiveness. According to the characteristics of the DEA model, the workload of data collection and processing will increase when the input and output indicators is excessive, and at the same time, the effectiveness of DMUs will be closer because of the association of quotas. It will affect the results of DEA (Wu 1992). We take the in-situ urbanization of unitary Yang County as a whole system to set up the DEA model— the first three indexes as input index, the others as output index. We can evaluate the sustainable development of in-situ urbanization by analyzing the input and output efficiency of the system. So, it is feasible to apply the DEA method in the evaluation of the sustainable development of in-situ urbanization.

5.2 DEA model

DEA model include different models such as C2R, C2GS2, C2WH, C2W and so on. In this paper, we use DEA method with the Archimedes infinitesimal in the C2R model, this model is:

$$\max h_{j0} = \mu^T Y_0$$

$$s.t. \begin{cases} \mu^T Y_j - \omega^T X_j \leq 0, j = 1, 2, \ldots, n \\ \omega^T X_0 = 1 \\ \omega^T \geq \varepsilon \cdot \hat{e}T \\ \mu^T \geq \varepsilon \cdot eT \end{cases}$$

Note:

$$\hat{e}^T = (1, 1, \cdots, 1) \in E_m$$
$$e^T = (1, 1, \cdots, 1) \in E_s$$

Then convert equivalent form of dual model into linear programming to solve the problem. The model is as follows:

$$\min \theta - \varepsilon \, (\hat{e}^T S^- + e^T S^+)$$

$$s.t. \begin{cases} \sum_{j=1}^{n} X_j \lambda_j + S^- = \theta X_0 \\ \sum_{j=1}^{n} Y_j \lambda_j - S^+ = Y_0 \\ \lambda_j \geq 0, S^-, S^+ \geq 0 \end{cases}$$

Among them, $\hat{e}^T = (1, 1, \cdots, 1) \in E_m, e^T = (1, 1, \cdots, 1) \in E_s, S^- = (S_1^-, S_2^-, \cdots, S_s^-)^T$ are the vectors composed by input corresponding slack variable, $S^+ = (S_1^+, S_2^+, \cdots, S_s^+)^T$ is the vector composed by produced remaining variables.

6 RESULTS ANALYSIS

According to the data collected in Table 3 and Table 4, using MATLAB tool, the computation results are as Table 5. By the MATLAB results, we can see that the index computation result of unitary Yang county in 2013 is $\theta = 1$, $s^- = 0$, $s^+ = 0$, this DMU is effective, and its efficiency is relatively the highest.

The result of 2008–2012 is DEA invalid. We analyze the data, and improve them. The results of 2008 are as follows:

$$\theta = 0.9080$$
$$\lambda_1 = \lambda_2 = \lambda_3 = \lambda_4 = \lambda_5 = 0, \lambda_6 = 0.44$$
$$s_1^- = 1.5237, s_2^- = 1.8430, s_3^- = 0$$
$$s_1^+ = 13.4867, s_2^+ = 7.8298, s_3^+ = 0$$

We should reduce the urban construction land area and the investment in fixed assets investment, at the same time increase the output of the fiscal revenue and the secondary and tertiary industries to make the DMU effective. The calculation results of 2009–2012 analyzed also have the same conclusion.

Table 5. The MATLAB results based on DEA model.

Year	2008	2009	2010	2011	2012	2013
Theta	0.9080	0.9068	0.9730	0.8291	0.9715	1.0000

7 CONCLUSIONS

According to the results of DEA model analysis, in order to promote the sustainable development of above-mentioned possibilities in-situ urbanization path, we should pay attention to improve the resources' efficiency along with the development of the local economy. On the one hand, pay attention to reduce the investment assets resources, on the other hand, pay attention to protect the environment. Don't put the development of economy based on the destruction of the ecological environment.

REFERENCES

Chao Xuan. 2014. The post-crisis era rural in-situ urbanization pattern analysis—take Henan province for example [J]. *Inquiry into Economic Issues*. 2014 (1).
Wu Guangmou. 1992. The relationship between the index characteristics and effectiveness [J]. *Journal of Southeast University*. 2 (5):124–127.
Yu Hongsheng. 2013. China's new urbanization development path selection and restriction factor research [J]. *The new urbanization*. 2013 (06).
Yang Yiqing. 2013. Foreign experience and mode of new urbanization development and China's path to choose [J]. *Research of agricultural modernization*. 2013 (7).
Zhang Yanming. 2009. The research on urbanization development model in urban fringe area village—take Jiangsu and Zhejiang economic developed areas for example [J]. *Journal of Zhejiang Normal University: Natural Sciences*, 2009 (9).

Manufacturing Engineering and Intelligent Materials – Lu & Abu Bakar (Eds)
© *2015 Taylor & Francis Group, London, ISBN 978-1-138-02832-6*

The vibration detection on slewing bearings of portal crane

X.P. Wang & C. Ye
Wuhan University of Technology, Wuhan, China

ABSTRACT: This paper takes the 80T portal crane as an example to show the process of the vibration detection on pillar-type slewing bearings of portal crane. After analyzing the vibration signal, this paper makes evaluation of slewing bearings system technical state and proves that this is of great significance because of its economic and practical application value.

Keywords: vibration detection; slewing bearing; portal crane; analysis

1 INTRODUCTION

The slewing device of portal crane is one of the most important institutions, how to test its running state is a complex engineering work. Equipment vibration signal analysis is one of the most important technical means of mechanical state detection and diagnosis. Portal crane slewing ring used to produce a variety of fault, such as wear and fatigue spilling, crack and deformation defects, which can cause abnormal vibration. The vibration acceleration sensor is retrieving turn bearing vibration signal, and makes in-depth analysis, so that it can evaluate slewing bearings system technical state.

2 DETECTION PROCESS

2.1 *The vibration detection of the upper bearing ring*

2.1.1 *The arrangement of vibration measuring points on the upper bearing ring*
There are 4 vibration measuring points in total, which are located in the axial and radial directions on the right driving pinion of the crane jib and the left wheel under the crane jib, as shown in Figures 1 and 2. The measuring point 1# (No. 1) is located in the axial direction on the right pinion, and the 2# (No. 2) is located in the radial direction on the right pinion, and the 3# (No. 3) is located in the axial direction on the left wheel, and the 4# (No. 4) is located in the radial direction on the left direction.

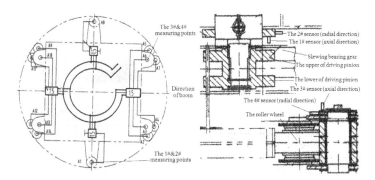

Figure 1. Arrangement of vibration measuring points on the upper bearing ring.

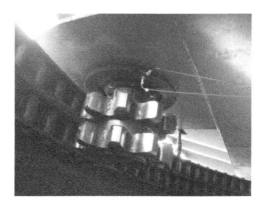

Figure 2. Arrangement of vibration measuring points in reality.

2.1.2 *Signal acquisition*
The data acquisition instrument records the vibration acceleration signals of the measuring points on the portal crane under different conditions.

2.1.3 *Sampling frequency*
The sampling frequency is 750 Hz.

2.1.4 *Testing process*
The testing process of the upper slewing bearing of the 80T portal crane consists of the following two conditions, no-load and load conditions:

1. The crane jib at the minimum amplitude, the crane rotates 2 circles clockwise under no-load condition;
2. The crane jib at the middle amplitude, the lifting hook bears 40t test load, the crane rotates 1 circle clockwise.

In the slewing process of the crane, from time to time, the noise that sounds like the strong discontinuous conflict comes out, especially in the no-load test, which is suspected to be caused by the deformation and releasing of the stress on the steel structure; at the same time, under no-load condition, at least 5 of 10 supporting rollers have no contact with the raceway, and 4 rollers on the counterweight side do the most tight contacting; and in the 40T test load and middle amplitude condition, 3 of 10 rollers still have no contact with the raceway, and two have light contact, and the four rollers under the counterweight of the crane still contact very tightly.

2.1.5 *The analysis method of vibration signal on slewing bearing*
Mainly, the strength analysis and signal frequency spectrum analysis method are used in the slewing bearing vibration.

Because the portal crane slewing bearing raceway has big diameter, whose rotating speed is slow and main feature is based on the vibration created randomly. Therefore, the slewing bearing is divided into several regions to analyze the vibration intensity. The Root-Mean-Square (RMS) and average of the values of 10 peaks appearing firstly in each period of time are calculated on 32 periods of time (16 segments) under different conditions of each measuring point, and then observe the fluctuation trends of the vibration intensity and the energy of the vibration, as shown in Figures 3 and 4.

The important vibration signals are analyzed with the method of frequency spectrum analysis, combined with related special frequency.

During the rotary process, the RMS and average peak values of each measuring point in different sections of the vibration signal fluctuate wildly; the RMS and average peak values of each measuring point under the no-load condition are more than those under

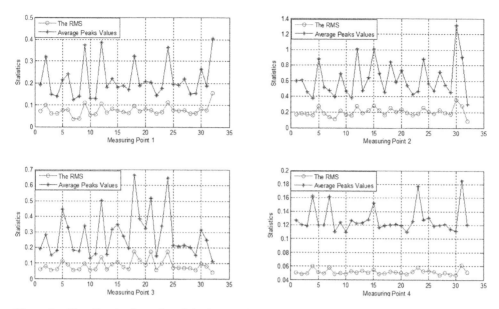

Figure 3. The vibration intensity of the 4 measuring points on the upper slewing ring and 16 segments in no-load and the crane jib at middle amplitude condition.

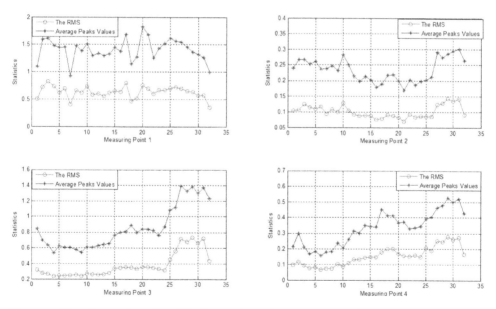

Figure 4. The vibration intensity of the 4 measuring points on the upper slewing ring and 16 segments in the 40T test load and the crane jib at middle amplitude condition.

test load condition; the vibration intensity of the measuring point 2 is biggest, then the measuring point 3; finally, the measuring point 1 and 4 whose vibration intensity are similar and relatively weak. According to the theoretical calculation, the meshing frequency of the driving gear is 0.75 Hz, while the impact frequency is 0.732 Hz, so we can get a preliminary conclusion of big meshing impact caused by the deformation of the raceway.

2.2 *The vibration detection of the lower supporting ring*

2.2.1 *The arrangement of vibration on the lower supporting ring*
There are 4 vibration measuring points in total, arranged on the surface of the rotary supporting ring in the vertical direction, as shown in Figure 5.

2.2.2 *Signal acquisition*
The data acquisition instrument records the vibration acceleration signals of the measuring points on the gantry crane within 2 circles.

2.2.3 *Sampling frequency*
The sampling frequency is 750 Hz.

2.2.4 *Testing process*
The testing process of the lower slewing bearing of the 80T portal crane consists of the following conditions:

1. The crane jib at middle amplitude, loading 40t, clockwise, rotation of 2 laps;
2. The crane jib at middle amplitude, loading 40t, anticlockwise, rotation of 2 laps;
3. The crane jib at middle amplitude, no-load, anticlockwise, rotation of 2 laps.

During the crane slewing, the noise that sounds like strong discontinuous impact comes out from the upper structure, which is suspected to be caused by the deformation and release of the stress on the steel structure.

2.2.5 *The analysis method of vibration signal on the lower supporting ring*
The analysis method of vibration signal on lower supporting ring is same as the upper. But the diameter of lower slewing bearing is small, so the slewing bearing is divided into 8 regions to analyze the vibration intensity. The RMS and average peak values are calculated on 16 periods of time (8 segments) under different conditions of each measuring point, and then observe the fluctuation trends of the vibration intensity and the energy of the vibration, as shown in Figures 6 and 7.

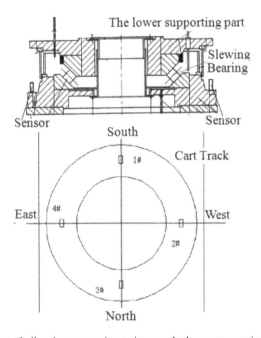

Figure 5.   Arrangement of vibration measuring points on the lower supporting ring.

118

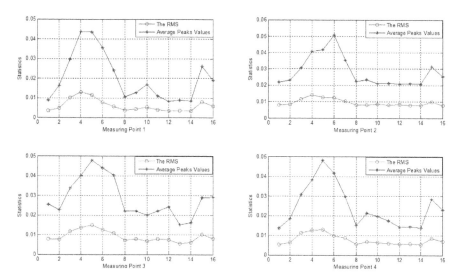

Figure 6.   The vibration intensity of the 4 measuring points on the lower supporting ring and 8 segments in the 40T test load and the crane jib at middle amplitude condition.

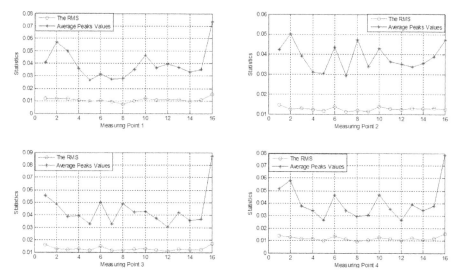

Figure 7.   The vibration intensity of the 4 measuring points on the lower supporting ring and 8 segments in no load and the crane jib at middle amplitude condition.

Compared with the upper bearing, the vibration intensity of each point on the lower supporting ring is far small, and the fluctuations are also much gentler. The RMS and average peak values of each measuring point under no-load condition are bigger than those under load condition.

3   CONCLUSION OF THE VIBRATION ANALYSIS

After the in-depth analysis, we have concluded the problems as follows:

1. When rotating, it has strong intermittent vibration impact, about 5 of 10 supporting rollers have no contact with the raceway, and 4 rollers under the counterweight have the

most tight contact with raceway, which should result from unbalanced force of the steel structure, and the impact caused by the deformation and release of the stress on the steel structure.

2. The teeth of driving gear wear heavily, and large impact is produced by meshing.
3. The center of the whole no-load crane gravity tends to lie to back of the crane, and there are much impact noise when rotating.

According to these problems, we give suggestions as follows:

1. About the 80t crane, repair correction of the supporting roller raceway deformation should be made, or make adjustments of the clearance between the rollers and raceway, to ensure that all rollers and raceway normally contact, eliminating abnormal impact noise.
2. Strengthen the lubrication management, timely on schedule lubrication, and do the spectral analysis and ferrography analysis once half-yearly to see how terrible the wear is developing.
3. Check the gear teeth thickness half-yearly.

On the whole, the portal crane slewing bearings after 15 years using have entered the late tour of duty. There will be more terrible fatigue, wear and cumulative problems. The contact condition between the rollers and raceway needs to be adjusted to eliminate the abnormal sound in the rotation. In the future, to strengthen, lubrication management and testing is extremely necessary.

4 CONCLUSION

In order to know the technical state of the portal crane slewing bearings, we take an 80T portal crane as an example, and do the vibration detection on the slewing bearings including the upper bearing ring and the lower supporting ring. For the upper bearing ring, the RMS and average peak values of each measuring point in different sections of the vibration signal fluctuate wildly, and the meshing frequency of the driving gear is 0.75 Hz, while the impact frequency is 0.732 Hz, so we can get a preliminary conclusion of big meshing impact caused by the deformation of the raceway. For the lower supporting ring, the vibration intensity of each point is far small, and the fluctuations are also much gentler. According to the result of the vibration analysis, we give some suggestions about what to do for daily management. The process has some degree of reference significance and practical application value.

REFERENCES

Gong qunxie, Liu zhijun, Li huabiao. 2003. Vibration and Diagnosis Application Technology Research of Large Rotating Bearing. *Noise and Vibration Control*.

Guo yunfei. 2013. Strength Analysis of Slewing Bearing System of Port Crane. Wuhan University of Technology.

Ji guoyi, Zhao chunsheng. 2010. Summary of Vibration Testing and Analysis. Machine Building & Automation.

Liu zhijun, Chen Jie. 2011. Monitoring and diagnosis technique on slewing bearing. *Modern Manufacturing Engineering*.

Ye yixi, Xiao hanbin. 2004. Extraction of Vibration Signal of Large-scale Bearing Based on Wavelet Analysis. *Journal of Wuhan University of Technology (Transportation Science & Engineering)*.

Manufacturing Engineering and Intelligent Materials – Lu & Abu Bakar (Eds)
© *2015 Taylor & Francis Group, London, ISBN 978-1-138-02832-6*

Synthesis and electrical conductivities performance of SmBa$_{0.5}$Sr$_{0.5}$Co$_{2-x}$Ni$_x$O$_{5+\delta}$ as cathode materials for IT-SOFC

S.X. Yuan, G.Y. Liu, G.Y. Liang, F. Fang, H.L. Chen, Y.K. Liu & X. Kong
College of Science, Honghe University, Mengzi, China

ABSTRACT: SmBa$_{0.5}$Sr$_{0.5}$Co$_{2-x}$Ni$_x$O$_{5+\delta}$ (x = 0, 0.3, 0.6) were synthesized by combined EDTA citrate complexing method. The effect of Ni substitution on the crystal structures, chemical compatibility, and electrical conductivities performance of the layered perovskite SmBa$_{0.5}$Sr$_{0.5}$Co$_{2-x}$Ni$_x$O$_{5+\delta}$ oxides has been investigated. XRD analysis indicates that the SBSCNO and SBSCNO30 are layer perovskites single-phase and the level of substitution with solubility is less than 0.6. There are no observed interfacial reactions or apparent secondary phases between SmBa$_{0.5}$Sr$_{0.5}$Co$_{2-x}$Ni$_x$O$_{5+\delta}$ and SDC electrolyte at 1000°C. The electrical conductivity shows metallic conducting behavior and it decreases with Ni content increasing. The conductivities of all the SmBa$_{0.5}$Sr$_{0.5}$Co$_{2-x}$Ni$_x$O$_{5+\delta}$ samples are more than 100 S/cm, which can meet the demand of the performance as cathodes in IT-SOFC.

1 INTRODUCTION

The world is faced with energy crisis and researchers devote more attention to Solid Oxide Fuel Cells (SOFC) which have clean energy conversion and high-energy efficiency. The operating temperature of the traditional SOFC is above 1000°C which leads to problems such as serious interface reaction, possibility of crack formation due to the mismatch of Thermal Expansion Coefficient (TEC) of the cell component, and so on (Kong & Ding 2011). Reducing the operating temperature to intermediate temperature (500–800°C) would solve these problems. However, the traditional La$_{1-x}$Sr$_x$MnO$_3$ cathode show poor oxide-ion conductivity, inadequate catalytic activity, and much higher Thermal Expansion Coefficient (TEC) compared to those of electrolyte materials at the intermediate operating temperatures (Kim & Manthiram 2009). Developing new mixed conductors with high catalytic activity is an approach for improving stable cathodes at intermediate temperature. Layered perovskite oxides LnBaCo$_2$O$_{5+\delta}$ (Ln = La, Pr, Nd, Sm, Gd and Y) have acquired immense attention as cathodes for IT-SOFC recently. The oxygen vacancies are localized in the LnO layer and show an ordering of the vacancies along the b-axis (Kim & Manthiram 2008, Kim et al. 2009). The diffusion coefficient D and the surface exchange coefficient K in the layered perovskite oxides are much higher than that of ABO$_3$ perovskite. LnBaCo$_2$O$_{5+\delta}$ oxides showed a decrease in TEC, conductivity, and electrochemical performance with decreasing size of the Ln^{3+} ions from Ln = La to Y (Kim & Manthiram 2008). SmBaCo$_2$O$_{5+\delta}$ show good cathode performance but it still suffers from a high TEC value of 21.2 × 10^{-6} within the temperature range of 50–800°C. In this paper, new cathode materials based on SmBa$_{0.5}$Sr$_{0.5}$Co$_2$O$_{5+\delta}$ by Ni substitution on B-site were investigated. Three SmBa$_{0.5}$Sr$_{0.5}$Co$_{2-x}$Ni$_x$O$_{5+\delta}$ compositions with different Ni substitution (x = 0, 0.3, 0.6) were synthesized. We studied the influence of Ni substitution on B-site in terms of structural characteristics, chemical stability with electrolyte Sm$_{0.2}$Ce$_{0.8}$O$_{2-\delta}$ (SDC), and electrical properties.

2 EXPERIMENTAL

The SmBa$_{0.5}$Sr$_{0.5}$Co$_{2-x}$Ni$_x$O$_{5+\delta}$ perovskite oxide powers were synthesized by a combined citrate and EDTA complexion method. Sm$_2$O$_3$ was first dissolved in nitric acid. Co(NO$_3$)$_3 \cdot$ 6H$_2$O,

Ni(NO$_3$)$_3$·6H$_2$O, Ba(NO$_3$)$_2$, Sr(NO$_3$)$_2$, and Sm(NO$_3$)$_3$, all in analytical grades were used as the metal sources. EDTA powder and citric acid were used as the complexion agent. The mole ratio of EDTA and citric acid to metal ions was fixed at 1:1.5:1. EDTA powder was added to ammonia water. Stoichiometric amounts of metal nitrates were dissolved in distilled water and then mixed with the NH$_4$-EDTA solution under stirring condition. The mixed solution then was heated at 80°C with stirring until a gel was formed. The gel was calcined in air at 1100°C for 5 h. NBCO and NBSCO oxide powers with PVA agglomerant were pressed into a strip of 60 mm × 5 mm × 5 mm, and then was heated at 1150°C and 1200°C for 2 h to form specimens. The shrinkage rate of specimens was 10% when heated at 1200°C.

The electrolyte powder of Sm$_{0.2}$Ce$_{0.8}$O$_{2-\delta}$ (SDC) was composed by the conventional solid reaction method with Sm$_2$O$_3$ and CeO$_2$ as initial materials. The as-synthesized SDC powders were calcined in air at 1200°C for 5 h. The chemical reactivity between SmBa$_{0.5}$Sr$_{0.5}$Co$_{2-x}$Ni$_x$O$_{5+\delta}$ cathodes and SDC electrolyte examined by sintering the mixture powders in 50%:50% weight ratio at 1000°C for 2 h.

The phase identification of synthesized SmBa$_{0.5}$Sr$_{0.5}$Co$_{2-x}$Ni$_x$O$_{5+\delta}$ powder and the mixture powders between cathodes and electrolyte were characterized by X-Ray powder Diffraction (XRD) using Cu-Kα radiation (D/MAX-III Japan) in the range of 20°<2θ<80°. The electrical conductivities of SmBa$_{0.5}$Sr$_{0.5}$Co$_{2-x}$Ni$_x$O$_{5+\delta}$ specimen were surveyed with a four-probe dc method using RTS-8 four-point probes meter in the temperature range of 100–800°C.

3 RESULTS AND DISCUSSION

X-ray diffraction patterns of SmBa$_{0.5}$Sr$_{0.5}$Co$_{2-x}$Ni$_x$O$_{5+\delta}$ (x = 0, 0.3, 0.6) (SBSCNO, SBSCNO30, SBSCNO60) sintered at 1100°C for 5 h are shown in Figure 1. The XRD characteristic peaks indicate that SBSCNO and SBSCNO30 are layer perovskites without any secondary phases (Kim & Manthiram 2009, Hu et al. 2013). On the contrary, the secondary phases are observed in the XRD patterns of SBSCNO60 indicating the Ni substitution limit of ~0.6 in SmBa$_{0.5}$Sr$_{0.5}$Co$_{2-x}$Ni$_x$O$_{5+\delta}$ with increasing Ni content. It is observed the peaks of SBSCNO shift toward lower angles which indicates an expansion in the unit cell volume. This is attributed to the difference in B-site substitution ions size between Ni^{3+} (0.056 nm for low spin and 0.060 nm for high spin), Co^{3+} (0.0545 nm for low spin and 0.061 nm for high spin), and Co^{4+} (0.053 nm) (Shannon 1979). The peaks broadening can be found in the XRD patterns with an increasing Ni content. It could be due to the orthorhombic distortion with a decreasing oxygen content. The same phenomenon was observed in NdBaCo$_{2-x}$Ni$_x$O$_{5+\delta}$ (Hu et al. 2012).

In general, the phase reaction between electrode and electrolyte can cause formation of an undesired insulating layer at the interface, which obstructs the oxide–ionic and electronic transport and deteriorate the electrochemical properties (Jun et al. 2014). The chemical

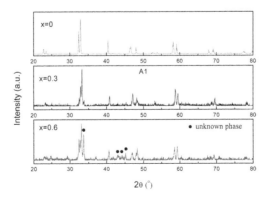

Figure 1. X-ray diffraction patterns of SmBa$_{0.5}$Sr$_{0.5}$Co$_{2-x}$Ni$_x$O$_{5+\delta}$ (SBSCNO, SBSCNO30, SBSCNO60) sintered at 1100°C for 5 h.

reactivity between SBSCNO30, SBSCNO60 cathodes, and SDC electrolyte is therefore examined by sintering the mixture powders in 50%:50% weight ratio at 1000°C for 2 h. The XRD spectra of SBSCNO30–SDC and SBSCNO60–SDC are illustrated in Figure 2. There are no observed interfacial reactions or apparent secondary phases between them which confirms $SmBa_{0.5}Sr_{0.5}Co_{2-x}Ni_xO_{5+\delta}$ are the chemically stable cathode materials for SOFCs based on SDC when the operating temperature is below 1000°C.

There are two primary conduction mechanisms that are relevant to MIEC oxides, electronic, and ionic conduction, both electronic holes and oxygen vacancies present into those materials. Because ionic conductivity is much lower than electronic conductivity, the conductivity measured can be well approximated to the electronic contribution. The temperature dependency of electrical conductivity of $SmBa_{0.5}Sr_{0.5}Co_{2-x}Ni_xO_{5+\delta}$ is illustrated in Figure 3. $SmBa_{0.5}Sr_{0.5}Co_{2-x}Ni_xO_{5+\delta}$ show a decrease in electrical conductivity with an increasing temperature indicating metallic conducting behavior. It can be explained due to the loss of oxygen from the lattice at higher temperatures. The electron holes are the major charge carriers and they are annihilated during the generation of oxygen vacancies (Zhou et al. 2009). The defect reaction (in accordance with K–V notation) can be represented as expressed in Eq. 1.

$$O_O^x = 1/2 O_2(g) + V_O^{\bullet\bullet} + 2e'$$

(1)

where O_O^x and $V_O^{\bullet\bullet}$ represent oxygen vacancies and oxygen ions respectively. Besides the generation of oxygen vacancies perturbs the (Co, Ni)–O–(Co, Ni) periodic potential and leads the carrier localization (Kim & Manthiram 2009). The electrical conductivity decreases with an increasing Ni substituted content at a given temperature which is due to the Ni–O bond

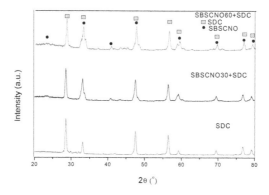

Figure 2.   The XRD spectra of SBSCNO30–SDC and SBSCNO60–SDC.

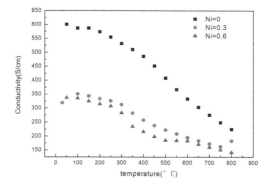

Figure 3.   The temperature dependence of electrical conductivity of $SmBa_{0.5}Sr_{0.5}Co_{2-x}Ni_xO_{5+\delta}$ from 50 to 800°C.

123

which is smaller when compared to the Co—O bond. The decreased bond results in reducing the electron holes. However, the conductivity of all the $SmBa_{0.5}Sr_{0.5}Co_{2-x}Ni_xO_{5+\delta}$ samples are greater than 100 S/cm (McIntosh & Gorte 2004), which can meet the demands of the performance as cathodes in IT-SOFC.

4 CONCLUSIONS

$SmBa_{0.5}Sr_{0.5}Co_{2-x}Ni_xO_{5+\delta}$ (x = 0, 0.3, 0.6) were synthesized by combined EDTA citrate complexing method and investigated for the potential application as novel IT-SOFC cathode material. The structure and electronic conductivity of $SmBa_{0.5}Sr_{0.5}Co_{2-x}Ni_xO_{5+\delta}$ composites have been studied with XRD and four-probe dc method. The XRD characteristic peaks indicate that SBSCNO and SBSCNO30 are layer perovskites single-phase without any secondary phases and is allowed obtaining a level of substitution with solubility less than 0.6. With an increasing Ni content, it is observed that the peaks of SBSCNO shift toward lower angles which indicate an expansion in the unit cell volume. The peaks broadening with an increasing Ni content could be due to the orthorhombic distortion with decreasing oxygen content. $SmBa_{0.5}Sr_{0.5}Co_{2-x}Ni_xO_{5+\delta}$ are the chemically stable based on SDC at an operating temperature of 1000°C. $SmBa_{0.5}Sr_{0.5}Co_{2-x}Ni_xO_{5+\delta}$ show a decrease in electrical conductivity with increasing temperature which is due to the loss of oxygen from the lattice. At a given temperature, the electrical conductivity decreases with an increasing Ni substituted content which causes a decrease in the (Co, Ni)—O—(Co, Ni) covalency bond. The conductivity of all the $SmBa_{0.5}Sr_{0.5}Co_{2-x}Ni_xO_{5+\delta}$ samples are more than 100 S/cm, which ensures their potentials as cathodes for IT-SOFC application.

ACKNOWLEDGMENTS

This work was financially supported by the National Natural Science Foundation of China (No. 51362011 and 51362012), the Chemistry Discipline Master's Site Construction Open Foundation of Honghe University of Yunnan Province (No. HXZ1309) and the Innovation and Entrepreneurship Training Program (DCXL1319).

REFERENCES

Hu, Y., Bogicevic, C., Bouffanais, Y., Giot, M., Hernandez, O. & Dezanneau, G. 2013. Synthesis, physicalechemical characterization and electrochemical performance of GdBaCo2–xNixO5+δ (x = 0–0.8) as cathode materials for IT-SOFC application. J. *Power Sources* 242:50–56.

Jun, A., Lim, T.H., Shin, J.Y. & Kim, G. 2014. Electrochemical properties of B-site Ni doped layered perovskite cathodes for IT-SOFCs. Int J. *Hydrogen Energy* 39:20791–20798.

Kong, X. & Ding, X.F. 2011. Novel layered perovskite SmBaCu2O5+δ as a potential cathode for intermediate temperature solid oxide fuel cells. Int J. *Hydrogen Energy* 36:15715–15721.

Kim, J.-H. & Manthiram, A. 2009. Layered NdBaCo2–xNixO5+δ perovskite oxides as cathodes for intermediate temperature solid oxide fuel cells. *Electrochim. Acta* 54:7551–7557.

Kim, J.-H. & Manthiram, A. 2008. LnBaCo2O5+δ oxides as cathodes for intermediate temperature solid oxide fuel cells. J *Electrochem Soc* 155:385–390.

Kim, J.-H., Mogni, L., Prado, F., Caneiro, A., Alonso, J.A. & Manthiram, A. 2009. High temperature crystal chemistry and oxygen permeation properties of the mixed ioniceelectronic conductors LnBaCo2 O5+δ (Ln = Lanthanide). J *Electrochem Soc* 156:1376–1382.

McIntosh, S. & Gorte, R.J. 2004. Direct hydrocarbon solid oxide fuel cells. *Chem Rev* 104:4845–4865.

Shannon, R.D. 1979. Revised effective ionic radii and systematic studies of interatomic distances in halides and chalcogenides. *Acta Cryst* A32:751–767.

Zhou, L., Shen, J.C., He, B.B., Chen, F.L. & Xia, C.R. 2009. Synthesis, characterization and evaluation of PrBaCo2–xFexO5+δ dascathodes for intermediate-temperature solid oxide fuel cells. Int J *Hydrogen Energy* 34:2416–2420.

Manufacturing Engineering and Intelligent Materials – Lu & Abu Bakar (Eds)
© *2015 Taylor & Francis Group, London, ISBN 978-1-138-02832-6*

Preparation and optical properties of porous alumina thin films

S.M. Yang, J.J. Gu & Y.K. Qi
Department of Physics, Hebei Normal University for Nationalities, Chengde, Hebei, China

ABSTRACT: The alumina thin films with brilliant structural colors have been fabricated by means of electrochemical oxidation in acid electrolyte. The relationship of the structural colors with the anodization time and the angle of incidence of illuminating light are discussed. The microstructures of the films have been characterized as well. The effective refractive index of film is calculated according to Maxwell-Garnett theory. The maximum reflection wavelength of films is calculated by Bragg's equation, and the generation mechanism structural color has also been discussed. In the end, organic-assist covering method has been applied to obtain the intricate multicolor pattern on the same alumina thin film. This thin film with brilliant colors may have potentials in color displays, decoration and anti-counterfeiting technology.

Keywords: alumina films; porous material; structure color; optical properties

1 INTRODUCTION

Structural color universally exists in natural world, for instance as that of feather of bird (Kolle M et al. 2010, Noh H et al. 2010) and wings of insects (Rassart et al. 2009, Parker et al. 2003). Recently, the manmade structural color systems have attracted wide attention, for example, multilayer structure (Philips & Bleikolm 1996), orderly porous surface structure (Liu et al. 2010).

Porous Anodic Alumina (PAA) membranes have attracted considerable attention due to the relatively simple equipment and technology required to produce them, the ease of controlling the process, and the many potential applications. Therefore, many researchers have focused efforts on PAA fabrication techniques and the properties of the membranes, including the optical properties (Losic et al. 2009, Li et al. 2010, Kustandi et al. 2010, Thompson et al. 2005, Stojadinovic et al. 2009). The research about the structural color of PAA thin film has already begun with the progress in the study of photonic crystal. In 1969, Diggle et al. (Kolle M et al. 2010, Noh H et al. 2010) reported that within the range of visible light, alumina thin films show bright color when their thickness is less than 1μm with the interference of light. In 2007, Wang et al. (Wang et al. 2007) reported that they have successfully prepared alumina thin film with comparatively high color saturation by depositing carbon nano-tubes in hole of alumina with CVD technique. Shortly after that, in 2010, Zhao et al (Zhao et al. 2010) realized the subtle control of the colors of alumina thin film by depositing carbon nano-tubes. In 2011, Sun et al (Xu et al. 2012) prepared alumina thin film with iridescence, using multiple oxidation method.

This paper has concentrated on the preparation technique of single structural color of PAA thin film by applying anodization method; the theoretical explanation and support have also been given. This research is significant in the bionic study of structural color of birds' feather. Since structural color has the merits as being environmentally friendly, never fades in color and has the property of rainbow effect, it may have wide potentials in color displays, decoration and anti-counterfeiting technology.

2 EXPERIMENT

High-purity aluminum foils (99.999%) were annealed at 400°C for 2 hours in an annealing furnace, the annealed Al foils were electropolished in a mixture of ethanol and perchloric acid (ratio by volume 4:1) for 5 min to smoothen the surface. After electropolishing, the foils were cleaned in acetone and deionized water and dried. After the anodization at the voltage of 40 V in oxalic acid electrolyte and constant temperature 5°C, alumina thin films have been prepared at the different oxidation time of 70 s, 80 s, 90 s, 100 s, 105 s, 110 s.

The structural and optical features of alumina thin film were characterized by an optical digital camera (Canon-EOS 600D) and field-emission scanning electron microscopy (SEM, Hitachi S-4800).

3 RESULT AND DISCUSSION

Figure 1 shows the structural colors of the six samples (with the angle of incidence of 20°) of the different oxidation time of 70 s, 80 s, 90 s, 98 s, 100 s, 110 s under the voltage of 40 V. It shows that the structural color turns from purple to red gradually with the increasing oxidation time.

To analysis the reasons for the above phenomenon, we give the FE-SEM images of the PAA films in Figure 2. It can be seen from the images that the forming of pores of PAA thin film is only at beginning stage, the influence of pores at their forming stage can be nearly neglected. Effective refractive index of PAA thin film is equal with that of pure alumina thin film. The average effective refractive index of PAA thin film is about 1.65 according to the calculation. The thickness of PAA thin film is about 290 nm according to the cross section image. The growth rate of PAA thin film in the process of oxidation is about 2.90 nm/s. The corresponding thickness of the PAA thin films prepared at the oxidation time from 70–110 s

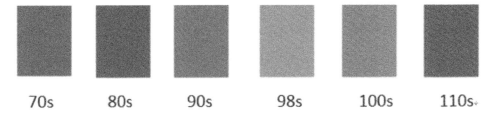

Figure 1. The structural colors of samples under the oxidation voltage of 40 V, oxidation time of 70 s, 80 s, 90 s, 98 s,100 s, 110 s.

Figure 2. FE-SEM images of an PAA thin film anodized for 100 s (surface image and cross section image).

are shown in Table 1. t is oxidation time and d is the thickness of film. Considering that phase difference (half-wave loss) exists between the reflected lights of upper surface and lower surface, the Bragg's equation can be written as:

In this equation λ is refraction wavelength; is refraction angle; m the order of the interference. The order of the interference can be estimated through film.

Thickness and the corresponding maximum wavelength obtained by the experiment; here, the order of the interference was one. The maximum reflective wavelengths of this series of sample are illustrated in Table 1. The data of the Table 1 indicates that the thickness of thin film increases with the increasing oxidation time, and the interference wavelength of reflective lights also increases. The calculation results of the wavelength according to Bragg equation are consistent with the colors of samples observed.

When we observe the color changes of samples in the natural light, we have found that the colors of samples change with the changing viewing angle, as shown in Figure 3. From Figure 3, we see that when viewing angle increases from 20° to 45°, the color of thin film turns from green to purple. To analyze the physical mechanism of this phenomenon, we have calculated the maximum reflective wavelengths (λmax) of corresponding thin films of different viewing angles by Bragg formula. The result is shown in Table 2, which is consistent with the images of digital photos. This indicates that this change of colors also obeys Bragg formula.

From the experimental results above, we know that the different colors can be achieved by changing the thickness of PAA thin film. We have also considered if the changing color could be achieved at a same PAA thin film, thus the ideal colorful patterns could be achieved for the practical needs. If we can do this, then the colorful PAA thin film will be very useful in decoration and anti-counterfeiting technology. Following the above idea, we have prepared the different colorful patterns (as shown in Fig. 4) with the organic-assist covering method. Figure 4 shows the patterns of the same PAA thin film being prepared at different areas of different oxidation time.

This indicates that the colorful thin film can be achieved by controlling the oxidation time to the same aluminum foil, and our idea is proved to be correct. The steps are as follow: firstly, we cover parts of surface of PAA film by printing ink with colorful patterns designed earlier, and then we anodize the other parts and repeat the process until the needed pattern is finished. In the end, we cleanse off the printing ink with acetone. The result above indicates

Table 1. The parameters of samples at the oxidation voltage of 100 V, oxidation time of 70 s, 80 s, 90 s, 98 s, 100 s, 110 s.

t (s)	70	80	90	98	100	110
d (nm)	203	232	261	284	290	319
λ (nm)	420	480	540	588	600	660
Colors of PAA thin film	Purple	Blue	Green	Yellow	Orange	Red

Figure 3. Photographs of the PAA films from different incident angles from 20° to 45°.

Table 2. The character of PAA thin films from different incident angles from 20° to 45°.

t (s)	90		
d (nm)	261		
Observed angles (°)	20	35	45
λ (nm)	540	470	408
Colors of PAA thin film	Green	Blue	Purple

Figure 4. The patterns of the PAA thin films different colors from different incident angles.

that with the organic-assisted covering method for controlling the thickness of PAA thin film is a simple and practical approach to realize the controlling of colors. And this will be no doubt useful for its application in anti-counterfeiting and decoration material.

This paper has described the process of the one-step preparation of alumina thin film with structural colors, which can be controlled by oxidation time. On this basis, the color's change can be realized at the same alumina thin film for achieving the ideal colorful pattern. This paper also discusses the microstructure, optical features and forming mechanism of alumina thin film with structural color. This is significant in the bionic study of structural color of birds' feather. Since structural color has the merits as being environmentally friendly, never fades in color and has the property of rainbow effect, it may have wide potentials in color displays, decoration and anti-counterfeiting technology.

REFERENCES

Diggle J.W., Downie T.C., Goulding C.W. 1969 Chen. Rev. 69 365.
Kolle M., Pedro M.S., Maik R.J.S., et al. 2010. [J]. *NaturNanotechnology*, 5:511–515.
Kustandi T.S., W.W. Loh, H. Gao, H.Y. Low, *ACS Nano 4* (2010) 2561–2568.
Liu X.Y., Zhu S.M., Zhang D., et al. 2010. [J]. *Mater. Lett.* 64:2745–2747.
Losic D., Lillo M., D. Losic Jr., *Small 5* (2009)1392–1397.
Li Y, Ling Z.Y., Chen S.S., Hu X., He X.H., Chem. Commun. 46 (2010) 309–311.
Noh H., Liew S.F., Saranathan V., et al. 2010. [J]. *Adv. Mater.* 22:2871–2880.
Parker A.R., Welch V.L., Driver D., et al. 2003. [J]. *Nature*, 426:786–787.
Philips R.W., Bleikolm A.F. 1996. [J]. *Appl. Opt.* 35:5529–5534.
Qin Xu, Yu-Hua Yang, Li-Hu Liu. 2012 *Journal of The Electrochemical Society*. 159 (1) C25.
Rassart M., Simonis P., Bay A., et al. 2009. [J]. *Phys. Rev. E*, 80:031910.
Stojadinovic S., Z. Nedic, I. Belca, R. Vasilic, B. Kasalica, M. Petkovic, Lj. Zekovic, *Appl. Surf.* Sci. 256 (2009) 763–767.
Thompson D.W., Snyder P.G., Castro L., Yan P., Kaipa J.A., Woollam J. *Appl. Phys.* 97 (2005) 113511.
Wang X.H., Akahane T., Orikasa H. 2007 *Appl. Phys. Lett.* 91 011908.
Zhao X.L., Meng G.W., Xu Q.L. 2010 *Adv. Mater.* 22 2637.

Manufacturing Engineering and Intelligent Materials – Lu & Abu Bakar (Eds)
© *2015 Taylor & Francis Group, London, ISBN 978-1-138-02832-6*

Research on automatic production control transformation of gas generator

J. Liu
Guangxi Vocational and Technical Institute of Industry, Nanning, Guangxi, China

ABSTRACT: In this paper, the methods of gas producing automation control transformation and related experiences of a gas plant were investigated. During the project, the equipments for instrument testing were transformed, the valve control system was improved, a blower converter was set up, and the DCS (Distributed Control System, which is a control system for a process or plant, wherein control elements are distributed throughout the system) was applied to test and control the flow of gas production. In addition, the plant developed the automatic coal-refilling system, the blower frequency conversion control system and the gas load balance adjustment control system according to the characteristics of productive technology and equipments. After the project, the automatic control capability of gas production of the full material bed was improved. The gas production was stabilized and the operators' labor intensity was reduced to a new low. Various technical and economic indicators were greatly improved.

Keywords: gas generator; full material bed; testing; automation control

1 INTRODUCTION

The project was implemented in a certain gas plant belonging to an aluminum industry company with an annual output of 1.4 million tons per year. Being restricted to the level of its automation equipment, it still followed the earlier tradition, monitoring and detecting instruments manually, thus failing to realize centralized control. It suffered from large energy consumption, poor stability, low quantity and quality of products, repeated equipment maintenance, and high frequency of accidents. Neither the quantity nor the quality of gas fuel could meet the requirements of baking furnace to produce aluminum. In 2009 it carried on an innovation to the operation mode of its gas generator and changed it into full material bed operation. However, the original manual control means of gas generator greatly hindered the improvement of its production capability.

Recently, in order to adapt to requirements for modern enterprises, improve gas production controlling technology, lessen the infection factors of handlers and reduce fluctuation of production indexes, the plant started the transformation project of gas generator controlling technology optimization which had a great significance on the development of the whole company.

2 ANALYSIS OF PRE-TRANSFORMATION CONDITION

There were 22 TG-3MI gas generators in this plant to produce gas fuel for 2 baking furnaces and residential consumption. In October 2009, it carried on the innovation project to replace the original lower layer operation with full material layer operation, which helped stabilize gas calorific value, reduce impurities and lower gas temperature. However, on the other hand, it raised many new problems, such as requirements for frequent coal replenishment, fire detecting and equipment inspection. Not only the stokers were required to be more responsible and standardized, but also their labor intensity multiplied. In addition, in the mode of full material layer operation, it was required to keep low-load balance for all furnaces (Zhang et al. 2003). However, the original manual control operation greatly affected the stability of production. In the production process, coal refilling, firing, air blowing, discharging, and other operations depended on central dispatching room to coordinate producing, which meant making arrangement on telephone was the main communication way. This way of production and management caused the following problems.

1. When the amount of gas consumed by furnaces decreased, the after-discharging pressure on the discharging position would rise sharply. Then order of load-adjusting would be sent to all discharging positions, dispatching room, central controlling room, air blowing room and stokers would be required to implement the order immediately. It cost at least 30 minutes for the load to be completely readjusted. Before that, in order to ensure the stability of the discharging machine, the high-pressure gas must be forced to discharge. The discharging rate reached over 6% per month. For the purpose of energy conservation, the plant tried to reduce the quantity of gas discharged, which caused the problem of high after-discharging pressure, resulting in vibration of the discharging machine. It would bring about not only frequent maintenance but also higher costs.
2. In the case of gas supply could not be increased timely, the discharging machine was forced to have electric backflow for the purpose of stable operation, increasing the electrical power consumption. Index showed that the electrical consumption in eight months in the year of 2009 was higher than the planned value of $45.4/Km^3$.
3. Delay of load regulation might lead to gas refilling into the air pipe because of over high gas system pressure, which might cause cracking of the explosion-proof plate Bai & Hu (2003).

Due to lack of automated recording of the production inspection system, all the operation indexes were written down by the operators at every two hours. If accident happened, it was difficult to make accidental analysis for lack of a comprehensive data before and after the accident. Operators depended on experience to control and regulate the production process, lacking of accurate calculation. As a result, the strategic optimization of producing control could not be realized.

Comprehensively speaking, the problems like backward gas control technology, ineffective production regulating system, untimely command, artificial influences on quality of production and indexes, greatly restricted the improvement of quantity and quality of gas products as well as the process of "reducing costs and increasing effectiveness". All above reasons called for the reform to automatic control of the gas generator.

3 RESEARCH ON PRODUCTION PROCESS CONTROL

3.1 Automatic control of gas generator

Gas generator is a device using coal to produce gas fuel, adopting air and water vapor as the gasification agent. In order to ensure normal gasification reaction in the furnace, there should be a reasonable and stable material layer distribution and high reaction temperature.

As daily operations in gas generator Li (2009), coal-refilling and ash discharging have a direct impact on the distribution, location, height and other factors of the material layer. The thickness of ash layer is affected by the amount of ashes while the height of the material layer is influenced by the quantity of coal refilled.

Under the condition that gas generator runs stably, the furnace temperature changes regularly with a regular coal-refilling operation. The time of refilling could be calculated by testing the furnace temperature. It has been proved theoretically and practically that if production conditions keep stable, the quantity of raw materials consumed is proportional to the output amount while the quantity of ashes discharged is proportional to the quantity of raw materials consumed. That is to say, if the ratio between these two does not change, a certain amount of coal would result in the same amount of ashes. In line with this rule, the time periods of the system of coal refilling and ash discharging could be calculated on the basis of this certain amount. In addition, DCS system is applied on the basis of standard operational procedures so as to realize automatic coal refilling and stable production. The automatic control process is shown in Figure 1.

The three time periods, say, T, T1, T2, greatly influence the stability of the material layer in the gas generator. A too long T period would result in the phenomenon of low material layer, over high outlet gas temperature, burning out of the nozzle button, and discharging of burning coals. It would also greatly reduce the quality of gas fuel produced as a too long T period intensifies the fluctuation of gas calorific value and increases the difficulty of baking furnace control. It is necessary to adjust coal-refilling period on the basis of referring to gas temperature at the outlet of gas generator as a parameter. As a result, an automatic-regulating relation is formed between coal-refilling period T and the outlet gas temperature. If the outlet gas temperature is high, T should be reduced. Conversely, if the outlet gas temperature lowers down, T should be increased.

From Figure 1, it could be assumed that actually there are two parameters exerting influences on the control process of coal refilling: the furnace temperature and refilling interval. The reason for taking coal refilling interval as an auxiliary condition is to prevent the situation that when the furnace operates unstably, the workers keep on refilling coal rather than poking the furnace to control the furnace temperature which would deteriorate problems inside the furnace.

3.2 *Automatic adjustment control of gas load*

The automatic control on gas load is realized mainly through inlet and outlet of the gas generator. In other words, the blowing system and discharging system should be rightly adjusted.

The blower fan speed is adjusted on the basis of the outlet pressure of the blower and the inlet pressure of the discharging machine (pre-discharging pressure). The speed should be turned down when pre-discharging pressure is higher than 2000 Pa while pressure of main

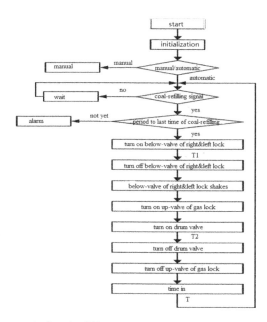

Figure 1. Automatic control of coal refilling.

blowing pipe is higher than 4000 Pa. The speed should be quickened when pre-discharging pressure is lower than 1000 Pa. As the outlets of four blowers (generally two for operation and two for backup) are connected to a same pipe, there might be phenomenon like uneven speed or reflux pour string in the process of frequency converter regulation. In this project, two frequency converters were applied. One converter was used as a major regulation device and the other one was used as output follow-up, to ensure the output power of the two frequency converters to keep in conformity.

However, the usage of frequency converter might cause the problem of blower surge. Especially in the case of furnace stops while discharging machine stops transportation, the speed of frequency converter becomes lower and the surge problem appears due to the "hump phenomenon". In order to prevent that problem, the output power of blower should be higher than the hump value. When the output power of the frequency converter decreases to a certain extent, the trend of decreasing should be stopped. Meanwhile, reflux and relief valve should be rightly controlled to ensure a stable operation of blower and prevent any damage to the equipment.

In this project, it was found out that although there were differences among different blowers, the lowest frequency most likely causing surge problem was between 30–32 Hz. So, we set the lowest frequency at 34 Hz. In addition, the frequency regulation speed should keep at a certain interval otherwise it would cause drastic vibration of the whole production system. When some problems turned to appearance, for example, pre-discharging pressure suddenly drastically changed, the frequency of blower should be adjusted immediately. Hence the pulse adjustment method was adopted. In the normal condition, the frequency of blower was regulated at 0.5 Hz. If it was proved the whole production system still operated stably at this frequency after 1–3 seconds, fine-tune regulation should be continually made till the frequency achieved to the lowest/highest. Through this way, the speed of blower could automatically self-adjust in accordance with the producing conditions. From the research, it was found out the best frequency was 35–38 Hz.

Load adjustment of the discharging system was mainly to adjust the inlet valve, relief valve and the large backflow valve. Both the inlet and outlet pressure of the discharging machine were controlled by regulating its inlet valve according to the pressure signals. The inlet pressure should be controlled at 1000–3000 Pa and the outlet pressure should be controlled at 34000–36000 Pa.

Generally the pre-discharging pressure is adjusted by the blower frequency converter. However, in certain conditions, for instances, when the pre-discharging pressure is lower than 500 Pa or electrical flow value of one discharging machine is too low (59 A in experiment), surge problem would exist and its frequency would reach the limits. In order to ensure a safe production, the large reflux valve should be turned on to regulate. Here the pulse switch mode was adopted: turn on t1 seconds and wait for t2 seconds. When the production system was stabilized, the counter current was turned down. In some cases, such as gas generator suddenly stops consuming gas, after-discharging pressure could ascend to over 36000 Pa. In order to ensure a safe production, relief valve should be opened to discharge gas till the pressure turns back to normal.

4 EFFECTS

Since the automatic control system was fully put into operation in 2011, real-time data collection, display, record, information sharing of the gas production process were realized.

4.1 *Gas producing automation level was enhanced*

1. Technological parameters of coal-conveying system, gas generator system, purification system, discharging system were timely recorded. It was easy and convenient for operators to get and analyse accurate real-time data and get full grasp of production condition.

2. The production process of gas generator was in accurate automatic control and kept in best condition. The labor intensity of workers was greatly reduced. The quality of gas fuel was improved. In addition, the problem of gas leaking was solved so that a considerable quantity of gas could be saved each year. The possibility of gas poisoning accidents decreased.
3. The gas load regulation system helped ensure the blower and discharging machine operate well and safely and realize the purpose of reducing gas leak and saving electricity. Meanwhile, the time to regulate production process was reduced.
4. Discharging machine, blower, circulating oil pressure and axle bearing temperature were in coordinated control. Timely alarm was realized to ensure the production safety.

4.2 *Technical and economic indicators were improved*

1. Gas calorific value stayed at a high level. The gas calorific value rose by 0.08 MJ/km^3. In 2009 the average value was 5.43 MJ/km^3. In 2011 the average value achieved 5.55 MJ/km^3.
2. Optimal control on production system and automatic regulation of gas load were realized. The gas leakage rate was reduced from an average 6.4% in 2009 to an average 0.09% in 2011. The gas leakage quantity was reduced by 2500 km^3 per year.
3. Electricity consumption decreased greatly. After the project, electricity consumed by the gas generator declined from an average 48.37 KWh/km^3 per year to an average 44.61 KWh/km^3 per year, which meant 1.3 million degrees of electricity were saved each year.

5 CONCLUSION

This project successfully realized automatic production control of gas generator, improved quantity and quality of gas products, ensuring the stability and safety of gas production, having a remarkable significance to the future reform of the traditional industry.

ACKNOWLEDGEMENTS

It has been approved as a Science and Technology Research Project of Guangxi Universities, Research on Automatic Control System of SBR Three-phase Fluidized Bed Processing Sugar Wastewater (Project No.: 2013 LX203).

REFERENCES

Bai Xumiao, Hu Junqing. 2004. Research and Exploration on Improvement of Gas Generator Gasification Efficiency [J]. *Non-ferrous Metals and Energy Saving*, 2004 (4).
Li Yuming. 2009. *Chemical Instrumentation and Automation (The Fifth Edition)*. Chemical Industry Press.
Zhang Guoquan, Gao Jinxuan, Liu Guoliang, Li Huicun. 2003. Microcomputer Automatic Control System of Gas Generator [J]. *Automation and Instrumentation*, 18 (02).

Manufacturing Engineering and Intelligent Materials – Lu & Abu Bakar (Eds)
© 2015 Taylor & Francis Group, London, ISBN 978-1-138-02832-6

Research of the fault diagnosis expert system of a Ship's Electric Propulsion System with an asynchronous motor

J.Y. Liu & H.P. Su
Inner Mongolia Technical College of Mechanics and Electrics, Hohhot, Inner Mongolia, China

ABSTRACT: In order to improve the intelligence and accuracy of a Ship's Electric Propulsion System with an asynchronous motor, introduced expert system approach in the fault diagnosis. The asynchronous motor vessels' electric propulsion system was discussed in the paper, and the induction motor vessels' electric propulsion system failure was analyzed in this paper. The fault tree analysis was introduced to asynchronous motor vessels of electric propulsion system, and the knowledge representation method based on rules was realized. Secondly, the structure of the asynchronous motor electric propulsion system fault diagnosis expert system was designed in this paper. Expert system is designed based on electric propulsion system fault diagnosis system.

Keywords: ship electric propulsion; asynchronous motor; expert system; fault diagnosis; fault tree

1 INTRODUCTION

In the fault diagnosis of ship's electric propulsion system, the experience and professional knowledge of ship's electrical engineers play a decisive role in the rapid diagnosis of faults. Currently, with the continuous development of ship electric propulsion system modernization in China and the incessant improvement of facility upsizing and automation, power generation and distribution equipment and control systems of these equipment are more and more complicated. It is hard to solve the problem timely only by several experts if a fault occurs in these large equipment. Meanwhile, because of the space limitation, the common diagnostic method of multi-expert, multi-specialized person is limited, therefore, it is quite necessary to establish a set of fault diagnosis system with multi-expert knowledge.

Based on the characteristics of ship electric propulsion system, the fault diagnosis expert system is designed in this paper, which diagnoses the faults for the ship electric propulsion system by adopting an expert system method integrating knowledge from multiple experts in one. With the combination of the best experience of a plurality of experts, its functional level can achieve or even exceed expert level.

2 THE ASYNCHRONOUS MOTOR VESSELS ELECTRIC PROPULSION SYSTEM AND THE STRUCTURE OF THE EXPERT SYSTEM

At present, the electric propulsion system has played an extremely important role in the development of ship; the selection of motors has become the key problem of the electric propulsion system. Due to its good operating performance, the three-phase asynchronous motor has excellent mechanical properties, which can realize the direct starting, braking and speed regulating function, perfectly solving the problems existing in current electric propulsion system, such as the higher cost of electric propulsion apparatus itself and serious energy loss. Thus, the asynchronous motor will be well applied in the ship electric propulsion system. The structure of asynchronous motor-based ship electric propulsion system is shown in Figure 1.

The parts between the propeller and power plant is selected as the research object of this system; for the fault analysis part, the fault phenomena, reasons and solutions of asynchronous motor, AC frequency converter, and control equipment will be listed down and summarized, respectively.

The structure of the expert system is shown in Figure 2. It mainly consists of six parts: man-machine interface, inference machine, knowledge base, comprehensive knowledge base, interpretation mechanism and knowledge acquisition, among which the man-machine interface is responsible for the interaction of users or domain experts with this expert system; inference machine implements the search, matching and conflict resolution of diagnosis rules; interpretation mechanism is in charge of the interpretation during the inference process of fault diagnosis; the knowledge base is used to store the professional knowledge needed by the system; the comprehensive knowledge base is a database mainly used to store the temporary data generated during the whole process from accepting the user information to offering the result of the problem; while the knowledge acquisition process is mainly used to achieve the self-learning function of the expert system.

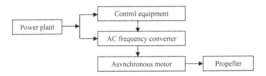

Figure 1.   The structure of asynchronous motor-based ship electric propulsion system.

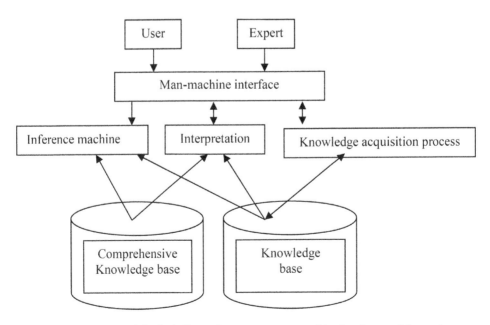

Figure 2.   The structure of the fault diagnosis expert system to a ship electric propulsion system.

3 THE KNOWLEDGE BASE OF FAULT DIAGNOSIS EXPERT SYSTEM OF ELECTRIC PROPULSION SYSTEM

At present, the diagnostic knowledge of electric propulsion system mainly comes from fault diagnostic trees of each equipment of the system, the process of establishing the knowledge base is also the process of converting the knowledge of fault trees into the rules of expert system. The relations between fault trees and knowledge base of diagnosis expert system are as follows: the top event of fault trees corresponds to the task that should be analyzed and solved by the expert system; each minimum cutset of fault trees is the failure mode of this system, corresponding to the final result that should be inferred by this expert system, fault trees' logical relationship from top to bottom corresponds to the inference process of expert system, fault trees' branch corresponds to the rule of knowledge base in expert system, the number of its branches corresponds to the number of rules. Figure 3 shows part of fault trees under too high asynchronous motor bearing temperature.

4 THE INFERENCE OF THE EXPERT SYSTEM

This system adopts the rule-based inference method. In the inference based on rules, the fault knowledge of asynchronous motor-based ship electric propulsion system is stored both in the rule set of the system and the knowledge base simultaneously. The system solves the practical problems mainly by using these rules and the information stored in the database. When IF statement of rules matched with the information of job content, the system executes the program of THEN rule. At this time, the rule is used and meanwhile THEN statement will be

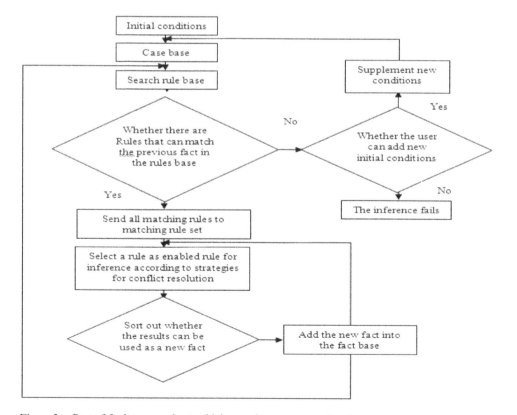

Figure 3.   Part of fault trees under too high asynchronous motor bearing temperature.

introduced into the working memory. While new statements that have been added to job, contents will be seen as IF part in rules, which will be used to match with other rules to activate the other rules. If the temperature of asynchronous motor bearings in asynchronous motor-based ship electric propulsion system is too high, the inference process is as follows:

IF bearing temperature is too high
AND grease is normal
AND No impurities in grease
THEN the match of bearing with shaft neck or the match of bearing with end cover is abnormal
OR THEN There is eccentricity inside bearing bore
OR THEN There is gap in the end cover of motor or the assembly of bearing cover
OR THEN The distance of bearing and shaft is abnormal
OR THEN the spindle of motor curves.

The rule-based inference offers its inference technology mainly by adopting a method that combines the initial evidence and the rules in rule base. The inference machine is mainly used to control the whole inference process of the expert system, which uses the relevant rules stored in the rule base according to the data the system inputs, namely the information stored in the database, solves problems according to specific inference models, and stores the diagnostic results in database. After the rule-based forward inference starts its inference, the inference machine will search for the matching knowledge in the knowledge base according to the initial data provided by users, and store the inference results into the comprehensive database as a fact, and then carry out the follow-up inference, and so forth, until the whole design results exist. The entire inference process flow is shown in Figure 4.

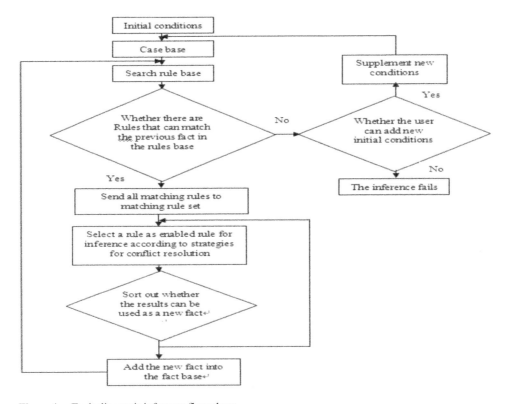

Figure 4.   Fault diagnosis inference flow chart.

5 CONCLUSION

Fault diagnosis expert system of electric propulsion system acquires knowledge by adopting expert experience, organizes the knowledge in the form of inference rule based on fault trees, and achieves inference machine of the system by establishing database and using forward inference strategy. Based on the Windows XP operating system platform, the system adopts Visual C++6.0 as the software development environment, uses SQL to establish the knowledge base, and adopts ADO database connection technology as the main connection mode of application interface and knowledge base.

REFERENCES

Cong wang, Ji Xiaohui & Liu Jingbo. 2002. The Design of the Fault Diagnosis Expert System to Ship Electric System. *Techniques of Automation and Application* 2(1):13–16.
Heung-Jae Lee, Bok-Shin Ahn &Young-Moon Park. 2000. A Fault Diagnosis Expert System for Distribution Substation. *IEEE Transaction on Power Delivery* 15(1):92–97.
Ji Xiaohui, Cong Wang & Zhang Helin. 2001. Research of Fault Diagnose Technology to a Ship's Electric System. *Journal of harbin engineering university* 22(6):1–3.
Wang Menglian, Ma Dan & Shen Feng. 2010. Research of Fault Diagnose Technology to a Ship's Electric Propulsion System. *Marin Electric* 30(12):23–25.
Wen-Hui Chen, Chih-Wen & Men-Shen Tsai. 2000. On-Line Fault Diagnosis of Distribution Substations Using Hybrid Cause-Effect Network and Fuzzy Rule-Based Method. *IEEE Transaction on Power Delivery* 15(2):710–716.

Manufacturing Engineering and Intelligent Materials – Lu & Abu Bakar (Eds)
© *2015 Taylor & Francis Group, London, ISBN 978-1-138-02832-6*

Calculation of induced voltage and current of double-circuit high-voltage line on the same tower lines based on EMTPE 202 kV and selection of grounding knife switches

J.Y. Liu & S.S. Lu
Inner Mongolia Technical College of Mechanics and Electrics, Hohhot, Inner Mongolia, China

L.F. Li
Inner Mongolia Electric Power Design and Research Institute, Hohhot, Inner Mongolia, China

ABSTRACT: The article, which takes EMTPE (Electro Magnet Transients and Power Electronics) as a development platform, calculates induced voltage and current on 220 kV double-circuit high-voltage line on the same tower. For providing theoretical basis on selection in grounding knife switches and safety work, numeral model is established and calculation results are also analyzed based on the influences of electric power operation mode, line parameters and other factors.

Keywords: EMTPE; 220 kV; double-circuit high-voltage line on the same tower; induced voltage; induced current; grounding knife switches

1 INTRODUCTION

With the acceleration of grid construction and the intensity improvement of "west-east power transmission", large land resources are occupied by overhead power lines. Though we have a vast territory, the available land resource in our country is quite limited. The conflict between line corridor used for grid construction and local land utilization aggravates. Lack of land resource limits the construction of overhead power transmission lines. During the 12th Five-Year Plan, the problems of the new transmission and transformation electric project are more and more prominent. As a result, the development of grid is restricted. For solution, the requirement of sustainable growth in load and increasing the transmission capacity of line corridor per unit area, the construction of double-circuit high-voltage line on the same tower is one of the most effective methods.

Simulation research, which is based on EMTPE 220 kV double-circuit induced voltage and current on the same tower, can not only help the analysis and improvement in operating type, but also can save the maintenance personnel. What's more, it also can provide reliable basis of reasonable selection for grounding devices. Once the number of double-circuit high-voltage line on the same tower increases, the induced voltage and current of overhauling circuit will increase greatly because of the electromagnetic coupling and electrostatic induction between circuits.

There are many accidents in our country and abroad. The maintenance staff on pylons will feel discomfort for weak stimulation, and dead for strong stimulation from a high-altitude falling. Furthermore, induced current may be large in some conditions, As a result, it will be bad for the operation of grounding switches. To be sure of the safe operation of the electric system, correct electric system simulation software will be used correctly.

EMTPE (Electro Magnet Transients and Power Electronics) is a digital simulation software package for electro magnet transients and power electronics, which is developed by Chinese Electric Power Research Institute on EMTP. It can calculate induced voltage and current.

2 SELECTION OF LINE PARAMETERS FOR CALCULATION

2.1 *Line parameters*

220 transformer substations are connected with 500 kV transformer substations by double-circuit 220 kV lines, and the length of line is about 90 km.

2.2 *Wire parameters*

The conductor type is LGJ-2 × 400. Parameters of wire and ground wire are shown in Table 1.

2.3 *Typical tower shape of double-circuit high-voltage line on the same tower*

Drum-shape towers of double-circuit lines are selected, typical poles and towers are shown in Figure 1. Parameters of typical poles and towers are shown in Table 2.

Table 1. Parameters of wire and ground wire.

Wire parameters	Circuit No.	Circuit I	Circuit II
	Conductor type	JL3/G1 A-400/35	In common with Circuit I
	Split root number	2	In common with Circuit I
	Intrabundle spacing (mm)	400	In common with Circuit I
	Section area of Al wire (mm²)	391	In common with Circuit I
	Section area of steel core (mm²)	34.4	In common with Circuit I
	Outside diameter of conductor (mm)	26.8	In common with Circuit I
	Diameter of steel core (mm)	7.5	In common with Circuit I
	20°C direct-current resistance (Ω/km)	0.0722	In common with Circuit I
	Phase sequence	Up, middle and lower: CBA (see phase sequence figure)	Up, Middle and Lower: ABC (See phase sequence Figure)
Ground wire Ground wire parameters	Ground wire number	Circuit I	Circuit II
	Conductor type	OPGW	In common with Circuit I
	Split root number	1	In common with Circuit I
	Intrabundle spacing (mm)	/	In common with Circuit I
	Al wire section area of Al wire (mm²)	140 (Full-Al steeling wire)	In common with Circuit I
	Section area of steel core (mm²)	/	In common with Circuit I
	Outside diameter of conductor (mm)	16	In common with Circuit I
	Diameter of steel core (mm)	/	In common with Circuit I
	20°C of direct-current resistance (Ω/km)	0.4	In common with Circuit I

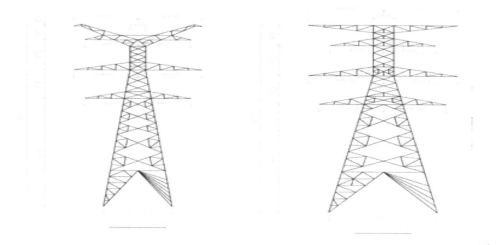

Figure 1. Typical poles.

Table 2. Typical parameters of pole and tower.

Suspension height of ground wire $h1$ (m)	40
Suspension height of the upper phase conductor $h2$ (m)	37
Suspension height of in-phase conductor $h3$ (m)	30.3
Suspension height of the next phase conductor $h4$ (m)	24
Horizontal distance of ground wire $l1$ (m)	11.6
Horizontal distance of the upper phase conductor $l2$ (m)	9.6
Horizontal distance of in-phase conductor $l2$ (m)	11.6
Horizontal distance of the next phase conductor $l2$ (m)	9.6
Insulator chain length $h5$ (m)	3.2
Ground wire sag (m)	6
Conductor sag (m)	10
Span (m)	400

Note: Data in the above table is the data of SZC1 type tangent tower, which has a few differences from the data of other tangent towers and resisting-tensile towers.

3 CALCULATION CONTENT

The outgoing line side of line spacing on 220 kV voltage class in transformer substation is connected with opening and closing induced current capacity of ground knife switch (Type A/B), i.e. whether ground connection induced current is above 80 A, and the induced voltage is above 5 kV.

4 CALCULATION CONDITION

The conductor type of double-circuit high-voltage line between 220 kV and 500 kV transformer substations is LGJ-2 × 400. And it chooses double-circuit high-voltage line on the same tower. Non-transposition type is used for the circuits. The height of tower body over the ground is in the pole and tower: The upper phase is 37 m, the in-phase is 30.3 m and the next phase is 24 m. The horizontal distance in the upper phase conductor is 9.6 m, the in-phase conductor's is 13.6 m and the next phase conductor's is 11.6 m. The conductor sag is

10 m, and the insulator chain length is 3.2 m. The section area of steel core is 34.4 mm^2, the Al wire's is 391 mm^2 and the 20° C direct-current resistance is 0.0722 Ω/km.

Both ground wires' type is OPGW, the horizontal distance of ground wire is 11.6 m, and the sag is 6 m. The section area of ground wire is 140 mm^2, the 20° C direct-current resistance is 0.0722 Ω/km.

Under the 200 kV voltage class, the transmission capacity of LGJ-2×400 wires is 610 MVA. Such type circuit's active capacity is not suitable for over 510 MW with 0.95 power factor and 0.88 allowance. According to the above current limitation, the data of circuit, pole and tower, EMTPE in Version 2.0 calculates induced voltage and current.

5 CALCULATION RESULT

When Current I (in the double-circuit high-voltage line between 220 kV and 500 kV transformer substations) is in line outage, and the current of Current II is considered as 510 MW+j63.7 Mvar, the calculation results are shown in Table 3.

Based on the calculation result in Table 3, and compared with relevant industrial standard (DL/T486-2010, see Table 4), electrostatic induced voltage of double-circuit high-voltage line is below 6.99 kV and is beyond the limitation of electrostatic induced voltage of Type A grounding knife switches, but under that of Type B switches.

The electrostatic induced current is below 3.01 A, which is beyond the limitation of electrostatic induced current of Type A switches, but within Type B switches. The electromagnetic induced voltage is below 4.53 kV, which is beyond such limitation of Type A switches, but within that of Type B; the electromagnetic induced current of double-circuit high-voltage line is below 158.38 A, which is beyond such limitation of Type A switches, but within that of Type B switches.

Table 3. Computation of interaction in transmission circuits between 220 kV transformer substation and 500 kV transformer substation (effective values).

Site	Voltage class	Phase	Electrostatic induction		Electromagnetic induction	
			Voltage/kV	Current/A	Voltage/kV	Current/A
Location of 220 kV transformer substation	220 kV	A	4.49	1.91	2.57	83.08
		B	3.01	0.47	0.51	5.37
		C	12.55	3.74	2.62	89.48
Location of 500 kV transformer substation	220 kV	A	5.15	1.77	2.60	82.52
		B	3.51	0.43	0.54	5.08
		C	13.20	3.59	2.67	88.42

Table 4. Standard table of rated current and rated voltage (Effective values) in grounding knife switches.

A Type grounding knife switches	Electromagnetic coupling	Induced current	80 A
		Induced voltage	1.4 kV
	Electrostatic coupling	Induced current	1.25 A
		Induced voltage	5 kV
B Type grounding knife switches	Electromagnetic coupling	Induced current	160 A
		Induced voltage	15 kV
	Electrostatic coupling	Induced current	10 A
		Induced voltage	15 kV

6 CONCLUSION

Electrostatic induced voltage of 220 kV double-circuit high-voltage line on the same tower below 6.99 kV, the electrostatic induced current below 3.01 A, the electromagnetic induced voltage below 4.53 kV, and the electromagnetic induced current below 158.35 A.

While non-transposition Circuit I on the same tower is in operation and the other circuit is in outage for overhauling, the electrostatic induced current through grounding knife switches is in direct proportion with line length, and has no relation with the transmission power of operation line. The length of such line is 90 km, the electrostatic induced current above 3.01 A, thus Type B grounding knife switches should be chosen. While non-transposition Circuit I on the same tower is in operation and the other circuit is in outage for overhauling, the electrostatic induced current through grounding knife switches is in direct proportion with line transmission power, and has no relation with the line length of operation line. The line transmission power between 220 kV and 500 kV transformer substations is 510 MW, thus Type B grounding knife switches should be chosen. Consider the above calculation results, the design suggestion shows that Type B grounding knife on 220 kV voltage class with both sides of double-circuit high-voltage line switches should be chosen.

REFERENCES

Hu Yi, Zhang Junlan & Zhang Lihua. 2002. Research on Hot-line Work of 500 kv Pylon with Double-circuit Lines. *power equipment*.

Li Bin, Cao Rongjiang & Gu Nihong. 1995. Study on making and behavior of the earthing switched on the common tower double circuit lines. *Power Systems Technology* 19(1): 4246.

Shi Wei. 1988. *Overvoltage Calculation of Electrical Power System*. Xi`an: Xi'an Jiaotong University.

Manufacturing Engineering and Intelligent Materials – Lu & Abu Bakar (Eds)
© *2015 Taylor & Francis Group, London, ISBN 978-1-138-02832-6*

Reed cellulose dissolves in ionic liquid [Amim]Cl/co-solvent system

F. Xie & J. Zhong
School of Chemistry and Chemical Engineering, Tianjin University of Technology, Tianjin, China

H. Liang
Key Laboratory of New Technology of Chemical and Biological Transformation Processes, Nanning, Guangxi, China

ABSTRACT: Highly effective cellulose solvents for the dissolution of reed cellulose have been designed by adding aprotic polar solvent DMSO to 1-allyl-3-methylimidazolium chloride ([Amim]Cl). In this work, we studied the dissolution and regeneration of reed cellulose in [Amim]Cl/co-solvent system. The effects of molar ratio of DMSO to [Amim] Cl have been studied in detail. Both the original and regenerated cellulose samples were characterized with wide-angle X-ray diffraction, thermogravimetric analysis and scanning electron micrograph. It was found that aprotic solvent DMSO can enhance dissolution of reed cellulose.

1 INTRODUCTION

Sustainability, industrial ecology, eco-efficiency, and green chemistry are directing the development of the next generation of materials, products, and processes. Cellulose, as the most abundant biorenewable resource in the world, is regarded as a promising material that could replace synthetic polymers and reduce global dependence on fossil fuel sources (Moutos et al. 2007, Xie et al. 2007). Cellulose and its derivatives are extensively used in industries such as food casing (Edgar et al. 2001), membranes and plastics etc (Azubuike et al. 2012). However, there are numerous intermolecular and intramolecular hydrogen bonds in cellulose, which results in that it's a challenge to find suitable solvents for its dissolution (Henriksson et al. 2008). The conventional solvent systems, such as N-Methylmorpholine-N-Oxide (NMMO) (Xu et al. 2010), LiCl/DMAc, DMSO/TBAF and NaOH/urea, have defects like high toxicity, high cost, harsh conditions, and difficulty in solvent recovery (Wu et al. 2005). Recently, Ionic Liquids (ILs) have been introduced as green solvents for cellulose with high solubility, and 1-allyl-3- methylimidazolium acetate has also been employed as efficient solvent for cellulose (Xu et al. 2013). However most of these solvent systems suffer some defects like high viscosity or high dissolution temperature. Thus, it is still imperative to develop potential cellulose solvents and to find the physiochemical properties of the regenerated cellulose material (Idris et al. 2013, Ma"ki-Arvela et al. 2010).

Here, we report our new reed cellulose solvent system: 1-allyl-3-methylimidazolium chloride/dimethyl sulfoxide ([AMIM]Cl/DMSO). The solubility of reed cellulose in this co-solvent system was studied and the physiochemical properties of the regenerated cellulose material were investigated by XRD, FTIR, TGA measurements.

2 EXPERIMENTAL

2.1 *Materials and characterization*

The reed cellulose used in the experiments was extracted from the dried wild reed (growing in Dongying city, Shandong province) by soaking in 12% sodium hydroxide (NaOH) aqueous solution for 2 h, and boiling for 3 h. The Degree of Polymerization (DP) = 514. All reed cellulose were cut into small pieces, and dried at 100°C for 12 h without activation treatment before use. N-methylimidazole (99%) and allyl chloride were obtained from Qingdao Chem. Co. Activated charcoal (40–60 mesh) was purchased from Sigma-Aldrich. DMSO (98.0%) was purchased from Tianjin Chem. Co. All other reagents were obtained from Tianjin Kewei Chem. Co. Original and regenerated reed cellulose were characterized by X-ray diffractometry (D/MAX-2500PC, Rigaku, Tokyo, Japan) and thermogravimetric analysis (TG209F3, NETZSCH Co. Ltd, Germany). SEM images of fibers were obtained by scanning electron micrograph machine (JSM-6700F, JEOL Co., Japan).

2.2 *Dissolution and regeneration of cellulose*

Dried reed cellulose was added into a glass vial which contained 4.0 g [Amim]Cl/DMSO. The glass vial was then immersed in an oil bath under an inert atmosphere of N_2. The mixture was heated and stirred at 100°C. Additional cellulose was added until the solution became completely clear under optical microscope. Complete dissolution was assumed to have occurred when no undissolved material was observed under microscope (Vitz et al. 2009). After cooling to room temperature, the solution was coagulated by adding water, and regenerated cellulose was precipitated. The precipitate was then washed twice with ethyl alcohol, separated by centrifugation for 20 minutes at 3000 rpm, and then dried under vacuum at 70°C for 2 days (Idris et al. 2013, Mäki-Arvela et al. 2010).

3 RESULTS AND DISCUSSION

3.1 *Dissolution of reed cellulose in ionic liquids*

Figure 1 showed the dependence of cellulose solubility on the molar ratio of DMSO to [AMIM]Cl in the [AMIM]Cl/DMSO co-solvent system at 100°C. Remarkably, the solubility of cellulose increases with increasing DMSO concentration in the molar ratio range from 0 to 2.5, then decreases with further increase of DMSO content. This can be explained by the assumption that concentration of the "free" Cl^- increases with addition of DMSO before the maximum solubility, and then it is diluted owing to the further increase of DMSO content in the IL. Furthermore, Figure 2 shows the rate of dissolution of reed cellulose in [AMIM]Cl/DMSO. Dissolution in co-solvent was visually slightly more rapid in the first 300 minutes, maybe because of the relatively low viscosity of the solution.

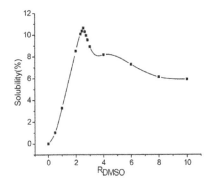

Figure 1. The dependence of cellulose solubility on the molar ratio of DMSO to [AMIM]Cl/in the co-solvent at 100°C.

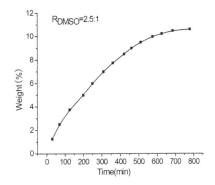

Figure 2. Rate of dissolution of reed cellulose in the [AMIM]Cl/DMSO co-solvent (R_{DMSO} = 2.5:1) at 100°C.

3.2 SEM studies

Scanning Electron Microscopy (SEM) was used to observe the bulk structure of cellulose. Figure 3 showed SEM images of original reed cellulose and cellulose after dissolution in [AMIM]Cl/DMSO with different ratio and regeneration into water. The reed cellulose shows fibers at × 1000 magnification in the SEM. Compared to raw material (a), the morphology of the regenerated material was significantly changed, displaying a rough, but conglomerate texture (b, c) in which the fibers are fused into a relatively homogeneous macrostructure. With increasing the ratio of DMSO, the uniformity of the regenerated cellulose also increased.

3.3 XRD studies

Figure 4 illustrated the X-ray diffraction patterns of raw material and regenerated cellulose material dissolved in [AMIM]Cl/DMSO (R_{DMSO} = 2.5:1). As shown in Figure 4, the diffraction curve of raw material was typical cellulose I structure. It had crystalline peaks at $2\theta = 16.33°$ and $22.07°$. After dissolution and regenerated with water, the regenerated cellulose material exhibited the typical diffraction crystalline patterns of cellulose II at $2\theta = 16.49°$, $20.41°$ and $22.18°$. These results showed that the transformation from cellulose I to cellulose II occurred after the ILs dissolution and regeneration. The other regenerated cellulose materials, which dissolved in other molar ratio of DMSO to [AMIM]Cl have the same situation but their diffraction peaks were weaker compared to regenerated cellulose material dissolved in [AMIM]Cl/DMSO (R_{DMSO} = 2.5:1).

3.4 TGA studies

Figure 5 displayed the TGA curves of raw material and regenerated cellulose material dissolved in [AMIM]Cl/DMSO (R_{DMSO} = 2.5:1). As shown in Figure 5, the thermal

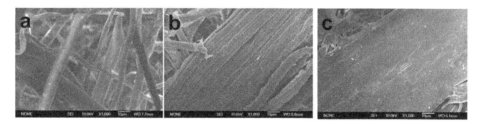

Figure 3. SEM micrographs of regenerated reed cellulose dissolved in [AMIM]Cl/DMSO at different molar ratio (a) R_{DMSO} = 0:0, (b) R_{DMSO} = 1:1, (c) R_{DMSO} = 2.5:1.

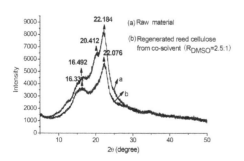

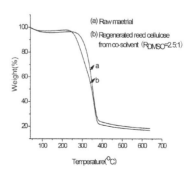

Figure 4. XRD of raw material and regenerated reed cellulose from co-solvent (R_{DMSO} = 2.5:1).

Figure 5. TGA curves of the original and regenerated reed cellulose.

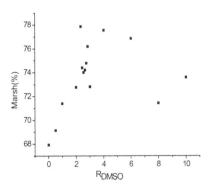

Figure 6. The marsh of the regenerated reed cellulose from co-solvent (R_{DMSO} = 2.5:1) at second stage.

decomposition process of original and regenerated cellulose had two stages. The small initial drop occurring near 100°C in all cases was due to the evaporation of incorporated water. For the original reed cellulose, the second stage started at 320.1°C and continued up to 376.4°C with a weight loss of 68.92%, which happened because of cellulose degradation. The second stage of weight loss of regenerated cellulose sample started at 304.7°C, and continued up to 378.6°C, with weight loss of 74.01%. The other regenerated cellulose materials have the same properties, but their decomposition temperatures are higher than R_{DMSO} = 2.5:1. E.g. For the cellulose samples regenerated from co-solvent (R_{DMSO} = 1:1, 2.4:1, 2.6:1, 3:1), their decomposition temperature started at 308.9°C, 305.9°C, 307.8°C, 311.2°C, and continued up to 378.4°C, 380.9°C, 378.6°C and 377.9°C, with weight losses of 71.39%, 77.86%, 74.78% and 72.81%. These results show that the thermal stability of regenerated cellulose is lower than original material, which is attributed to the partial destruction of crystalline part and hydrolysis. Figure 6 shows the char yield (nonvolatile carbonaceous material) of the regenerated cellulose stage in all cases were different which is due to the different molar ratio of DMSO to [AMIM]Cl.

4 CONCLUSIONS

The co-solvent in this work can directly dissolve reed cellulose in high concentration at 100°C, and especially reed cellulose is soluble in [AMIM]Cl/DMSO (R_{DMSO} = 2.5:1) up to 10.56 wt%. A regenerated reed material was obtained by precipitation from water. After dissolution and regeneration in ionic liquid, the crystalline structure of cellulose was converted from cellulose I to cellulose II. The thermal stability of cellulose regenerated from this ionic liquid was lower than the original cellulose. Generally, The co-solvent [AMIM]Cl/DMSO (R_{DMSO} = 2.5:1) showed high solubility for reed cellulose and this study will provide the basis for better development and utilization of reed cellulose.

REFERENCES

Azubuike, C., Rodrígue, H., Okhamafe, A. & Rogers, R. 2012. Physicochemical properties of maize cob cellulose powders reconstituted from ionic liquid solution. Cellulose 19(2): 425–433.
Edgar, K.J., Buchanan, C.M., Debenham, J.S. & Rundquist, P.A. 2001. Advances in cellulose ester performance and application. Progress in Polymer Science 26(9): 1605–1688.
Henriksson, M., Berglund, L.A., Isaksson, P. & Lindstroem, T. 2008. Cellulose nanopaper structures of high toughness. Biomacromolecules 9(6): 1579–1585.
Idris, A., Vijayaraghavan, R., Rana, U.A. & Fredericks, D. 2013. Dissolution of feather keratin in ionic liquids. Green Chemistry 15(2): 525–534.

Mäki-Arvela, P., Anugwom, I., Virtanen, P. & Sjöholm, R. 2010. Dissolution of lignocellulosic materials and its constituents using ionic liquids-A review. *Industrial Crops and Products* 32(3): 175–201.

Moutos, F.T., Freed, L.E. & Guilak, F.A. 2007. Biomimetic three-dimensional woven composite scaffold for functional tissue engineering of cartilage. *Nature Materials* 6(2): 162–167.

Vitz, J., Erdmenger, T., Haensch, C. & Schubert, U.S. 2009. Extended dissolution studies of cellulose in imidazolium based ionic liquids. *Green Chemistry* 11(3): 417–424.

Xie, H., King, A., Kilpelainen, I. & Granstrom, M. 2007. Thorough chemical modification of wood-based lignocellulosic materials in ionic liquids. *Biomacromolecules* 8(12): 3740–3748.

Xu, A.R., Wang, J.J. & Wang, H.Y. 2010. Effects of anionic structure and lithium salts addition on the dissolution of cellulose in 1-butyl-3-methylimidazolium-based ionic liquid solvent systems. *Green Chemistry* 12(2): 268–275.

Xu, A.R., Zhang, Y.J., Zhao, Y. & Wang, J.J. 2013. Cellulose dissolution at ambient temperature: Role of preferential solvation of cations of ionic liquids by a cosolvent. *Carbohydrate Polymers* 92(1): 540–544.

Zhang, H., Wu, J., Zhang, J. & He, J.S. 2005. 1-Allyl-3-methylimidazolium chloride room temperature ionic liquid: A new and powerful nonderivatizing solvent for cellulose. *Biomacromolecules* 38(20): 8272–8277.

Manufacturing Engineering and Intelligent Materials – Lu & Abu Bakar (Eds)
© 2015 Taylor & Francis Group, London, ISBN 978-1-138-02832-6

Experimental research on loading device of the magneto-rheological retarder

Y.D. An, W. Wang, X.J. Qi, Y. Bao & D.G. Lv
Heilongjiang Institute of Technology, Harbin, Heilongjiang Province, China

ABSTRACT: The paper designs a loading device to provide different work loads to the magneto-rheological retarder. In order to analyze magneto-rheological retarder performance under different loads, the loading device uses a planetary gear train loading, with no-load startup operation variable load, applied load angle unlimited travel and other properties. In this paper, some experimental research of the loavcbnbvnding device designed were studied, the results show that the magneto-rheological retarder loading device proposed completely meets the loading requirements.

1   INTRODUCTION

A retarder is an auxiliary device of modern heavy-load commercial vehicles (coaches or heavy trucks). When heavy-load commercial vehicles are normally arrested, the retarder will provide extra braking and this will improve vehicles' braking effectiveness and braking steadiness, which greatly improve usage longevity of braking system. Magneto-rheological retarder is a kind of new device which makes use of magneto-rheological liquid as working medium. Magneto-rheological liquid is a kind of intelligent material with wide usage and good performance whose physical properties present Newton fluid behavior with low viscosity under the condition of zero magnetic field and Binghan property with high viscosity and lazy flow under the condition of strong magnetic field. Moreover, rheology is instantaneous and reversible under the action of this kind of magnetic field, and cutting field strength after rheology will steadily match with magnetic field strength. Thus, it is feasible to use magneto-rheological onto retarder and helps to develop new-style braking retarder with low cost, high efficiency, high sensitivity, and energy saving. This paper carries out the design and experiment of the loading device of magneto-rheological retarders. Loading devices can make static and dynamic loading onto magneto-rheological retarder and satisfy all kinds of loading demands needed by magneto-rheological retarder (Tian et al. 2014).

2   STRUCTURE COMPONENTS AND WORKING PRINCIPLES OF MAGNETO-RHEOLOGICAL RETARDER LOADING DEVICES

2.1   *Retarder experiment table*

A retarder experiment table is an experiment system that is used to test retarder working performance design, in which transmission shafts of retarders are loaded through the loading device, operating condition of transmission shaft is simulated to test decreasing speed effect that the retarder impacts on the transmission shaft, and working performance of the retarder can be evaluated. The basic components of experiment table are as following in Figure 1.

The closed experiment table consists of reducer, retarder and main components of loading device. There are two ports of A and B in loading device output whose B port is connected to

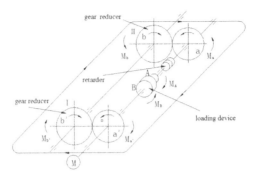

Figure 1. Components of the retarder experiment.

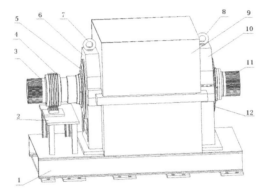

Figure 2. General structure of the loading device.
1-base 2-group ring frame 3-input shaft 4-group ring 6-input port bearing frame 7,8-flying ring 9-protection guard 10-output port bearing frame 11-output shaft 12-output port labyrinth seal.

driving machine and A is connected to transmission shaft. The loading device produces loading torque between output port A and B under the action of interior loading driving motor, then brakes loading device by arresters. That is, that closed torque is put on transmission shaft, then the whole experiment table derived by driving machine forms closed power and simulates actually operating situation of transmission shaft (Tian et al. 2014, Li et al.2014).

2.2 *Structure components of the loading device*

The designed loading device mainly is made up of group rings, permanent magnetism DC torque motor, and the first and second-order reducing mechanism considering magnetorheological retarder experiment table's demand on loading device. The first-order reducing mechanism is harmonic wave gear mechanism and the second is NGWN (3 K) planetary gear reducing device.

The permanent magnetism DC torque applied loading motor of loading devices and two-order reducers are mounted in interior devices. The applied motor gets power from the group rings on the shaft port, drives the first and second-order reducing mechanism, and performs applied loading. The closed port A is connected to loading outer shell and the closed port B is connected to output shaft of the second-order reducing mechanism. Thus, huge torque between closed port A and B will be produced through the action of reducing speed and adding torque of two-order reducing mechanism. The torque is closed in the experiment table by taking advantage of locking mechanism of the applied loading motor. The construction of the loading device is shown in Figure 2.

3 STABILITY TEST ON THE LOADING DEVICE OF MAGNETO-RHEOLOGICAL RETARDER

3.1 Contents of stability test

The loading device can realize dynamic applied loading. As a non-steady-state rotor, it has to experiment for non-steady-state stability test, in order to ensure operation stability of the loading device to get to G6.3. The two ends of maximal outside diameter of the loading device are used as balanced surface I and balanced surface II, and the whole circumference is divided into 26 test points by 14.4°. As the closed end A is set, the applied loading motor drives the closed end B to every point and locks them, and stability of the rotor is tested by the speed of 230 r/min. Thus, unbalanced units and unbalanced angles on the balanced surface I and the balanced surface II are gotten, and this is shown in Figure 3.

3.2 Analysis of stability test

Figure 3 shows that the balanced surface I and the balanced surface II bring about unbalanced unit, and discreteness of phase angle is greatly obvious. However, unbalanced unit and unbalanced phase of the balanced surface Iand the balanced surface II possess similar changing regularity achieving from curvilinear motion situation, which helps stability balance of the loading device. Stability precision grade of fore-and-aft balance can be compared in Figure 4.

Stability precision grade of the loading device greatly increases after balance through analyzing the experiment, the top point rises nearly 77%, precision grade after balance get to G6.3 and is about between 43.2° and 86.4°, and stability grade is considerably high. In actual application, ideal loading area can be decided according to working condition range of actual loading angle (Guo & Liao2012).

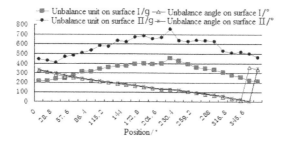

Figure 3.   Curves of the unbalanced.

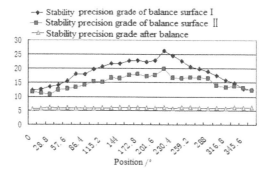

Figure 4.   Stability precision grade curve.

4 STATIC STATE LOADING EXPERIMENT OF LOADING DEVICE OF MAGNETO-RHEOLOGICAL RETARDER

4.1 *Experiment analysis of static state loading angle*

Static state loading experiment on the loading device studied will be processed, in order to check whether the loading device works according to the rules or not, whether applied loading angle at the two closed ends comes into being or not, whether the loading device checked can meet demands of design or not apply torque needed or not by the built check platform.

To check output port of the loading device under the control of control system to be able to realize any applied loading angle and operation of group rings to be out of normal.

The loading device is fixed onto experiment platform by using ground screws. The experiment proves that when torque shaft of A end of the loading device is fixed on the experiment platform, operation of conduction slip ring controlled by control system breaks many times, and normal operation and break can perform.

The needed applied loading angles are input on the control platform powered by the loading system in interior of the loading device. Forward and reverse direction operation are realized by controlling the applied loading motor, and any trip angles of forward and reverse direction on out put end B also are realized.

To check whether turning angles on dial of output end turn onto position of the set applied loading angle from original zero position.

The difference of the applied loading angles increases with turning angles adding known through experiment data, and appears the trend of accumulation. In fact, the loading device only need to turn a little angle to produce torque wanted, thus the difference within scope of the applied loading angles can be accepted.

4.2 *Experiment analysis on static state loading torque*

To check control ability of the loading device control system to loading mechanism, and verify accuracy of the applied torque by the experiment of static state applied loading torque.

The loading device is set on the experiment platform and two housings used by the experiment are individually put on A end torque shaft of the loading device and output shaft end of B end. The two housing lever arms are both adjusted to horizontal position by controlling the applied loading motor and then the four ends on the two housing lever arms are restricted by the experiment platform.

In order to check amount of the applied torque, a pressure sensor is installed on two ends of A end lever arm of the applied loading device closing end, whose one end is mounted on the top of lever arm and the other end is mounted under the lever arm.

The applied loading torque value can be gotten by support counterforce of the two ends of lever arm detected through the pressure sensor multiplying by the individual arm force, and the summation of torque is the value, when the control system starts the applied loading motor in the loading device. The check process is shown in Figure 5.

Table 1. The applied loading angle data (°).

The set applied loading angle	Values shown on the dial	Difference
50	49.8	0.2
100	99.1	0.9
150	148.9	1.1
200	198.1	1.9
250	247.8	2.2
300	296.9	3.1
360	355.9	4.1

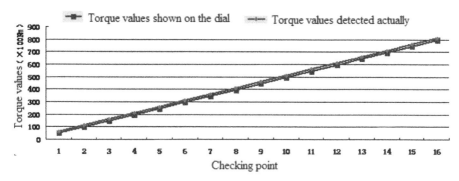

Figure 5. The applied loading torque is checked.

Table 2. Experiment data of applied loading torque (100 Nm).

No.	Values shown on pressure sensor		Torque values detected actually	Torque values shown on the control table	Difference ratio
	Left	Right			
1	380	310	62.8866	61	3.00%
2	710	540	113.925	111	2.57%
3	1020	770	163.1406	159	2.54%
4	1340	1020	215.0904	211	1.90%
5	1630	1250	262.4832	252	3.99%
6	1950	1500	314.433	306	2.68%
7	2240	1730	361.8258	346	4.37%
8	2540	1980	411.9528	397	3.63%
9	2850	2240	463.9026	446	3.86%
10	3150	2490	514.0296	497	3.31%
11	3450	2740	564.1566	546	3.22%
12	3740	3000	614.2836	596	2.98%
13	4030	3250	663.4992	646	2.64%
14	4300	3520	712.7148	696	2.35%
15	4600	3790	764.6646	746	2.44%
16	4869	4060	813.7891	796	2.19%

Experiment data of the applied loading torque detected is shown in Table 2. Loading torque of the applied loading can be calculated according to length of lever arms. Length of the left lever arm is 0.89 m and length of the right lever arm is 0.99 m.

Comparative analysis on actually detected value of static state applied loading torque detected by experiment. It is obvious in the table that the torque value applied and the actual torque value detected by experiment on each checking point is nearly close, and accuracy of static state applied loading torque is verified.

In experiment, the loading device realizes that to load any torque value, and the applied loading value can be limitlessly adjusted; moreover, exceptional fluctuation and unloading doesn't appear during applied loading process, and the applied loading doesn't unload, after maximum rating load 79200 has been loaded for one hour (Tian et al. 2011, Yue et al. 2011).

5 CONCLUSION

This paper does some research on the loading device and verifies correctness of the loading device theory design and rationality of physical design. Performance of the loading device can satisfy design demands of technique index, realize loading of any torque value at static and dynamic state, and keep stable.

The ideal loading area of the loading device is clarified on the basis of guaranteeing G6.3 balance quality through analyzing stability experiment of the loading device which guarantees that retarder experiment platform simulates actual vehicle operation working condition.

Accuracy of static applied loading angle and applied loading torque of the loading device is checked by analyzing static loading experiment of the loading device.

ACKNOWLEDGEMENTS

Fund Project: Scientific and technological projects in Heilongjiang Province (GC12A412).

REFERENCES

Guo H.T., Liao W.H. 2012. A novel multifunction rotary with magnetorheological fluid. *Smart Materials and Structures*. 21(6):1–9.

Li Guofa, Zhao Pu, Liu Chang, Gao Wei, Shan Cuiyun, Zhang Jiyuan. 2014. Development and experiment of layered magneto-rheological torque transmission devices. *Journal of Jiangsu University (Natural Science Edition)*. Vol. 35 No. 1:20–24.

Li Xia, Fu Junfeng, Zhang Dongxing, et al. 2012. Study on traction resistance based on the experiment analysis of a vibration ripper. *Transactions of the CSAE*. 28(1):32–36.

Tian Z.Z., Hou Y.F., Wang N.N. 2011. Effect of wall characteristics on transmission properties of magnetorheological transmission device. *Journal of Functional Material*. 42(11):1962–1964.

Tian Chao-yuan, Zhang Jiang-tao, Huo Zhi-jun, Wang Zhi-wei. 2014. Theoretical Design and Simulation Analysis of Disc MRF Brake Tractor & Farm Transporter. Vol. 41 No. 3:19–21.

Yue E., Tang L., Luo S.A., et al. 2011. The study on modifying suspended phase of high-performance magnetorheological fluid. *Journal of Functional Material*. 42(8):1433–1435.

Manufacturing Engineering and Intelligent Materials – Lu & Abu Bakar (Eds)
© *2015 Taylor & Francis Group, London, ISBN 978-1-138-02832-6*

Detection of debonding defect in CFRP reinforced concrete using infrared thermal imaging

L.N. Liu

Jincheng College, Nanjing University of Aeronautics and Astronautics, Nanjing, China

ABSTRACT: This paper presents an experimental study of quality inspection and defect detection for CFRP reinforced concretes. An experimental infrared thermal imaging inspection system was established to inspect CFRP reinforced concrete specimens with different types of internal debonding defects. During the experiments, transient thermal excitations were applied to the specimens and thermal images of the surfaces of the specimens under different conditions were obtained. The changes of a sequence of the thermal images over time were observed and analyzed. The results have demonstrated that the idea of using the infrared thermal imaging technology to detect debonding defects in CFRP reinforced concrete is completely feasible, and image analysis on the thermal images of defects can accurately reflect the status of defects.

Keywords: debonding detection; CFRP reinforced concrete; infrared thermal imaging; image analysis

1 INTRODUCTION

Carbon Fiber Reinforced Polymer (CFRP) is a kind of ideal material for repair and reinforcement; it has been widely used in various engineering fields Guo (2012). In this paper, an experimental study for inspection the quality of bonding of the CFRP reinforced concretes by infrared thermal imaging is presented (Feng et al. 2014). An experimental infrared thermal imaging inspection system was established, and transient thermal excitations were applied to the specimens with several type of internal debonding to obtain their thermal images. Image analysis technique was applied to analyze the sequences of thermal images to obtain information about the internal debonding. Experimental results have demonstrated the effectiveness of the infrared thermal imaging technique for detection of debonding in CFRP reinforced concrete.

2 PRINCIPLE OF INFRARED THERMAL IMAGING INSPECTION

Infrared is a kind of electromagnetic wave, with the same nature as the radio waves and visible lights. By using special electronic devices, the temperature distribution on the surface of an object can be converted to visible images, and different temperatures are displayed in different colors. Such a technology is called infrared thermal imaging technology. When an object is heated, in the process of tending to thermal equilibrium, the change of surface temperature field in space and time not only relates to the kind of material, but also be affected by the internal structure and inhomogeneity. In most cases, the local defect makes the thermal wave transmission non-uniform, the thermal wave experiences scattering and reflection. This can be reflected on the surface temperature field. By controlling thermal excitation and using infrared thermal imaging technology to record the temperature of the surface infrared radiation as visible image (as shown in Fig. 1), the non-uniformity of the

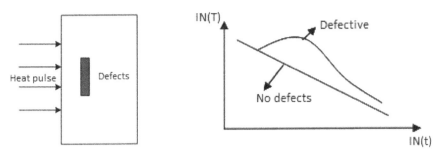

Figure 1. Principle of infrared thermal imaging inspection.

material can be obtained and the defect under the surface can be found, thus the purpose of NDI is achieved.

3 EXPERIMENTAL STUDY

3.1 *Infrared thermal imaging inspection system*

As shown in Figure 2, the experimental infrared thermal imaging inspection system established in this study mainly consisted of an infrared thermal imaging camera, a thermal excitation source, an excitation time controller and a computer processing system. The infrared thermal imaging camera was used to detect thermal radiation and record thermal images. The infrared imaging camera used in the experiments was FLIR S65, it had an acquisition frequency of 50 Hz, the acquisition time interval was 20 ms, and the temperature resolution was up to 0.06°C. The transient thermal excitation source used in the experiments was Photo Light-1000 W heat lamp. Meanwhile, the JSS48 A-1Z digital time controller was employed to control the duration of thermal excitation.

3.2 *Experimental specimens*

The specimens used in the experiments were concrete specimens bonded with CFRP sheets Li (2010). To simulate the CFRP debonding caused by the spalling on the concrete surface with different size and depth, artificial circular shape holes with different diameters and depths were made on the specimens. The specific shapes and sizes of the concrete specimens and debonding defects are shown in Figure 3. Two kinds of defect forms were considered: specimen 1 had defects with the same diameters but different depths, and the specimen 2 had defects with the same depths but different diameters. By using epoxy resin adhesive, CFRP sheets were bonded on the surface of the concrete specimens. The thickness of the CFRP sheets was 0.167 mm.

3.3 *Results of transient thermal excitation method*

By using Photo Light-1000 W heat lamp, the two specimens were applied with transient thermal excitations. The excitation durations were 3 to 5 seconds. The infrared thermal imaging camera started to record thermal images from the beginning of heating, while stopped a while later after the end of heating. It contained the heating process and cooling process. The thermal images with maximum contrast were selected and analyzed. Grey linear transformation were applied to these images to enhance the contrast Tian (2006). Figure 4 shows the thermal images with maximum contrast for specimen 1 and specimen 2 before and after grey linear transformation, respectively. Figure 5 shows the change of contrast with time for specimen 1 and specimen 2 under transient thermal excitation.

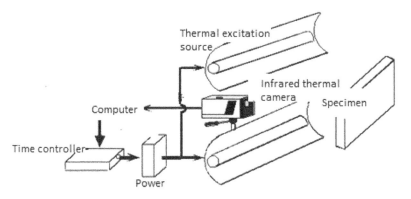

Figure 2.   Illustration of infrared thermal imaging inspection system.

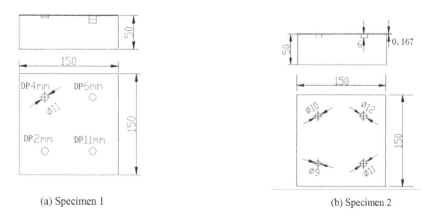

(a) Specimen 1   (b) Specimen 2

Figure 3.   Dimensions of CFRP reinforced concrete specimens and defects.

(a) Specimen 1 change before and after transformation   (b) Specimen 2 change before and after
                                                                                    transformation

Figure 4.   Thermal images with maximum contrast for specimen 1 and specimen 2 before and after grey linear transformation.

From Figures 4 and 5, it can be seen that, for defects with the same diameters, the deeper the hole, the greater the contrast; for defects with the same depths, the larger the size, the greater the contrast. Under the same thermal excitation, the difference between the defects and other parts in the specimens is more obvious (namely greater contrast) with deeper, larger defects and longer thermal excitation. And the contrast reaches a maximum at a certain time, then decays constantly until vanish. So there is an optimal observation time window, which is related to the distance between the surface and the defects. In this experiment study, both specimens bonded with the same CFRP sheets, theoretically the distances between the surfaces and the defects should be consistent. However, due to factors such as different adhesive

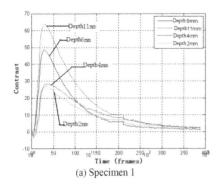

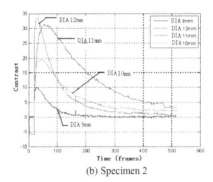

| (a) Specimen 1 | (b) Specimen 2 |

Figure 5. Change of contrast with time for specimen 1 and specimen 2 under transient thermal excitation.

Table 1. Number of defected pixels for specimen 1 and specimen 2.

Depth (mm)	Diameter (mm)			
	9	10	11	12
2			139	
4			154	
6			159	
8	35	100	166	255
11			212	

thickness, the actual distances may not be exactly the same, which is reflected in the contrast change curve with time that the contrast maximums appear in different time positions.

3.4 Quantitative analysis of the defects

In the image, the size of the defects can be characterized by the number of pixels (Yu et al. 2007). For binary images, the number of pixels with value of 1 is used to represent the size of the defect. And the actual size of the defect can be obtained by the proportion relationship between the pixel size and the actual size of the specimens. This proportional relationship can be determined by the imaging distance of the infrared thermal camera and the lens ratio formulated as:

$$PIX = SIZE \cdot \frac{Mrow \times Mcol}{Length \times Width} \tag{1}$$

in which $Length$, $Width$ represent the actual length and width of the specimen, respectively; $Mrow$ and $Mcol$ are the number of rows and number of columns of pixels, respectively; SIZE is the actual size of the defect. After opening an thermal image, through gray level transformation, denoising and enhancement processing, image segmentation are used to get the binary image, and the number of pixels with value of 1 is counted as PIX, then the actual size of the defect can be obtained by the proportional relationship defined in equation (1).

Table 1 shows the number of pixels for the defect in specimen 1 and specimen 2. From Table 1, it can be seen that, for defects with the same depths, the trend of size increment identified by image analysis are consistent with that of the actual defect sizes; for defects with the same sizes but different depths, the sizes identified by image analysis increase with the increase of depth, which can be explained that the deeper the hole is, more serious the heat dispersion.

162

4 CONCLUSION

This study carried out an experimental study to detect different types of internal debonding defects in CFRP reinforced concrete specimens with an experimental infrared thermal imaging inspection system. Transient thermal excitations were used to heat the specimens, and the experimental results have demonstrated that it is completely feasible to use infrared thermal imaging technology to detect the debonding defects in CFRP reinforced concrete. This technology has the advantages of fast, non-contact, large area scanning. The image analysis applied to the thermal images can directly obtain information that reflects the status of the defects, which can be provided for the repair of the CFRP reinforced concrete structure for reference.

REFERENCES

Feng Liqiang, Wang Huanxiang, Yan Dawei, Liu Jilin. 2014. Experimental Analysis of Infrared Thermography for Detecting Internal Defects of Decorative Layer on Exterior Wal [J]. Journal of Civil, Architectural & Environmental Engineering, 36(2): 57~61. (in Chinese).

Guo Wenyang, Liu Jingtao. 2012. Application of FRP in engineering structure strengthening [J]. Science & Technology Vision, 12: 179~180. (in Chinese).

Li Liang feng. 2010. Experiment Study About Infrared Thermography Detecting on Bonded Quality of Strengthened Structure of Bonded Steel Plates [J]. Fujian Architecture & Construction, 144(6): 48~50. (in Chinese).

Tian Yupeng. 2006. Infrared detection and diagnosis technology [M]. Beijing: Chemical Industry Press, 2006: 157~216. (in Chinese).

Yu Manli, Pan Wei, Zhu Ruohan. 2007. Application of IR thermograph and its development in engineering [J]. Science and Technology of Overseas Building Materials, 28(3): 51~54. (in Chinese).

Manufacturing Engineering and Intelligent Materials – Lu & Abu Bakar (Eds)
© *2015 Taylor & Francis Group, London, ISBN 978-1-138-02832-6*

The architecture and implementation of a knowledge-based engineering design verification system

J. Lin, Y.M. Hou, J. Cheng & L.H. Ji
State Key Laboratory of Tribology, Tsinghua University, Beijing, China
Department of Mechanical Engineering, Institute of Engineering Design, Tsinghua University,
Beijing, China

ABSTRACT: Verification is one of the major steps of design process, and it is also the most heavily calculation-relied part in many engineering design cases. This paper proposes a computerized verification system for engineering design, which is developed based on a theory framework called the design verification model. This model looks inside the verification cycles and tries to formulate the containing activities and their organization. With these understandings, a knowledge-based general-purpose verification system is developed to automate various design verifying cases. By implementing functional units of the system as D-Robots, the system has acquired extreme flexibility, expandability, and efficiency. An IC process chamber design case is also presented as an example of applying the system and the method.

1 INTRODUCTION

Engineering design plays a very important role in modern society, and numerous researches have been devoted to this field. Although great modernization have been achieved nowadays, higher level of automation is still demanded by many in engineering design cases, including from fully customizing products to building extraterrestrial colonies. Those requirements demand much more designing workload than the present industrial capability. The technology gap continually propels the development of higher level intelligent system to perform the design works automatically and spontaneously.

The automation of design must be based on the understanding of design. There are many theories of engineering design, like the Systematic Design Approach (Pahl & Beitz 2007), the FBS theory of design (Gero 1990, 2004), the Axiomatic Design theory (Suh 1990, 2001), TRIZ (Altshuller 1999), the Embryo Design theory (Hou 2005), etc. These theories have shown how to perform design, but most researches believe that the engineering design is so creative that it is still impossible for artificial intelligence to fully take over. Meanwhile, some of the design activities are less creative and repetitive, like the design verification.

The verification is so important to the modern industry that the infrastructure for verification must be gathered ahead of any design coming into form. After all, how can a company be called automobile designer without a wind tunnel?

As turning to the biology field, verification gets even more universal that all the species are designed through verification. For billions of years, Mother Nature has changed the design of DNA arbitrarily and occasionally, and then put the outcome organism directly to the trial, verifying whether it can represent certain level of surviving and self-duplication capability. In this case, verification could even be called passive design, which means doing design by verifying the result.

Inspired by the fact that all the varieties of creatures are "designed" with such a simple logic, it implies that there could be a simple and beautiful rule for all the verification problems. If it does, then it will be possible to make a general verification system to conclude all the verifying cases. The automation of verification could be achieved ahead of other parts of Design Automation (DA).

In this paper, a framework of design verification is established and a general verification system for engineering design is developed. This system is developed to contain product knowledge and process knowledge used in verifying engineering design. And the contained knowledge is systematically organized and can be invoked spontaneously so as to achieve automatic verifying. Meanwhile, the system is developed to tackle the massive calculation duties simultaneously. The parallel calculating capability could be expanded according to instant requirements with extreme flexibility.

To build a system like this, several elements are essential. First, a general model for all kinds of engineering design verification must be established to represent their common principle. Secondly, the system should have enough intelligence of planning how to organize the massive calculations, and performing automatically in a parallel manner. And thirdly, the system should have enough flexibility and expandability to not only integrate various kinds of calculation module, but also allow the domain experts to inject knowledge into the system. And it will be better if the knowledge injection work can be done in those experts' familiar manner.

In the next section of the paper, a brief review of modern KBE technology and KBE system will be represented, along with some concerns of KBE. And then in Sections 3–5, those essential aspects mentioned above will be discussed, respectively. By showing these aspects, the architecture of this verification system will gradually become clear. If not, an actual example is represented in Section 6. And in the last section, some conclusions and future expectations are provided.

2 RELATED WORKS

2.1 The design and verification

There are many theories of engineering design, for instance, the systematic design theory which is the lead theory in the industry (Paul & Beitz 1977), or the Function-Behavior-Structure Framework which is the most famous in understanding the essence of engineering design (Gero 1990), or the axiomatic design theory which is very influential in guiding the designer to make right decision (Suh 1990), or the TRIZ theory which helps logically searching for the solution (Altshuller 1946, 1999), and the Embryo Design theory which divides the engineering design into six phases inspired by the development of embryo (Hou 2005). All these theories described how design goes, with their own point of view, and all of them separates verification from other activities of design.

Paul & Beitz (2007) described the activities of designers as "applying scientific and engineering knowledge to the solution of technical problems, and then optimizing those solutions within the requirements and constraints set by various considerations". This definition implies that the full design process is composed of two different parts, the applying knowledge part, so-called design, and the optimizing part belongs to what we are discussing here, the verification. Verification is to verify whether the solution can actually satisfy the requirement. Meanwhile, solution optimizing can also be achieved during verification. The optimization is a kind of special case of verification, it is composed of multiple verifying process with shrinking available behavior arranges.

Since the verification is regarded as repetitive and non-creative activity here, and obviously most of the engineering design verification requires Multi-Disciplinary Optimization (MDO), it meets both the applicable conditions of employing Knowledge-Based Engineering (KBE) technology (La Rocca 2012).

2.2 Commercial software

Nowadays, many types of commercial engineering software are employed in the verification and optimization of engineering design. Most of them are performing as certain tools to the designers, like geometry modeling tools, FEM modeling tools, workflow tools, optimizing tools, and so on.

As the idea of **KBE** gets more and more practical, modern commercial software also get influenced. The intelligence level get improved, many jobs get more and more automated, and many workflow software have integrated the functions of invoking other software. But most of these automations are achieved by human interference, rather than absorbing knowledge into the systems. And the achieved automatic actions cannot be transformed into knowledge for accumulation or reuse.

By using workflow integrate software, like Optimus, ModelCenter, ANSYS Workbench, etc. the users need to define a working sequence of invoking different working tools, modeling or analysis software. And a pre-defined script of operation list must be provided to those working software. Within the script file, only parameters which can be matched with position or regular expressions mechanism could be changed, no matter through a software specified API or directly editing the ASCII format script file.

The modern commercial workflow integrating tools do have some similarities with **KBE** system. But since both the workflow sequence and the specific working script are given by the designers, rather than the knowledge formulized in the system, they can still only be considered as offering tools to the users but not a knowledge-based system.

2.3 *KBE technology*

Knowledge-based engineering is a relatively young technology. In recent years, **KBE** has gradually showed great efficiency especially in dealing with repetitive and non-creative works. Add that, the boundary between creative and non-creative are not that solid, but highly depend on the abstract generalization. Once the hidden principle was found, the creative work could fall into non-creative one.

The traditional technique and system used in the engineering design could be called operation-based. A major difference of operation-based and knowledge-based system could be found by seeing how the system is developed. While developing traditional engineering software, the developers concerns most about what kind of functions they would provide to the end users, or the designers. These systems provide a working environment where the designer should decide whatever operation to be performed or whatever tool to be used.

Meanwhile, the developer of **KBE** system concerns most about the knowledge involved in the system, what kinds of knowledge are needed to inject into the system, and how to systematically store and invoke the knowledge. With this knowledge, the users of a **KBE** system could execute a series of operations automatically so as to achieve certain level of automation.

This is the essential difference between today's dominant commercial software and **KBE** system, and tells where they miss the new-generation design and engineering system standard of cognitive, collaborative, conceptual and creative (Goel et al. 2012, La Rocca 2012).

2.4 *KBE systems*

The software entity which is designed under certain systematic framework and developed with various knowledge based are called **KBE** System. And when the knowledge about a specific use case is injected into the system, it would be instantiated into a **KBE** Application.

Among all the published **KBE** systems we have read, the DEE and IDEE systems developed by T.U. Delft for aerospace or automobile industry in recent years (Van Dijk et al. 2011, Cooper 2009) are most inspiring to us. And they are surely among the state-of-the-art **KBE** systems.

The DEE and IDEE systems do have extreme advancement, that they have been reported to tremendously reduce the cost and time consumed. But the approach that how the product knowledge was formulized in the systems seem a little bit inconvenient.

DEE employs an Object-Oriented program language (GenDL, superset of Lisp) which integrated with a geometry modeling kernels to contain the product knowledge (Verhagen et al. 2012). But considering the efficiency of geometry information describing, using a program language is not compared with modern GUI modeling software. Not to mention the process of grasping a programing language is time consuming. For a KBE system, the geometry knowledge providers are not developers, such as the verification system introduced here, it maybe impractical for them to use a program language. Instead, a method of building parametric geometry model by combining the commercial modeling software with extra product knowledge will be addressed later in this paper.

2.5 *KBE concerns*

In a Delft recent critical review of KBE technology, several higher level concerns have been stated (Verhagen, 2012).

- Case-based development of KBE applications
- A tendency toward development of 'black-box' applications
- A lack of knowledge re-use
- A failure to include a quantitative assessment of KBE costs and benefits
- A lack of a quantitative framework to identify and justify KBE development

The author has mentioned some suggestions to remedy these shortcomings, both from methodology and method level.

In addition, the case-base and lack of re-use situations for KBE are quite linked together. And while developing the verification system, we found that they all come from the different object s of system building. The traditional engineering softwares offer tools to the user, and tools are supposed to be case irrelevant. But KBE systems or applications offers the user with the knowledge, which is highly case related. It is natural to find that case-based knowledge much more difficult to be reused than traditional engineering softwares.

But look again at the knowledge formulated in the KBE system, they are the knowledge about actually operating those tools, with the knowledge, the users of KBE application can systematically reuse the operations. And the operations could be reused because of the hierarchy of knowledge and operations, the knowledge are located above operations and the principles of operation are described within the knowledge. So, if we want to make those case-based knowledge to be reused, we must build another hierarchy above those use cases. And that leads to the idea of developing this general-purpose knowledge-based verification system. With the general model introduced later, the verification of various engineering design cases are kind of similar, which supports more knowledge reusability.

The re-use of knowledge relies on the hierarchy of knowledge, how to re-use the knowledge is itself a kind of knowledge which is located at higher level than the knowledge being re-used. But the order of knowledge does not spontaneously develop with the increase of knowledge (Horvath 2004). In order to reuse the knowledge, the KBE system must have a more complex hierarchy of the KBE cases or KBE applications.

3 MODELING AND ARCHITECTURE OF VERIFICATION SYSTEM

The verification system is developed based on the understanding of engineering design verification. The verification model will be represented here to explain how the system works from the conceptual level.

3.1 *Design verification model*

The design verification model in Figure 1 shows the position of verification among other design activities. The elements of this model mostly come from John Gero's FBS theory, and the model could also be seen as the FBS theory, but be presented in a special way so that the relationship between verifying and other design activities are focused on. The verifying in this model is equivalent to the analysis and evaluation of FBS theory (Gero 1990).

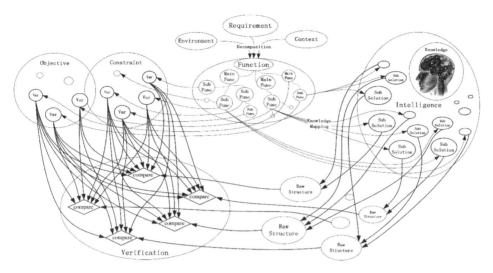

Figure 1.   Model of engineering design verification.

The Embryo Design theory has proposed a Growth Form Design Model (GFDM), which classified the activities of design into 6 stages: Function, Surrogate, Property, Specification, Feature, and Structure (Hou 2005). GFDM described the mapping process of engineering design from function to structure in a bionics point of view other than the classical FBS model. The feedback route of GFDM called measurement is corresponding to the verification of design, but the Embryo Design theory did not describe in detail how verifying works. The Design Verification Model in Figure 1 shows the relationship between verification and other design activities, and the frameworks showed in Figure 4 proposes the idea of how verification works. These two models are the fundamental ideas which the verification system is built on.

The verification system is developed to perform fully automatically, because the verifying activities of engineering design are considered highly repetitive. Some activities in the GFDM are also considered as repetitive. A new concept software entity, called Design-Robot (D-Robot), was proposed to handle the repetitive duties of design and verification. The Embryo Design and the verification system are well linked together, expecting to contribute to enhancing the Design Automation.

3.1.1   *Requirements and functions*

The activities of engineering design begin with the requirements. The requirements that need to be fulfilled inducts into functions, along with the environment and the context. The context means the form or latter requirement related to this one.

Most requirements cannot be fulfilled by a single function of the to-be-designed object. It always needs to be decomposed into a group of sub-functions. The procedure that proceeds from requirement into decomposed function group is considered to be the most difficult part for artificial intelligence to accomplish, for the environment and context contain knowledge of such a wide range and that boundary of it cannot be explicitly defined. Like the systematic design theory or any classical design methodology suggests, identifying the requirement always plays a key role in the whole design circle.

As the design procedure reaches the decomposed functions, it splits into two branches: one to the quantitative variables, the other to the sub-solutions.

3.1.2   *Behavior branch*

The functions that the designers want to realize need to be verified before any conclusion can be made, and they must be represented as quantitative variables, which is equivalent to

169

the Behavior Variable concept in FBS theory, for the computer to deal with. So each of the decomposed functions must be mapping to a corresponding variable.

Those mapping variables could have been further classified into constraints or objective, nominates from optimization. The constraint is a variable with available range, upper or lower or both, and the objective is a variable with an expected trend, either the designer wants to maximize it or minimize it or as close as possible to a certain level. And a variable could be constraint and objective at the same time, for instance the designer wants to maximize a variable, but only bigger than a certain level could make the design work. But in this system, we treat this situation as two variables with same sampling strategy (will be discussed later), but one constraint and one objective. The constraints and objectives which a design task must fulfill are aggregated into the behavior, a term from FBS theory. Figure 2 shows a UML model of the behavior related classes in the system.

3.1.3 *Structure branch*

The other direction that comes out from the sub-functions is mapping to sub-solutions, which is the trickiest part of engineering design. For every single sub-function, the designer must find a method to tackle with it. Sometimes, there will be several possible solutions there that can fulfill one function, but only after every single function has mapping solutions could the design go on. Today, the solution mapping is still mainly dependent on human brain, but the cortex are not very good at mapping work. There are mainly two reasons for it. The first is the information gap: there is just no corresponding solution there. In order to ensure the knowledge reserve, designer teams always be found. But that could fall into the second drawback even deeper, the low efficiency of information searching of cortex. Occasionally, the solution is there, but you just cannot reach it. Only after the designer is done, and you are seeing it, one realizes, how I could not have thought of that. Some methodologies put special effort on the solution mapping of engineering design. The most famous one is TRIZ (Altshuller 1999). Most of the TRIZ principles are about how to think systematically and comprehensively. And the framework of a next-generation patent searching and sorting system based will surely do great help to it (Li et al. 2012).

Each of the sub-function must map to one or more sub-solutions before the design goes on. And then the sub-solutions can be put in combinations into the raw-structures. The raw-structure, which means the topology of the product, is defined, but the parameters are still undetermined. The role of parameter here is equivalent to the Structure Variable in FBS

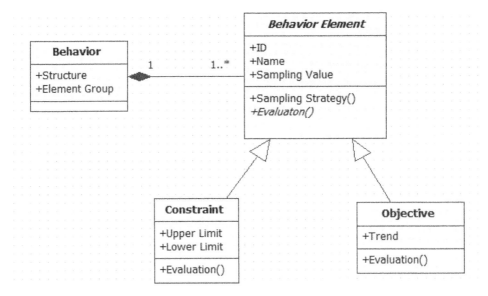

Figure 2.   UML model of behavior.

170

theory. The parameters could have possible range or possible values, due to the assemble relationship, the manufacture capability, or the supporting market, etc. Figure 3 shows the UML models of structure class and parameter class in the system.

The behavior and the raw-structure are the input for the verification system, what the system will do now is to predict the performance of the structure, and verifying if the performance of the candidate structure could satisfy the required behavior with the parameter stay in certain available range. And by converting the calculated value of an objective behavior into the constraint of next verification case, the system performs the optimization, the sub-set of verification.

The concept of function, behavior, and structure are inherited from the FBS model of design. For any case of engineering design, the designer always have to go from the function to the structure, and to a computer system, the behavior is acting as a quantifiable middle role, which can be derived from both sides, so as to make the function and the structure comparable. The verifying procedure is similar to the analysis and evaluation of the FBS model, but rather than reformulation while the evaluation do not pass, the verification system could theoretically explore the whole design space at same time. The reason for that comes from the different expertise of human designer and the computer, along with the extreme parallel capabilities of the system.

John Gero established the FBS theory in 1990, and extended it into a dynamic model called situated FBS framework (2002). Fundamental it is, the FBS model is still been criticized today. Some researchers questioned whether the FBS model explained the principles of design activities, or is just a guide for any digital design system (Dorst, 2005). The model shown in Figure 1 tends more to the latter. Obviously, a human designer would not follow those streams while doing design. More likely, from nowhere, the designer got a most appealing solution, and then other possible branches are cut down and put aside till after the intuitive idea has been verified. This picture shows the sequential character of human brain, which is hugely different from the logic of a computer.

The model of design verification explains the role of verifying among other design activities, and the boundary of the system is clarified. It takes two inputs: the raw-structure along with the undetermined parameters and the behavior which the structure should achieve, and the system output of the verification result of whether the structure is capable of performing as expectation and with what configuration (the value set of those parameters) will the structure perform that way.

Here, we see the reason why representing product knowledge with program language is not fit for this system, which is mentioned before. That is, because the structure is input to the system not generated within the system. The models provided by the user are in commercial geometry model format, and the system was developed to maintain them that way, and employ special knowledge module to extract information out and generate multiple analysis model.

What is more, the website of Genworks releases a new version of GenDL while this research is undergoing. It seems that the enterprise edition will support commercial CAD

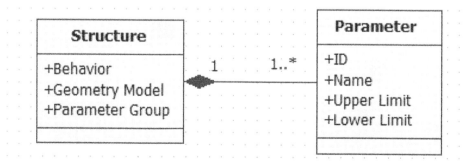

Figure 3. The UML model of structure and parameter.

171

model converting and translating. Will this improve GenDL with more users friendly or make it less different with the approach employing in this verification system is still under observation.

3.2. *System architecture*

Figure 4 shows the architecture of the verification system, and it also explains how the verifying process works. The input of the verification system is to-be-verified structure, required behavior, and undetermined parameters of the structure. The parameters are equivalent to the structure variables in FBS theory, and input to the system along with their possible value range. All arrangements are combined into the design space of the structure. A class named Explorer will manage to find out the possible behavior of the structure while the parameters changing within the design space. If the structure can perform required behavior, the verification is passed, and meanwhile, the available arrangement is discovered by the system. Otherwise, the verification has failed.

3.2.1 *Class explorer*

The arrangement of each parameter is aggregated to form the design space of the structure by the Explorer. Every possible value set of the parameters is called a "configuration" of the structure, and will act as a point in the design space. The purpose of Explorer is to explore the whole design space to verify whether the structure is capable of giving expected performance.

In order to achieve this, the Explorer needs to decompose the design space into sampling missions. Each mission is a mapping procedure from the design space to the behavior space. The reachable behavior of the structure will be gradually made clear while more and more sampling missions are accomplished. And the exploring will end whenever the verification result can be made.

The actual mapping process is performed by a class called Sampler, which is implemented as a Design Robot (D-Robot). The Sampler will be discussed later.

The algorithm which Explorer employed to decide which point should be sampled is fully flexible. Most of the existing optimization algorithm will do. However, considering the most advantage of the system lay on parallel calculation, those algorithms which can deploy multi sample points will fit the system best. For instance, the response surface based algorithms are better than the gradient-based algorithms, because building a response surface allows multiple sample missions to be performed simultaneously, but the gradient-based algorithm can only give the next sample point after the former one is done.

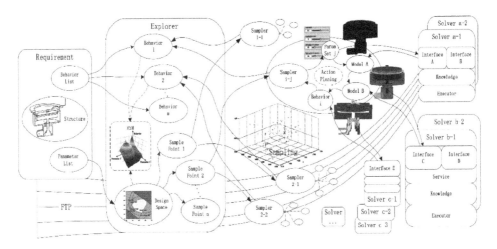

Figure 4.   Architecture of verification system.

172

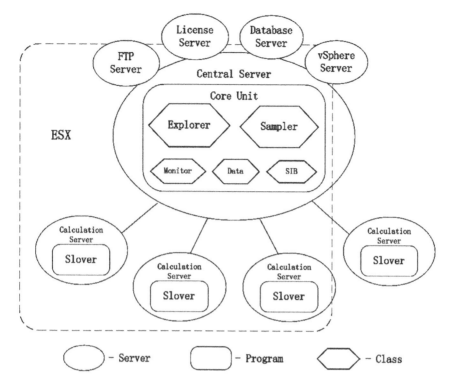

Figure 5. System architecture.

The class Explorer and the class Sampler are the major functional entities. Along with some supporting class, they form the core unit of the system at central server. The Solver are not class, each solver is an independent calculation unit located at distributed server, communicating with the central server over network. The core unit and the solvers form a tree-like hierarchy as shown in Figure 5.

4   THE IMPLEMENTATION: D-ROBOT

The verification system is developed to realize fully automatic running. The method of achieving this goal is to implementing this system with a new concept software entity called Design Robot (D-Robot).

Most of the sample missions, which need to measure the behavior of the structure with given parameters, cannot be done directly. The Sampler needs to invoke multiple calculating tools, called Solver, to accomplish the mission step by step. The action sequences are decided by the Sampler by its action-planing capability provided by the D-Robot architecture. And each of the Solvers is implemented as an independent web service accessible to the network.

The idea of D-Robot is brought out to handle the Gene Transcription mechanics by Hou's Embryo Design theory (Hou & Ji 2013). The role of D-Robots in engineering design is introduced in the literature, and the realization of D-Robot will be presented here. A D-Robot is like an agent for its automatically performing manner. But instead of the mental state of an agent, like beliefs, decisions, capabilities, obligations, etc. (Shoham 1993), the D-Robot employs the components of an industrial production line robot to organize its attributes and methods, such as sensors and effectors. Its job is to perform some repetitive activities of design automatically, which, according to the Embryo Design theory, is the process to extract the information from the gene and then make a transcription. Meanwhile, the pathway of doing it should be figured out by the D-Robot itself without human interference.

When agents are used in conceptual design, they generate problem solutions by integrating qualities of the human design process into computational algorithms based on Stochastic Optimization, Genetic Algorithms, Artificial Life, Multi-Agent Systems, Qualitative Physics, Bond Graphs and Utility Theory (Campbell 2000). However, some design activities are repetitive and can be executed based on discrete rules. In these cases, less smart entities than agents will reduce the complexity of the system.

4.1 D-Robot architecture

The idea of D-Robot is the major method of implementing the whole verification system. It helps the system run spontaneously and automatically, as well as clearing the boundary of each module.

4.1.1 Sensors and effectors

The idea of D-Robot originally came from the industrial robot. While in the production line, each robot accepts the raw input from the line, and performs a series of actions upon it before putting it back to the line. In order to accomplish the expected actions, industrial robots must have the sensors and effectors for measuring and operating, respectively. Those multifunction robots, which were designed to deal with different products, need extra sensors to detect the imported product so as to apply different operations.

For instance, a Solver in the system called Mesher is a D-Robot. It takes the geometry model, and transforms it into discrete model for FEM calculation. The input file is in IGES or STEP format, and should be inspected thoroughly before any action can be taken. The Mesher employs ANSYS ICEM to conduct the actual meshing job, and generate a journal file (the processing script) to guide the ICEM with every single operation. And the operations in the journal file are documented with the ID of the geometry elements (nodes, curves, etc.) involved in the operation. So the sensor of the Mesher, a file analysis module, must check each element of the input model, make sure every maneuver in the script points to the right object. After that, the effector takes its actions, ICEM been launched, and the mesh model will be generated.

4.1.2 Action planning

The word robot originally came from Czech 'robota', means servitude, forced labor (Wiki). The most significant purpose for human to build robots is that they can serf human with labor. And a good serf should know how to fulfill the demand by his own thought, and does not require his lord to give every order. It demands the D-Robot having an action-planning function to spontaneously figure out how to schedule its actions. This character will be discussed in detail while introducing the Sampler.

4.1.3 Multiple state and gene transcription

The demand that the lord gives to his serf could be a complex one. It may need the serf to accomplish it step by step.

In this verification system introduced in this paper, the D-Robot needs to transform the design models from one state to another, which is the Gene Transcription process defined in the Embryo Design theory.

4.1.4 Discrete rules and nested hierarchy

To fulfill any demand of his lord, the serf must have the corresponding knowledge. But it is impossible and unnecessary to have a cleverest serf to know all the required knowledge. He may have his subordinate serf to grasp the knowledge of discrete domain, each serf doing what he could, and together to fulfill the lord's demand.

The idea of discrete knowledge is also applied in the verification system. There are many D-Robots employed in the system, organized in a tree-like hierarchy. Each of them has a discrete and knowledge base, contains the rules and expertise corresponding to its role in the system. And those knowledge bases are designed in an expandable way, which enables the expertise's and the domain expertise to add new context into it.

4.2 Sampler, the D-Robot

The Sampler is one of the major functional classes of the system. It is developed as D-Robot, and works in the core unit of the system. Figure 4 shows the relationship between Explorer and Sampler. In a running system, the Sampler receives sample mission from the Explorer and returns with the sample result. Every time a sampling mission is deployed, a Sampler object get instantiation, and each object works with an individual thread.

Mapping from the parameters to the behavior, the Sampler has to firstly make action planning. With the help of Service Interfaces Board (SIB), the Sampler gets the information of all available Solver interfaces. The information includes the addresses, structures, input models, output models, and the working states. Those models are several of simplified state of the designed structures, and they bridge the path from the structure to the behavior.

The service interfaces board is provided by another object named Solver Manager, which is also running in an individual thread. The Solver Manager collects the information about all Solvers and puts them together. Thus, the Sampler can use that information to make action planning.

After the analysis path is decided, the Sampler could make its first move. Again, he looks at the SIB, and tries to find an idle Solver which correspond to the first step of the calculation path. Once found, the Sampler marks that solver as occupied on the SIB and then invoke that Solver to do the actual calculation. While calculating, the Solver monitors the state of the Solver periodically. Once the calculation is done, the quantitative or brief result will be reported to the Sampler, and those heavy-load output models will be uploaded to the

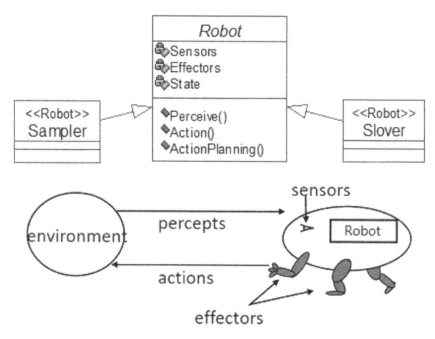

Figure 6.   D-Robot architecture.

file storage. To add that, if no idle Solver could be found, the request from the Sampler will trigger an eMotion cloning, a new copy of the demanded solver will be made. Section 5.4 will explain this in detail.

Typically, an unbranched calculation takes 3–5 steps. For instance, the first Solver take the parameters and the structure as input, and generate the analysis model, which is the geometry model that describes only the specific region that was involved in the analysis. And then, the second Solver takes the analysis model as input, and generates the discrete model, which enables the FEM calculation upon the model. And the third Solver applies the FEM calculation to the discrete model, and hoping to get a converged result of physical field within the analysis region. And the last Solver puts measurement upon it and retrieves the desired quantified behavior.

From the system's point of view, each of those simplified models represents a facet of the to-be-verified structure. The information contained in those models are not created but discovered by the Solvers. They are inherited by the structure ever since the values of those parameters are decided by the Explorer. Based on the idea of Embryo Design theory, those parameters are part of the structure's gene. With the parameters set, the structure's gene is solidified. The gene are transcribed by the sampler to different Solvers and developed into different model, just as the same chromosome can develop into different cells and organs by meeting different inducible factors.

5 TECHNOLOGY: FLEXIBILITY AND EXPANDABILITY

To build this system, all kinds of advanced IT technologies are employed, both of software and hardware. They have greatly ensured the extremely flexibility, expandability and efficiency of the system. Some distinctive ones are introduced here briefly, hoping to further contribute to the development of KBE. All of them are language irrelevant.

5.1 *Service Oriented Architecture (SOA)*

Service oriented architecture is a software design pattern for cooperating the distributed computers connected over network (Erl 2005). As shown in Figure 5, all the Solvers are packaged as web service, their offering functions and demanding inputs are announced with WSDL (Web Service Description Language), and the communicating messages are sent with SOAP (Simple Object Access Protocol) (WSDL 1.1, http://www.w3.org/TR/wsdl).

Web Service programming is fully language irrelevant. And Microsoft Visual Studio has provided the tools for generating all those components.

The formerly mentioned SIB registers all the services and interfaces by reading the WSDL of the Solvers so as to build contracts between the core unit and the Solvers.

5.2 *Dynamic multi-threading*

The thread pool pattern is very useful in computer programming to meet the requirement for parallelization. In the verification system, each object of Explorer and Sampler is instantiated with an individual thread taking from the thread pool. When the mission is accomplished, the object will be deconstructed with the thread returning to the thread pool. The thread pool does not require explicit coding from programmers, some advanced languages, like C# and Java, can manage it automatically in the background, and packages for other languages can also be found.

The maximum number for parallel thread is decided by the memory space. Compared to the maximum number of parallel Solvers, it usually is considered as infinite.

5.3 *Virtualization*

The parallelization for Solver is supported by Virtualization with an excellent function, called vMotion, offered by VMware ESX platform (Daniels 2009, Gleed 2011). The Solvers of the verification system are deployed with virtual machine in a VMware ESX server. When the Sampler cannot find an idle Solver from the SIB, it will send a request message to the ESX server to trigger a vMotion cloning operation. The ESX will make a new copy for the existing virtual machine containing the Solver without interrupting its current work. Cloning a virtual machine with a typical size of ten Gigabytes within the same RAID costs only several minutes. (But across storage vMotion is terribly time consuming.) When the cloning is done, the new virtual machine will be launched and registered to SIB, available to be invoked by Sampler. And the redundant virtual machines can be deleted at anytime to release the resource.

With the help of virtualization technology, the verification system can maintain only an instance for each Solver, and automatically raise new Solver according to instant demand. Meanwhile, the effort for knowledge injection is also reduced. All the contents of the Solver will be cloned.

5.4 *Capsulation decoupling*

Capsulation decoupling has been proved as an effective software engineering manner, which helps to clarify the duties for every programmer or developer group, and improve the cooperation and parallel developing. The main program of the system's core unit contains two types of classes, the functional class and the data class. Explorer and Sampler are functional class, at the same time, they are also acting as the boundary of separated module or working unit. Inside these modules, it shows a picture of traditional OOA and works with an individual thread, while outside them, they never directly communicate with each other. Who they do communicate with are the data classes. All the data classes in the system are generalized from the same parent class called Data, which has the methods to related variables in the memory with the database. But all children data classes have no

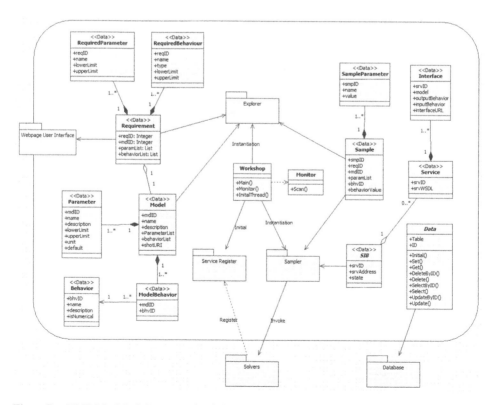

Figure 7.   UML Model of the core unit of the system.

177

methods, they are designed to offer input value and contain output value for the functional classes. Thus, the data classes form the boundaries of functional classes, and the functional classes themselves are decoupled. The UML model of the main program of the system is shown in Figure 7.

6   EXAMPLE: PECVD CHAMBER OPTIMIZATION

Here is an example of applying the verification system to engineering design. The SC300 is a vacuum chamber for conducting a kind of IC manufacturing process, called PECVD (Plasma Enhanced Chemical Vapor Deposition). Like all other IC manufacturing process, PECVD requires various fields in the chamber distributed uniformly. But exhaust outlet of SC300 is placed asymmetrically because of the motor of the rolling platform, as shown in Figure 8. And that will lead to a nonuniform flow field in the chamber. So the designers wanted to place a gas distributor at the exit of the chamber to balance the flow. Figure 8 shows the structure of the SC300. And the problem is: what is the best diameter of exhaust holes A and B which could lead to the most uniform flow field in the chamber?

This question is typical in IC equipment industry, and fits to the verification system very well. As the requirement comes to the system, the structure is known, but several parameters are undetermined. The behavior to be verified is the uniformity of the flow field, which can be quantified as the deviation of the flow speed close to the upper edge of the roller.

The structure of SC300 was provided by its manufacturer in Solidworks format, which was normally used by the design engineers. When the structure comes to the system, a CAD engineer from the KBE application developer team, who acting as a domain expert to the system, will transport the structure to a parametric driving geometry model, and put that model into the model base of Solver Modeler.

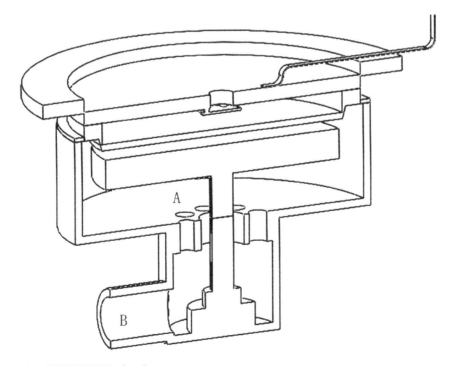

Figure 8.   SC300 PECVD chamber.

Compared to the original structure, the parametric model contains some more information, like the topology of the model, the mapping rules from the assembling to the parts, a set of parameter table, and some other companion information. This information should be explicitly declared along with the original Solidworks files.

Meanwhile, the operation for generating the analysis model must be recorded in Solid-Works macro file. In this case, the operation is to get the inner volume of the chamber which is occupied by the fluid.

The CAD modeling engineer is responsible for transforming the original structure offered by the design engineer into parametric model. And the CAD engineer need not understand the design of the structure as the design engineer does. And that extra information for parametric model is appended to the model with tools developed by KBE application developers. The CAD engineer is not required to know anything about program coding, he/she is acting as a domain expert, the knowledge injector, to the system.

Aggregate the jobs of the Modeler as a D-Robot. Firstly, receives invoking message from Sampler, and then perceives (parses) the required structure and parameter values, and next finding the parametric model from its knowledge base and generate operational script with given parameter, and then generates analysis model, and then uploads the model, and finally reports to Sampler.

The analysis model will then be sent to next solver called Mesher. What Mesher does is to transform the analysis model into a discrete model for FEM calculation usage. And the knowledge injection process is similar to that of Modeler. So does the FEM calculation Solver and the post-processing Solver.

Those knowledge-injecting process may seem laborious, but actually these works are done by different expertise with their already mastered skill, it's pretty easy for them to do it. Once the knowledge injection is complete, the Solvers can perform those operations automatically.

Now the Solvers are ready, after the Solver Manager collected and updated the interfaces of the Solvers, the Sampler is capable of planning a calculation path from the structure to the behavior.

Normally, a calculation case as the sale of SC300 will take an engineer at least half an hour to generate analysis model, and half an hour to generate mesh model, and another hour of FEM calculation to get convergence, and extra 20 minutes to quantify the uniformity, under the circumstance that no mistake happens. In the system, the Explorer employed a RSM algorithm and gave 25 sample points for the first sampling around, and another 25 samplings for the second around, and final 8 samplings to get the optimized result, and it takes about 6 hours to complete all 58 sampling missions with cloning 3 geometry modelers, 3 mesh modelers, 8 FEM calculators, and 2 post-processor Solvers to be invoked by Sampler. Without employing the system, this quest will take at least 1 week. The information of employed hardware is shown in Figure 9.

Server (Dell PowerEdge R910)		Solvers	
CPU	2.0GHz x32	CPU	2.0GHz x2-8
Memory	128GB	Memory	2GB-16GB
Disk	300GB	Operating System	Windows 7 Enterprise Debian 6.0.5
Virtualization System	VMware ESX 5.1		
Storage (Dell Compellent SC200)		Programming Language	C#, Python, Java, PHP
Volumn	24TB (Raid 5)	Applications	Dassault Solidworks 2013 ANSYS Fluent 13.0 ANSYS ICEM 13.0 ANSYS HFSS 13.0 CFD-ACE 2013 MathWorks Matlab 2010 Noises Optimus 10.7
Central Server			
CPU	2.0GHz x4		
Memory	4GB		
Operating System	Windows Server 2008 R2		
Programming Language	C#		

Figure 9.   Server configuration.

7 CONCLUSIONS AND FUTURE WORK

This research is focused on the nature of design and verification activities, and proposes a model describing the detail frameworks of how engineering design verification works. A knowledge-based verification system was developed based on these understandings. And the system is introduced from three dimensions. Firstly, the architecture level, the relationship between verification and other design activities is presented along with the full picture inside verification, so as to conclude the boundary and the architecture of the verification system. Secondly, the implementing level, the detail construction of D-Robot is presented with an instance of Sampler. D-Robots are developed specially to confront the repetitive and less creative activities of engineering design. And they are employed all over the verification system, since the whole verification is regarded as repetitive and less creative. Thirdly, the technical level, several advance IT technology which have been considered as the most important ones for supporting the verification system with flexibility, expandability and efficiency are briefly introduced, including service-oriented architecture, dynamic multi-threading, virtualization, and capsulation decoupling.

From the example of the SC300 optimization application, the verification system exhibited its high potential for time saving with parallel calculation and distributed architecture. And another distinct advantage of the system is that the discrete solver modules and the inner design of knowledge-base separate the process of knowledge acquisition and facilitate various of domain expertise to inject their knowledge into the system by the manner they familiar with, but without programming skills.

Although the KBE technology has enhanced the level of design automation, the engineering design is still too creative to be fully taken over by artificial intelligence nowadays. But a new theory called Memory Prediction Frameworks (MPF), which is about how cortex performing intelligent activities, gives the DA researchers some new hope. The theory claimed that no matter how intricate and complex human intelligence is, the most basic elements of cortex activities are only building memories and making predictions (Hawkins 2004). And some neuron network prototypes have been developed to mimic human intelligence and support Hawkins' theory (George 2008). If further work could really prove that the MPF theory shows the essential principle of human intelligence, it will be best news to DA researchers. For it will provide all the bricks of modeling human intelligence and make it possible to automatically perform all creative design activities.

For now, this research is only focussing on the verification part of engineering design, which is more understandable to researchers and more easy to be treated as repetitive and non-creative activities. And the significance of this verification system will break its limit only after being included as a part of a fully automatic design system.

ACKNOWLEDGEMENTS

This work was supported by National Science and Technology Major Project of China (Grant No. 2011ZX02403-004), National Science Foundation of China (Grant No. 51175284), and Tsinghua University Initiative Scientific Research Program.

The system was developed by Mechanical Engineering Department and Software School of Tsinghua University.

REFERENCES

Altshuller, G. 1999. *The Innovation Algorithm: TRIZ, systematic innovation, and technical creativity.* Worcester, Technical Innovation Center.
Cooper, D. 2009. Knowledge-based Techniques for Developing Engineering Applications in the 21st Century, *Journal of Aircraft,* Vol. 46, No. 6.
Campbell, M.I., Kotovsky, K. & Cagan, J. 2000. Agent-Based Synthesis of Electromechanical Design Configurations. *J. Mech. Des.* 122(1), 61–69.

Dorst, K. & Vermaas, P.E. 2005. John Gero's Function-Behaviour-Structure model of designing a critical analysis. *Research in Engineering Design*, 16, 17–26.

Daniels, J. 2009. Server virtualization architecture and implementation. *Crossroads,* 16(1).

Erl, T. 2005. *Service-Oriented Architecture: Concepts, Technology, and Design.* Prentice Hall PTR.

Gero, J.S. 1990. Design prototypes a knowledge representation schema for design. *AI Magazine* 11(4): 26–36.

Gero, J.S. & Kannengiesser, U. 2004. The situated function–behaviour–structure framework. *Design Studies,* 25, 373–391.

George, D. 2008. *How the brain might work—A hierarchical and temporal model for learning and recognition.* Stanford University.

Gleed, K. 2011. vMotion—what's going on under the covers? VMware vSphere Blog.

Goel, A.K. et al. 2012. Cognitive, collaborative, conceptual and creative—four characteristics of the next generation of knowledge-based CAD systems: A study in biologically inspired design. *Computer-Aided Design*, 44 (2012) 879–900.

Hou, Y.M. 2005. *Research on growth form design—an embryogenesis approach: model and application.* Tsinghua Univ.

Hou, Y.M. & Ji, L.H. 2013. Role of D-Robots in Designing. *Advanced Materials Research*, vols 694–697, 1717–1721.

Hawkins, J. & Blakeslee, S. 2004. *On Intelligence.* Henry Holt and Company.

Horvath, I. 2004. A treatise on order in engineering design research. *Research in Engineering Design* 15: 155–181.

La Rocca, G. 2012. Knowledge based engineering: Between AI and CAD. Review of a language based technology to support engineering design. *Advanced Engineering Informatics*, 26, 159–179.

La Rocca, G. & van Tooren, M. 2009. Knowledge-based engineering approach to support aircraft multidisciplinary design and optimization, *Journal of Aircraft*, Vol. 46, No. 6.

Li, Z. et al. 2012. A framework for automatic TRIZ level of invention estimation of patents using natural language processing, knowledge-transfer and patent citation metrics. *Computer-Aided Design,* 44, 987–1010.

Pahl, G. & Beitz, W. 2007. *Engineering Design: a Systematic Approach (3rd Edition),* Spring-Verlag, Berlin, Heidelberg.

Shoham, Y. 1993. Agent-oriented programming. *Artificial Intelligence*, 60, 51–92.

Suh, N.P. 1990. *The principles of design.* New York: Oxford University Press.

Suh, N.P. 2001. *Axiomatic Design: Advances and Applications.* Oxford University Press.

Van Dijk, R. et al. 2011. Multidisciplinary design and optimization of a plastic injection mold using an integrated design and engineering environment, *NAFEMS World Congress*.

Van Tooren, M. 2003. Sustainable Knowledge Growth. *Inaugural speech*, March 5th.

Win, J.C. et al. 2012. A critical review of Knowledge-Based Engineering: An identification of research challenges. *Advanced Engineering Informatics,* 26, 5–15.

Zeng, Y. & Horvath, I. 2012. Fundamentals of next generation CAD/E systems. *Compute-Aided Design,* 44(10).

Manufacturing Engineering and Intelligent Materials – Lu & Abu Bakar (Eds)
© *2015 Taylor & Francis Group, London, ISBN 978-1-138-02832-6*

Structural and electrical properties of high C-orientation GaN Films on diamond substrates with ECR-PEMOCVD

D. Zhang, Y. Zhao, Z.H. Ju, Y.C. Li, G. Wang, B.S. Wang & J. Wang
New Energy Source Institute of Shenyang Institute of Engineering, Shenyang, China

ABSTRACT: Preferred orientation GaN films are deposited on freestanding thick diamond films by Electron Cyclotron Resonance Plasma Enhanced Metal Organic Chemical Vapor Deposition (ECR-PEMOCVD). The TMGa and N_2 are applied as precursors and different N_2 flux is used to achieve high quality GaN films. The influence of N_2 flux on the properties of GaN films is systematically investigated by X-Ray Diffraction analysis (XRD), Atomic Force Microscopy (AFM) and Hall Effect Measurement (HL). The results show that the high quality GaN films deposited at the proper N_2 flux display a fine structural and electrical property.

Keywords: GaN films; diamond sunstrates; SAW devices; ECR-PEMOCVD

1 INTRODUCTION

Recently, Surface Acoustic Wave devices (SAW) have been realized to be core components of broadband digital communication and video systems. With the rapid development of mobile communications industry, the SAW devices with high frequencies and low insertion have become the focus attention of 3G communications. The SAW devices with operation frequencies higher than 1 GHz will be widely applied in the wireless communication devices. However, the development of SAW devices with operation frequency greater than 2.5 GHz has been greatly limited due to the restrictions in the conventional photolithography process. The use of piezoelectric materials based on diamond as SAW substrates can offer an attractive means for relaxing the lithographic criteria since diamond has the highest stiffness coefficient and sound velocity.

At present, ZnO/diamond/Si layered structures with a diamond layer of 10–100 microns grown by Chemical Vapor Deposition (CVD) technique have been reported (Assouar et al. 2001, Huang et al. 2002, Chen et al. 2005, Seo et al. 2002, Lehmann et al. 2001), however, the heat dissipation performance of the structure was considerably degraded due to the poor heat dissipation property of silicon substrate compared to the diamond. In order to achieve the SAW devices with higher frequency based on GaN/diamond structure, the GaN films with preferred orientation and smooth surface are expected. In this paper, GaN/diamond films were prepared by Electron Cyclotron Resonance Plasma Enhanced Metal Organic Chemical Vapor Deposition (ECR-PEMOCVD) system. The N_2 reactivity can be remarkably enhanced by ECR process, which was necessary for the formation of GaN films under the low temperature. The influence of N_2 flux on structural, morphological, componental and electrical prosperities of GaN films is systematically investigated.

2 EXPERIMENTS

In our experiments, we used substrates of freestanding thick diamond films with thicknesses of about 0.5–0.8 mm prepared by direct current glow discharge PCVD. (Jin et al. 2002,

Bai et al. 1998) Trimethyl gallium (TMGa) and N_2 were used as the source of Ga and N, respectively. The temperature of TMGa was kept at $-14.1°C$ with semiconductor well. The N_2 reactivity can be remarkably enhanced by ECR process, i.e., there are much more particles of reactive nitrogen over the substrate as a result of ECR enhancement, which was necessary for the formation of GaN films under the low temperature. In addition, the higher N_2 flux was used to prove an N-rich atmosphere. Subsequently, the flow rate of TMGa was 0.5 sccm, which controlled by mass flow controller. The deposition temperature is 400°C and the GaN films were grown for 180 min. To investigate the effect of N_2 flux on the structural and electrical properties of as-grown films, the N_2 flux varied in the range of 80 sccm to 120 sccm.

The crystalline quality and orientation of the samples were determined by X-Ray Diffraction (XRD) using a D/Max-2400 ($CuK_{\alpha1}$: $\lambda = 0.154056$ nm). The morphologies of GaN films were analyzed by Atomic Force Microscopy (AFM). Electrical properties of diamond substrate coated GaN films are investigated using Hall Effect measurement (HL5500).

3 RESULTS AND DISCUSSION

3.1 Structural properties

Figure 1 shows the XRD pattern of GaN films deposited on freestanding thick diamond at different N_2 flux. Among the peaks, the stronger peaks appeared at about 43.9° and 75.2°, which are characteristic of the diamond (111) plane and diamond (220) plane, respectively. It can be seen that with the N_2 flux increasing from the 80 sccm to 90 sccm, the preferred orientation of the GaN films become clear gradually and the diffraction intensity of the peak becomes stronger gradually. When deposition temperature further increasing from the 90 sccm to 110 sccm, the preferred orientation of the GaN films becomes poor gradually and the diffraction intensity of the peak becomes smaller gradually, which is close related with the poor orientation is somewhat ascribed to be that the reaction are not complete when the N_2 flux is too low or too high. Moreover, nitrogen donor impurity might induce more defects in the GaN films with the N_2 flux increasing, which lead to the degradation of crystal quality. From the experimental results, it is clear that high c-axis orientation of GaN films is perpendicular to the substrate at the N_2 flux of 90 sccm.

3.2 Electrical properties

Hall Effect measurements are applied to investigate the electrical properties of GaN films deposited at different temperature, and the experimental results show that the GaN films are n-type conductivity. As can be seen from the Table 1, when the N_2 flux increase from 80 scccm to 90 sccm, the carrier concentrations of GaN films gradually decrease and the

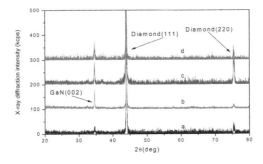

Figure 1.   XRD spectrum of the GaN films deposited at the different N_2 flux of 80 sccm (a), 90 sccm (b), 100 sccm (c) and 110 sccm (d), respectively.

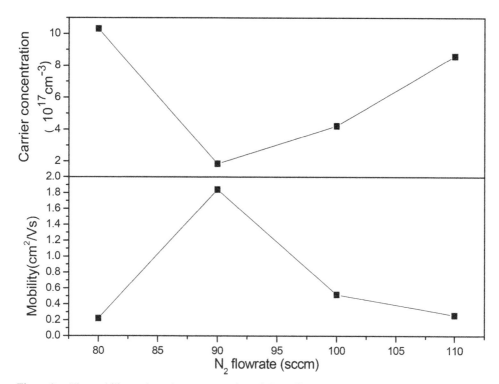

Figure 2.   The mobility and carrier concentration of GaN films deposited at the different N₂ flux.

mobility gradually increases. However, the carrier concentrations of GaN films gradually increase and the mobility gradually decreases when the N₂ flux increase from 90 scccm to 110 sccm. The reason with the poor mobility was ascribed to the scattering effects, which includes defect scattering, grain boundary scattering and so on. And the large background carrier concentrations of GaN films has most commonly been attribute to nitrogen vacancies, high partial pressure of nitrogen at the high growth temperature favors this assumption. And others suggest that the Oxygen incorporation has been shown to increase conductivity and the n-behavior is associated with intrinsic defects in the material. The reason with the samples of poor electrical property was assumed to be as follows: when the N₂ flux is too low, the reaction are not complete and show much defects scattering with the crystal quality which leads to the poor electrical properties. However, when the N₂ flux is too high, there are more N reactive particles over the substrate, accumulation of defects at the interface and results in the degradation of electrical properties under this condition. The results show that the N₂ flux plays an important role on the electrical properties with the GaN films deposited on diamond substrates and can help us to understand how to obtain the high electrical quality GaN films deposited on diamond substrates.

3.3   Morphological properties

A typical Atomic Force Microscope (AFM) image for the GaN films deposited at the N₂ flux of 90 sccm is shown in the Figure 3. As can be seen from the images, the surface islands of GaN films are uniform density and the GaN films have a smooth surface. Therefore, the uniformly and densely GaN films with minimum RMS of 4.5 nm were successfully grown on diamond substrate, which completely satisfies the requirements of surface roughness for high frequency SAW devices.

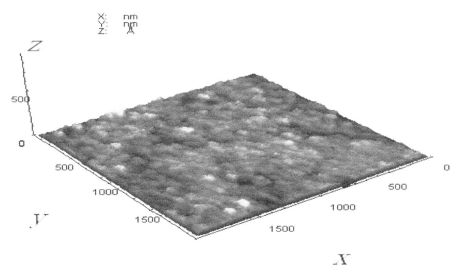

Figure 3.   AFM surface morphologies of the GaN films deposited at the N$_2$ flux of 90 sccm.

4   CONCLUSION

Highly c-axis-oriented GaN films are deposited on freestanding thick diamond films by Electron Cyclotron Resonance Plasma Enhanced Metal Organic Chemical Vapor Deposition (ECR-PEMOCVD). The results show that the high quality GaN films with small surface roughness and high c-orientation are successfully achieved and the Hall measurements indicate that the samples show n-type conductivity. This achievement is very important for the application of SAW device.

ACKNOWLEDGEMENTS

This work was supported by the Natural Science Foundation of China (Grant No 61304069, 61372195, 61371200); The Scientific Research Fund of Liaoning Provincial Education Department under Grant L2014516, The Dr. startup fund LGBS-1408, LGBS-1402 and LGQN-1406.

REFERENCES

Assouar M.B. Bénédic F. Elmazria O, et al., 2001. "MPACVD Diamond Films for Surface Acoustic Wave Filters," *Diam. Rel. Mater*, vol. 10, pp. 681–685.
Bai Yi Z, Jiang Zhi Gang, Wang Chun Lei, et al., 1998. "Effects of Alcohol Addition on the Deposition of Diamond Thick Films by dc Plasma Chemical Vapor Deposition Method," *Chin. Phys. Lett*, vol. 15, pp. 228–229.
Bi B, Huang W.S, Asmussen J, et al., 2002. "Surface Acoustic Waves on Nanocrystalline Diamond," *Diam. Rel. Mater*, vol. 11, pp. 677–680.
Chen J.J, Zeng F, Li D.M. et al., 2005. "Deposition of High-quality Zinc Oxide Thin Films on Diamond Substrates for High-frequency Surface Acoustic Wave Filter Applications," *Thin Solid Films*, vol. 485, pp. 257–261.
Jin Z.S, Jiang Z.G, Bai Y.Z, et al., 2002. "Synthesis of Thick Diamond Film by Direct Current Hot-Cathode Plasma Chemical Vapor Deposition," *Chin. Phys. Lett*. vol. 19, pp. 1374–1376.
Lehmann G, Schreck. M, Hou L, et al., 2001. "Dispersion of surface acoustic waves in polycrystalline diamond plates," *Diamond and Related Materials*, vol. 10, pp. 686–692.
Seo S.H, Shin W.C, Park J.S, 2002. "a Novel Method of Fabricating ZnO/diamond/Si Multilayers for Surface Acoustic Wave (SAW) Device Applications," *Thin Solid Films*, vol. 416, pp. 190–196.

Manufacturing Engineering and Intelligent Materials – Lu & Abu Bakar (Eds)
© *2015 Taylor & Francis Group, London, ISBN 978-1-138-02832-6*

A subset optimized based High Frequency direction finding cross location algorithm

X. Shen, W.L. Ji, R. Liu, D.Y. Liu & X.F. Zhang
The State Radio Monitoring Center, Beijing, P.R. China

ABSTRACT: This paper innovatively proposes a subset optimized based high frequency direction finding cross location algorithm. The proposed algorithm pre-processes the monitoring data to increase the accuracy of the input monitoring data. And it is accomplished by selecting optimized subset which is based on the extracted composite feature in the first localization results and then selecting the direction finding stations with high degree of confidence. And the final location results are gotten using these optimized subsets for different target areas. The measured results show that compared to the traditional location algorithm, the proposed one increases the location accuracy and improves the performance of the current high frequency location system.

1 INTRODUCTION

Based on the electromagnetic wave propagation, determining the location of a High Frequency (HF) emitter which has vital military uses is one of the fundamental functions of electronic warfare systems Poisel (2005). Over the last few years, the location of a HF emitter is a hot research point and many scholars have devoted their efforts to the research of HF emitter location method. In general, the location of a HF emitter is often estimated from two steps. Firstly, estimate the parameters which will be used in the location algorithm. Secondly, use these parameters to estimate the target location of a HF emitter. Depending on the classification of these needed parameters, the HF emitter location method can be divided into three main categories: Direction Of Arrival (DOA) based method, Time Difference Of Arrival (TDOA) based method and direction and time difference of arrival based method (Stansfield 1947, Gavish & Weiss 1992, Dogancay & Gray 2005, Liang et al. 2005). In the DOA based method, the emitter location is obtained by intersecting the direction bearing lines taken at different direction finding stations. In the TDOA based method, the arrival time of the same signals is measured between different Direction Finding (DF) stations. The direction and time difference of arrival based method needs the DOA and TDOA parameters in the meantime. But estimation of the arrival time is critical in strict time synchronization. Moreover the method using TDOA parameter increases the system complexity. Therefore over the last few years, the DOA based method with low system complexity requirements has been large-scale applied in the engineering fields. And in these applied DOA based system, the process of the DOA is a hot point and has been much more researched to increase the location accuracy (Chen 2002, Fu 2006, Abdual Matheen et al. 2012). However the location accuracy of the applied DOA based method is much lower than that of these other two methods. It requires efforts to increase the location accuracy of the DOA based method. The weighted average method, the least square method and the maximum likelihood method are usually used in the location technology to increase the location accuracy (Xu & Xue 2001, Liu et al. 2012, Song & Ma 2014). But these mentioned methods are almost post processing technology without concerning about the monitoring data pre-processing. As the volume of data increases, this input distortion monitoring data increase which will influence the location accuracy.

In this contribution, a subset optimized based high frequency direction finding cross location algorithm is proposed. As known to our knowledge, it is the first time that the proposed algorithm changes the focus to the first location results and pre-processes the monitoring data. The measured results show compared to the traditional location algorithm, the proposed one not only increases the location accuracy but also improves the performance of the HF location system.

2 PROBLEM FORMULATION

In DOA based location algorithm, the DF azimuth and the DF station coordinates are the input parameters. Assume that the HF direction finding cross location system is composed of m direction finding stations (i.e. DF station) with the coordinates DFn (Xi, Yi) (i = 1, 2, 3, ..., m) respectively. And the azimuth of each DF station is θi (i = 1, 2, 3, ..., m) respectively. In the absence of noise, the DF lines will intersect at a single point. Taking the influence of ionosphere changes into consideration, the errors in azimuth cannot be inevitable. This errors which is represented by Δθi (i = 1, 2, 3, ..., m) for each DF station will create a polygon-shaped area instead of a single point. And the target location of the emitter is within this polygon-shaped area. Here three DF stations are taken for example as shown in Figure 1. Considering the errors of azimuth, the target location is represented by the polygon-shaped area with thick line. Regardless of these errors, the target location is a single point represented by ★. It is worth noting that the number of DF stations and their geographical position distribution will influence the polygon-shaped area and then influence the final target location of emitter. In practical application, more than three DF stations are used to locate the target emitter. And in the situation with more than three DF stations, there would be much more intersections. Here the method applied to the situation with three DF stations can be applied to other situations.

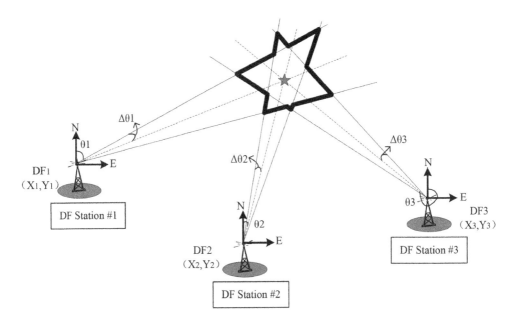

Figure 1.   DOA based location sketch for azimuth error case with 3 DF.

3 SUBSET OPTIMIZED BASED HIGH FREQUENCY DIRECTION FINDING CROSS LOCATION ALGORITHM

In this section, a subset optimized based high frequency cross location algorithm has been proposed. The basic idea of the proposed algorithm is to pre-process the monitoring data, extract the distance-stations location composite feature and create the optimized subsets based on the composite feature. The algorithm flowchart is presented in Figure 2. The main algorithm steps are as follows:

1. Get the first location results.

   Assume the number of direction finding stations is m. In this step, the first location results are gotten and would be used in the next step. Usually, we can use the traditional HF emitter direction finding cross location algorithm to get the first location results. Here the weight of azimuth lines level δ_j and the weight of crossing angle level α_{jk} are introduced to define the reliability of measured azimuths and crossing angles. The final target location is determined by these two weight variables. The best estimate latitude coordinates for many direction finding stations can be calculated by (1) and (2).

$$X_T = \frac{\sum_{j,k=1(j\neq k)}^{m}\left(X_{T_{jk}}\frac{1}{\delta_j\delta_k}\frac{1}{\alpha_{jk}}\right)}{\sum_{j,k=1(j\neq k)}^{m}\left(\frac{1}{\delta_j\delta_k}\frac{1}{\alpha_{jk}}\right)} \qquad (1)$$

$$Y_T = \frac{\sum_{j,k=1(j\neq k)}^{m}\left(Y_{T_{jk}}\frac{1}{\delta_j\delta_k}\frac{1}{\alpha_{jk}}\right)}{\sum_{j,k=1(j\neq k)}^{m}\left(\frac{1}{\delta_j\delta_k}\frac{1}{\alpha_{jk}}\right)} \qquad (2)$$

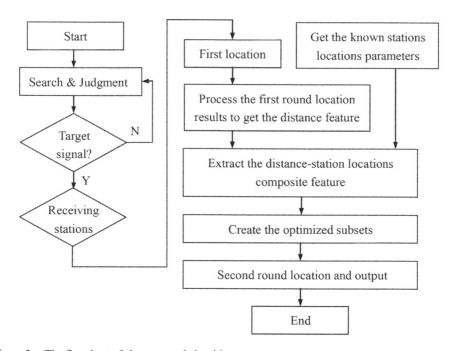

Figure 2. The flowchart of the proposed algorithm.

2. Process the first location results to get the distance feature.

Calculate the great-circle distance between the first location result (X_T, Y_T) and m direction finding stations respectively. And the measurement distance matrix D is $D = (D_1, D_2, ..., D_i, ..., D_m)$. The D_i means the great-circle distance between the first location results (X_T, Y_T) and the ith direction finding station.

Set the threshold value k_1 and k_2 in kilometres. If $k_1 < D_i < k_2$, the corresponding direction finding station is available. The available direction finding stations is M.

3. Get the known stations locations parameters.

Set the threshold value k_3 in kilometres. If the great-circle distance between the first location results (X_T, Y_T) and the known stations locations in the database is less than k_3, the corresponding station is available. The available stations in the database is N. And the measurement stations location matrix DF is

$$DF = \begin{pmatrix} DF_{11}, DF_{12}, \cdots, \cdots \\ \cdots, \cdots, DF_{ij}, \cdots \\ DF_{N1} \cdots, \cdots, \cdots \end{pmatrix}$$

The ith row means the needed direction finding stations which are used to locate the ith known station in the database. These needed direction finding stations are pre-stored in the database based on experience data. Assume the number of needed direction finding stations which are used to locate the total N known stations is n. It can be gotten by intersections of these needed direction finding stations.

$$n = \bigcap_{i=1}^{N} DF_{ij}$$

4. Extract the distance-station locations composite feature.

Concern the direction finding stations which are gotten in step (2) and step (3) with the number M and n respectively. Set the number of final needed direction finding stations is $Number$.

$$Number = \begin{cases} m, & M < 3 \text{且} n < 3 \\ M \cup n, & others \end{cases}$$

5. Create the optimized subsets.

Select the azimuth value corresponding to the $Number$ direction finding stations to generate the optimized subsets of azimuth A_{Number}.

$$A_{Number} = (A_1, A_2, ..., A_i, ..., A_{Number})$$

The A_i means the azimuth value of the ith direction finding station which is selected based on the distance-stations locations composite feature.

6. Second round location and output.

The final results are gotten using the optimized subsets of azimuths in step (5).

Table 1. Parameters of the proposed algorithm.

Parameters	Meaning	Values
δ_j	Weight of azimuth line level	
α_{jk}	Weight of crossing angel level	
k_1	Threshold #1	200/km
k_2	Threshold #2	1500/km
k_3	Threshold #3	200/km

The proposed algorithm uses the first location results and pre-processes the monitoring data. The extracted distance-stations locations composite feature is used to optimize the subsets. And the selected direction finding stations with high degree of confidence are used in the second round location. These threshold values in the algorithm debug are as shown in Table 1.

4 MEASUREMENT RESULTS AND DISCUSSION

To minimize the influence of the ionosphere changes and the man-made influence, we permit that these azimuths of different frequency signals for the same emitter are gotten at the same time every day for each DF station by a certain worker. And the measurements last several months to get adequate data. Here one emitter with the coordinate which located at 108.54° east longitude and 34.12° north latitude is taken for example.

The distribution of the measured location results gotten from the proposed algorithm and the traditional HF direction finding cross location algorithm is shown in Figure 3(a). The shape ★ represents the actual geographical location of the target emitter. It is clearly that the measured location results gotten from the proposed algorithm are much more concentrated around the actual location of the target emitter than that gotten from the traditional algorithm. Furthermore, the maximum latitude and longitude deviation of the proposed algorithm is about ±1°. On the other, the maximum latitude and longitude deviation of the traditional algorithm is about ±3°.

The great-circle difference between the measured result and the actual location of the emitter is shown in Figure 3(b). It is clearly on the whole, the great-circle difference between the measured result and the actual location of the emitter for the proposed algorithm is much lower than that for the traditional algorithm. The numerical results show that the great-circle difference for the traditional algorithm is about 265 km, and that for the proposed algorithm is about 87 km.

The RMS error of measured results for the traditional algorithm is about 87 km, and that for the proposed algorithm is just about 40.7 km. There is apparently a 53.2% drop in the RMS error. So the quantified measured results illustrate that compared to the traditional algorithm, the proposed algorithm increase the location accuracy and decrease the location error.

The subset optimized based high frequency direction finding cross location algorithm use the pre-processing technique. It is accomplished by selecting optimized subset which is based on the extracted distance-stations locations composite feature in the first location results.

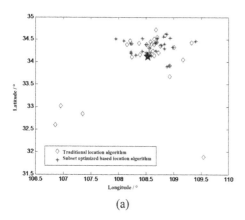
(a)

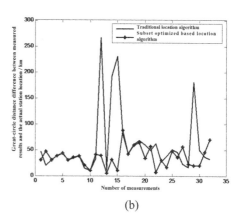
(b)

Figure 3.   (a) The distribution of the measured location results gotten from the proposed algorithm and the traditional algorithm; (b) The great-circle difference between the measured results and the actual location of the emitter.

And then select the direction finding stations with high degree of confidence. The final location results are gotten using these optimized subsets for different target areas.

5 CONCLUSION

In actual engineering application, the major issue to be solved is to increase the location accuracy of the current high frequency location system. In this paper, a subset optimized based high frequency direction finding cross location algorithm is proposed. The proposed subset optimized based high frequency direction finding cross location algorithm pre-processes the monitoring data to increase the accuracy of the input monitoring data. It is accomplished by selecting optimized subset which is based on the extracted distance-stations locations composite feature in the first location results. And then select the direction finding stations with high degree of confidence. The final location results are gotten using these optimized subsets for different target areas in the second round location. The quantified measurements indicate that the proposed subset optimized based high frequency direction finding cross location algorithm can increase the location accuracy and decrease the location error.

ACKNOWLEDGEMENT

This work was supported in part by the National Natural Science Foundation of China (Grant No. 61227801).

REFERENCES

Abdual Matheen, S. Padam Raju, K. Murali Prasad, G. 2012. Location fixing by 3 DF stations using triangulation method. *International Journal of Research in Computer and Communication Technology* 1(3): 75–81.

Chen, L. 2002. Study of passive location and tracking technique. Beijing: *Doctoral Dissertation of Beijing University of Aeronautics and Astronautics.* (in Chinese).

Dogancay, K. & Gray, D.A. 2005. Closed-form estimators for multi-pulse TDOA localization, Proceedings of the Eighth International Symposium on Signal Processing and Its Applications 2: 543–546.

Fu, X. 2006. The design and application of direction finding and location in the national wireless short-wave monitoring system. Sichuan: Master Dissertation of Sichuan University. (in Chinese).

Gavish, M. & Weiss, A.J. 1992. Performance analysis of bearing-only target location algorithms, *IEEE Transactions on Aerospace and Electronic Systems*, 28(3): 817–828.

LiangJian, S. Mueller, H.C. Marx, M. 2005. Pedestrian detection and localization using antenna array and sequential triangulation, *IEEE Proceedings of Intelligent Transportation Systems*: 126–130.

Liu, J.J. Gong, X.F. Yang, J.J. Du, L. 2004. Algorithm research based on DF and location. *The mod-ern electronic technology*, 171(4): 49–55. (in Chinese).

Poisel, R.A. 2005. Electronic warfare target location methods. Boston: Artech House.

Stansfield, R.G. 1947. Statistical theory of DF fixing. *Journal of IEEE 14 part IIIA*(15): 762–770.

Song, W.B. & Ma, X. 2014. A new target location algorithm with bearing-only measurements. *Telecommunication engineering*, 54(11): 1488–1492. (in Chinese).

Xu, J.R. & Xue, L. 2001. LS algorithm used in DF and locatation. *Chinese Journal of Radio Science*, 16(2): 227–230. (in Chinese).

Manufacturing Engineering and Intelligent Materials – Lu & Abu Bakar (Eds)
© *2015 Taylor & Francis Group, London, ISBN 978-1-138-02832-6*

An algorithm for solving multicast routing with delay and bandwidth constraints

C. Zhang
Network and Electronic Education Center, Luoyang Normal University, Luoyang, Henan, China

X.K. Chen
Academy of Information Technology, Luoyang Normal University, Luoyang, Henan, China

ABSTRACT: Lots of communication applications require a source to send multimedia information to multiple terminals through a communication network. More and more real-time multimedia applications require a network capable of satisfying QoS constraints such as delay and bandwidth so that the messages reach each terminal node at almost the same time within a certain specified time limit. The paper proposes a new algorithm for solving multimedia multicast routing, which find the low-cost multicasting tree with delay and bandwidth constraints. Computer simulation results show that the proposed algorithm is able to find a better solution for solving multicast routing, and it can meet the real-time requirement in multimedia communication networks.

Keywords: multicast communication; delay and bandwidth constraints; QoS; multicast routing

1 INTRODUCTION

Multicast transmission of multimedia data is a crucial service provided by the network layer; in fact, it allows the operator to spare a huge amount of network resources in many circumstances. Applications, such as video and audio conferencing, collaborative environments and distributed interactive simulations by large, involve a source node sending messages to a selected set of terminals nodes with varying Quality of Service (QoS) delivery constraints. In this work, we consider an additional criterion that can be used to characterize the quality of the multicast tree for interactive, real-time applications. In addition to end-to-end delay bounds, we assume that the multicast tree must also guarantee bounds on the variation among the delays along the individual source-terminals paths. This bound is needed as a provision of synchronisation among the various receivers, in order to ensure that the information is received by all receivers within an acceptable time difference. First works addressing this problem dealt with a single multicast session and focused on minimizing the transmission cost of each single tree. Many algorithms have been presented to solve the NP-complete unconstrained case, known as the Steiner tree problem. Other works, such as other algorithms, have extended this problem by introducing constraints on the resulting QoS, often evaluated in terms of end-to-end transmission delay.

In the real world, several multicast sessions occur simultaneously, a new and more complex optimization problem needed to be represented: the group multicast routing problem, which consists in the study of the best combination of trees for more sessions concurrently. So, our present work addresses the problem of designing multicast algorithms that take the delay and bandwidth variation as QoS constraint. Until now, only few papers have been published on this topic (Kim et al. 2006, Wang et al. 1999, Priwan et al. 1995, Chen & Yener 2000, Wang et al. 2002). In this paper, it was first studied in Rouskas & Baldine (1997), where

Badline and Rouskas proposed a delay variation multicast algorithm, referred to as DVMA. The authors of this paper have considered the problem of determining multicast trees that guarantee certain bounds on the end-to-end delays from the source to each of the terminal nodes, as well as on the variation among these delays. Their algorithm finds a multicast tree spanning the set of multicast nodes. The DVMA works on the principle of finding the dth shortest paths to all the nodes. If these paths do not satisfy the delay variation bound, longer paths are attempted. The DVMA algorithm does not attempt to optimize the tree in terms of cost. And the DVMA algorithm performs especially well when the number of terminal nodes is relatively small compared to the total number of nodes. Heberman considered the problem of generating cost-conscious multicast trees in Heberman (1997) that satisfy the delay and delay variation constraints imposed by the user process. He proposed a Cost Conscious Delay Variation Multicast Routing Algorithm (CCDVMA) having a high time complexity $O(klmn^4)$. Mokbel presented a Delay and Delay-Variation Constraint Shortest Path algorithm (DVCSP) in Mokbel (1999). This algorithm is divided into two separate phases. Stage I is responsible for yielding a set of paths for each terminal. Its main task is to keep P tokens in each node with reasonable delay difference between any pair of tokens such that every token T satisfies the relation Delay(T) $\in$ (0, Δ]. Stage II collects all the paths from stage I, and chooses one path from each terminal so that the delay variation constraint is satisfied. It is a centralized algorithm where all the information about the network is supposed to be kept at the source node.

Our work aims at presenting a new algorithm with lower cost and delay than what is proposed. The algorithms we proposed use cost information only from neighboring nodes as they proceed which makes it more practical from an implementation point of view. It also uses Dijkstra algorithm in the path selection process allowing terminal nodes to have priority and thus be selected over other nodes. In addition, it is based on the shortest-path tree and minimal spanning tree algorithms. At last, it combines the minimum cost and bandwidth.

2   THE NETWORK MODEL AND PROBLEM DEFINITION

Usually, a network is usually represented as a weighted directed graph $G = (N, E)$, where V symbolizes the set of nodes in the network, such as routers and stations that embody the set of communication links connecting network nodes. For E, every edge $e = (x, y)$ ($e \in E$, x, y $\in V$) defines cost weight-function $C(e):E{\rightarrow}R^+$ and delay weight-function $D(e):E{\rightarrow}R^+$, respectively, means the cost and the delay per link. For graph G, delay between the node u, v (u, v $\in V$) written as (u, v) is represented by

$$Delay(u,v) = \sum_{(x,y)\in p(u,v)} D(x,y),$$

cost by

$$Cost(u,v) = \sum_{(x,y)\in p(u,v)} C(x,y).$$

Bandwidth and delay are two important parameters in the high bandwidth computer network. Bandwidth is the remainder band; the bandwidth of a path is the minimum of the entire remainder bandwidth of the path. It is also called bottleneck bandwidth and given a function $B(P)$(Hypothesis is assert that b denotes bandwidth threshold, P is a path). For G, delay and bandwidth constraints Steiner tree is defined by giving a source node $s \in V$, a set of terminal nodes $D \subseteq V-\{s\}$ and delay δ(δ is a positive real number), a multicast tree $T = (V_T, E_T)$ ($T \subseteq G$, $D \subseteq V_T \subseteq V$, $E_T \subseteq E$) whose roots includes all nodes of terminal nodes D can be generated.

Assume the minimum bandwidth constraint of multicast tree is b, the maximum delay constraint id is δ, given a multicast demand R, then, the problem of bandwidth-delay constrained multicast routing is to find a multicast tree T, satisfying:

$$\min_{eu,v\in p}[B(u,v)\,|\,e_{uv}\in p]\ge b$$

$$\sum_{e_{uv}\in p} D(u,v)\le \delta \tag{1}$$

$$C(T)=\min(C(T_s),T_s\in S(R))$$

3 THE PROPOSED ALGORITHM

In this section, we introduce firstly a fast and efficient heuristic algorithm to solve the problem we proposed. The multiple constrained problems relate to the property of QoS, and it is NP complete problem. In this paper, path selected function will be constructed. This function will combine more parameters of QoS and the paths which have the minimum weight value can be calculated by giving value of e. Then, the synthetic value of p which has more parameters can be worked out. Because some conflict probably happens among some QoS parameters when all these are required for the optimal solution, the path named p probably does not exist. When this situation appears, a near-optimal solution p can be obtained in a subnet to satisfy the condition of multiple QoS. In the proposed algorithm, we consider the three components: A. Initial processes to select the nodes representing the ones of m shortest paths that satisfies bandwidth constraint between the source node and the terminal nodes B according to constrains to determine the norms of delay and bandwidth C. constructing tree T.

The algorithm of D_B_ MultiRouting(Delay and Bandwidth constraints routing algorithm) is shown by the following pseudocodes:

D_B_ MultiRouting(G, D, s, δ, b)
//Where, G represents a network topology, D is a set of terminal nodes, s is a source node,
//δ is an upper bound of end-to-end delay and b is the minimum bandwidth constraint.
Input: a network topology G = (V, E), a source node s, a set of terminals D, an upper bound of end-to-end delay δ, a positive real number b.
Output: a multicast tree T with the best multicast delay and bandwidth constraints
begin
$dv_{min}\leftarrow\infty$, T$\leftarrow\varphi$, $v_c\leftarrow\varphi$;
// dv_{min} records the minimum multicast delay variation, T records the present constructed multicast tree, v_c records the stand-by central node.
for i = 1 to |V| do
call Dijkstra's Algorithm to calculate P, the shortest path from s v to v_i, $p_j\in$ P;
Delay(p_j) = δ-ppd(s, p_j); Bandwith(p_j) = b-ppd(s, p_j);Max(p_j) = 0;Min(p_j) = ∞;
set Flag;//Flag = true;
for w = 1 to |D| do
calculating the shortest path from v_i to v_w to select p_k;
if ppd(p_k, v_w)>Delay(p_k)
    then {Flag = false; break;}
        else update Max(p_k) or Min(p_k);
end if
end for
if (Flag = = true) and dv(v_i) $\leftarrow$max{ppd(s, p_j)+ Max(p_j)}
    then $d_v(v_i)$ $\leftarrow$dv(v_i)-min{ ppd(s, p_j)+ Max(p_j)};
    if dv_{min} > dv(v_i) then {dv_{min} = dv(v_i); v_c = v_i;} end if
end if
end for

if $v_c = \varphi$ then break;
for each $v_w \in M$ do
    $T \leftarrow T \cup \{p \in \min(vc, vw)\}$;
$T \leftarrow T \cup \{p \in \min(s, vc)\}$;
return T;
edd.

4   SIMULATION EXPERIMENT

From the description, we can see that the algorithm is achieved by finding the actual inter-section node of the shortest path from the source to the candidate central node with the shortest path from the candidate central node to each terminal. All these intersection nodes will be on the shortest path from the source to the candidate central node. Since more than one terminal may share a common intersection node, the maximum and minimum delays from an intersection node to all its sharing terminals are calculated. These two maximum and minimum values, each plus the shortest path delay from the source to the intersection node, are, respectively, the actual maximum and minimum multicast delays from the source to all the terminals who share the intersection node. Therefore, the algorithm can achieve the actual multicast delay and bandwidth variation of the multicast tree constructed by using any network node as the candidate central node.

Thus, a theorem is given: The algorithm of D_B_ MultiRouting can construct a delay and bandwidth bounded multicast tree if such a tree exists and the tree is loop-free.

Proof Suppose, there exists a delay bounded multicast tree, there is at least one link from source node to every terminal node that satisfies delay constraint, because of the given δ and b, the algorithm can use the least delay path to construct a tree whose path satisfies δ, there-fore the tree exists. In the construction process, if some nodes were not selected at some steps of the algorithm execution due to their high cost, then they will be selected and connected during some other coming steps through other paths. Furthermore, every node is visited only once during the construction process. On the other hand, suppose, there are loops in the tree, then there will be two or more nodes connected with the tree, so the assumption is wrong. The experiment in this paper employs the modified Waxman to generate the network model. The model is very identical to the real network, and it is widely applied in the simulation experiments. Using this model, the network node randomly lies in a rectangle-shaped area of 400×400 and the probability of the existence of edge is:

$$p(u,v) = \beta \exp\left(\frac{-d(u,v)}{\alpha L}\right) \tag{2}$$

Different values in the range can determine the quality of the network generated and make it more identical to the real environment. The cost is defined as the distance between the two nodes, the delay is defined as the cost multiplying a random number between 0 and 1. In our experiment, $\alpha = 0.30$, $\beta = 0.42$, with two hundred experiments, we get an average as a result. DVMA, CCDVMA and D_B_ MultiRouting in this paper are made an analysis as far as the cost is concerned; the result is as shown in the following. In this experiment, $n = 20$, $m = 15$, $\delta = 10$, b takes different values. DVMA has a good success rate with time delay. The success rate of CCDVMA is low in less delay-variation but when delay δ tends to the success rate tends to 100%. Our success rate tends to 100% when δ equal to 9 and our algorithm has a fast convergence speed as compared to the algorithm we mentioned.

5   SUMMARY

We propose a new algorithm for solving multimedia multicast routing based on delay and bandwidth constraints. The proposed algorithm uses shortest paths algorithm to construct route set. By giving priority to the multicast member and shared edge, it effectively reduces the cost of the tree; therefore, it is more practical.

REFERENCES

Aida H, Saito T. 1995. The multicast tree based routing for the complete broadcast multipoint-to-multipoint communications. *IEICE Trans Commun*, E78-B: 720–8.

Chen S, Yener B. 2000. The multicast packing problem. *IEEE/ACM Trans Network*, 8:311–8.

Heberman. B.K. 1997. Cost, Delay, and Delay Variation conscious multicast routing. Master Thesis, Graduate Faculty of North Carolina State University, Raleigh.

Kim M, Bang Y.C., Lim H.J., Choo H. 2006. On efficient core selection for reducing multicast delay variation under delay constraints, *IEICE Transactions on Communications*, E89: 2385–2393.

Mokbel M.F. 1999. New Algorithms for Multicast Routing in Real Time Networks. Master Thesis, Faculty of Engineering, Alexandria University.

Priwan V, Aida H, Saito T. 1995. The multicast tree based routing for the complete broadcast multipoint-to-multipoint communications. *IEICE Trans Commun*, E78-B:720–8.

Rouskas G. and Baldine I. 1997. Multicast routing with End-To-End Delay and Delay Variation Constraints, *IEEE Journal on Selected Areas in Communications*.

Wang C, Lai B, Jan R. 1999. Optimum multicast of multimedia streams. *Comput Operat Res*: 461–80.

Manufacturing Engineering and Intelligent Materials – Lu & Abu Bakar (Eds)
© 2015 Taylor & Francis Group, London, ISBN 978-1-138-02832-6

The experimental study of orthogonal flexible automatic grinding for the small diameter blind hole based on turn-mill machine tool

Z.X. Li, X. Jin, Z.J. Zhang & W.W. Lv
School of Mechanical Engineering, Beijing Institute of Technology, Beijing, China

ABSTRACT: According to the mechanism of grinding process, an orthogonal flexible automatic grinding method for getting a small diameter blind hole is proposed based on orthogonal machining sports. By using the new method and the traditional processing technology of drilling, boring etc., managing steel 3J33 are studied in the experiment for getting a blind hole with small diameter. The comparative experimental results show that the blind hole surface roughness of the small diameter obtained by this method can reach the minimum, Ra0.2 µm, the edge of the blind hole has smooth surface and crack depth are minimum. The results show that the method can effectively improve the processing quality of small diameter blind holes and provide a new idea to study how to improve the machining precision of micro parts.

Keywords: processing technology; lapper process; blind hole; managing steel; micro parts

1 GRINDING MECHANISM

Grinding belongs to the free abrasive micro cutting machining; Figure 1 shows a grinding processing model for single abrasive particle. During grinding process, under the grinding pressure, many fine abrasive particles are cut; the abrasion generates heat and friction between the work-piece and lead to atomic disorder. So, micro deformation of the work-piece material or micro removal can be realized, the final formation of chip and a new surface can be obtained.

Grinding is the sum embodiment of material microscopic deformation or removal for all local processing points. The mechanism of grinding with different inhomogeneous degree of the grinding processing unit and work-piece material is different.

The superfine abrasive grain is less than the material defects and makes the ultra fine abrasive force less stress damage of materials than required. Superfine abrasive grain can get high quality machining surface by chemical effects of abrasive, grinding fluid in grinding process. Figure 2 shows the three stages of grinding process.

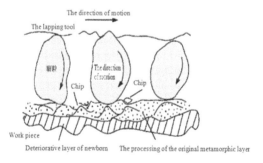

Figure 1. Free abrasive lapping model.

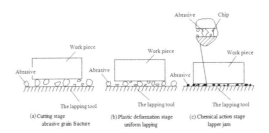

Figure 2. Schematic diagram of abrasive grinding process.

1. The cutting stage: free abrasive is rounded. Free abrasive size is often uneven, only large grinding abrasive works in the initial stage, local high pressure and high temperature of contact point crusts convex peak of the fine abrasive particles, rounds edge. Then, the more participate in cutting abrasive, the grinding efficiency is higher.
2. The stage of plastic deformation: abrasive is ground uniformly, grinding surface has small fluctuations. The abrasive edge is further ground and becomes blunt, local temperature of grinding point increases, plastic deformation of the grinding surface happens, the grinding surface's peak and valley of the work-piece flattens in the plastic flow and becomes micro chip under the repeated deformation, cooling and hardening.
3. The chemical action stage: lapping tool is blocked, active abrasive has chemical reaction. Abrasive particles and micro chip have a sliding friction effect on the work-piece, the oxidized film happen on the work-piece surface after active abrasive plays a chemical role, the oxidized film is easily rubbed off and does not hurt the matrix. A repeated process of rapid formation and oxide film quickly rubbed off can speed up the process of grinding and decreases the work-piece surface roughness values.

2 SMALL DIAMETER BLIND HOLES ORTHOGONAL FLEXIBLE AUTOMATIC GRINDING EXPERIMENT

Using micro turn-milling machine tools to design an orthogonal flexible automatic grinding method, as shown in Figure 3.

Figure 3 shows the orthogonal flexible automatic grinding; grinding process system is mainly composed of the work-piece (flexible joint, installed on the turning spindle), grinding rod (installed on the milling spindle), micro turn-mill machine tool (Kim et al. 1997, Kriangkrai).

The flexibility in grinding processing is realized depending on the grinding rod, small diameter blind holes, the grinding paste filling between them and trace interpolation processing method, as shown in Figure 4. Using orthogonal flexible automatic grinding as an automatic grinding processing method of small diameter blind hole to carry out experimental study and compare the effects of different processing methods on the blind hole machining quality.

2.1 *The experimental conditions of small diameter blind holes orthogonal flexible automatic grinding*

The experiment of small diameter blind holes orthogonal flexible automatic grinding uses the micro turn-mill machine tool CNKM25-I to study, the main parameters as shown in Table 1.

The experimental materials is the managing steel 3J33 with $\Phi 10 \times 100$ mm, experimental tool comprises a hard alloy drill bit, hard alloy cutter, cast iron grinding rod etc.

2.2 *Experiment parameters of small diameter blind holes orthogonal flexible automatic grinding*

Figure 5 shows the process of small diameter blind holes orthogonal flexible automatic grinding of 3J33 material, the experimental process parameters used are shown in Table 2.

Figure 3.   Small diameter blind holes orthogonal flexible automatic grinding method.

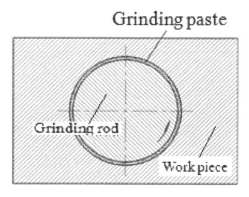

Figure 4.   Small diameter blind holes automatic grinding.

Table 1.   Main parameters of turn-mill machine tool CNKM25-I.

Serial number	Parameters	Numerical
1	Spindle motor power	3.7/5.5 kW
2	Spindle speed	0~6000 rpm
3	Milling spindle power	1 kW
4	Milling spindle speed	0~60000 rpm
5	Feed motor power	X, Y, Z: 1.2 kW, 0.5 kW, 1.2 kW
6	Milling motorized spindle which can clamp tool diameter	< Φ7 mm

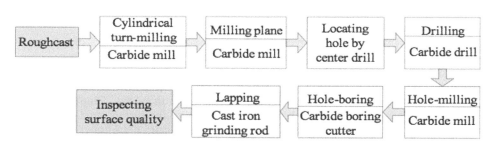

Figure 5.   Small diameter blind holes orthogonal flexible automatic grinding process.

201

Table 2. Main technological parameters of orthogonal flexible automatic grinding experiment.

Serial number	Machining process	Process parameters	Cutting tool	Processes target
10	Cylindrical turn-milling	v_i: 500 r/min, f: 5 mm/min, a_p: 0.1 mm;	Φ 5 mm, four edges carbide end mill	Φ 9.8 mm outer cylinder
20	Milling plane	v_2: 7800 r/min, f: 10 mm/min, a_e: 1.5 mm	Φ 5 mm, four edges carbide end mill	7 mm × 7 mm cross section beam
30	Drilling	v_2: 3000 r/min, hole depth: 2.5 mm, f: 3 mm/min	Φ 2 mm, carbide drill	Hole number: 7
40	Hole-milling	v_2: 10200 r/min, f:2 mm/min, hole depth: 2.3 mm, interpolation radius: 0.25 mm	Φ 2 mm, two edges carbide end mill	Hole number: 6
50	Hole-boring	v_2: 1500 r/min, hole depth: 2.3 mm, f: 1 mm/min	Φ 2.64 mm, carbide boring cutter	Hole number: 4
60	Hole-lapping	v_2: 300 r/min, f: 1 mm/min, lapping time h: 12 min, lapping depth: 2.0 mm, grain size: 6 μm, interpolation radius: 0.065 mm	Φ 2.497 mm, cast iron grinding rod	Hole number: 2

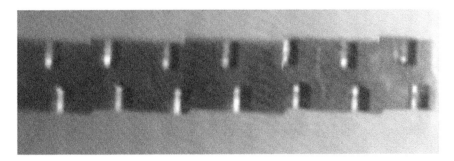

Figure 6. Small diameter of the blind hole by grinding.

The small diameter of the blind hole is processed, the test piece along the blind hole axis plane is cut by NC wire cutting machine tool and cleaned by kerosene, the cutting surface is ground with #600, #1000, #1500, #2000 sand paper, respectively. Using high pressure air gun to blow the specimen surface, the samples are obtained, as shown in Figure 6. The 7 holes in upper line from right to left are obtained by following: drilling, hole-milling, milling, boring, boring, grinding, grinding, machining. The 7 holes in low line from left to right are obtained by grinding, grinding, boring, boring, milling, milling, drilling.

2.3 *Experiment results and analysis*

1. The blind hole surface roughness obtained by different processing methods
   Using white light interfere meter to measure surface roughness of two groups of samples of the blind hole, the measurement results are shown in Table 3. The value of bore hole surface roughness is maximal, Ra = 0.8 μm, and surface roughness value of hole obtained by milling is lower than the value of hole obtained by drilling, about Ra = 0.4 μm.
   The superficiality of the hole obtained by boring is better than that of milling process, the surface roughness value is about Ra = 0.3–0.4 μm, the surface quality of the holes by grinding is best, the surface roughness value is about Ra = 0.2 μm. The results show that the surface finishing hole roughness value is the lowest and the best surface quality.

Table 3. Surface roughness of blind hole processing by different methods.

Processing method	Surface roughness of the first group hole/μm	Surface roughness of the second group/μm
Grind	0.2	0.24
Milling	0.474	0.465
Boring	0.37	0.31
Drilling	0.752	

Table 4. Surface residual stress of the blind hole with different processing methods.

Processing method	Residual stress of the first group hole/MPa	Residual stress of the second group hole/MPa
Grind	−254.9	−255.24
Milling	−218.3	−231.27
Drilling	−213.5	−204.85

(a) Drilling hole's surface crack(*5000) (b) Drilling hole's surface crack (*5000) (c) Milling hole's surface crack (*2000) (d) Milling hole's surface crack (*2000)

(e) Boring hole's surface crack (*5000) (f) Boring hole's surface crack (*5000) (g) Grinding hole's surface crack (*2000) (h) Grinding hole's surface crack (*5000)

Figure 7. The surface micro crack obtained by different technological means.

2. The surface residual stress of blind hole by different processing methods
   The surface residual stress of the sample blind hole are measured by using X-ray diffraction, the measurement results are shown in Table 4. The surface residual stress of hole-drilling is the lowest; the drilling hole is obtained by reaming, its residual stress value increased; the milling hole is obtained by boring, its surface residual stress decreased. Results indicated that there is little difference between the surface residual stress value obtained with the different methods, surface residual stress of grinding hole increased is mainly because of the direct extrusion between grinding rod and work-piece during grinding process, causes increase in surface compressible residual stress.

3. Surface crack of blind hole processed by different methods
   Using scanning electron microscope to test micro crack of the blind hole on the sample surface, the detection results are shown in Figure 7. Drilling hole's surface (a) and (b) have obvious defects, their defect width is up to 10 μm; the milling hole's surface (c) and (d) don't have obvious defects, but have obvious cracks, their depth is about 20 μm; Boring

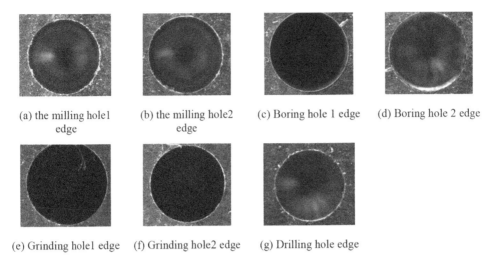

(a) the milling hole1 edge   (b) the milling hole2 edge   (c) Boring hole 1 edge   (d) Boring hole 2 edge

(e) Grinding hole1 edge   (f) Grinding hole2 edge   (g) Drilling hole edge

Figure 8.   The hole edge obtained by different approaches.

hole's surface (e) and (f) don't have obvious defects, but have obvious cracks, their depth is about 10 μm; the grinding hole's surface (g) and (h) have obvious cracks, their depth is about 5 μm. Overall, after grinding, the surface crack's depth is minimum, milling hole is longer, the obvious defects appeared in the boring hole's surface. Through grinding the boring blind hole, it can effectively reduce the surface crack.

4. The hole edge of blind hole processed by different methods

Using universal tool microscope19 JPC-V to test the edge of blind hole, the detection results are shown in Figure 8. It shows that the edge of drilling hole (g) has a lot of hard edges, its margin is irregular; the edge of milling hole (a) and (b) are better than drilling hole, milling hole is more regular than boring hole; boring hole (c) and (d) edge are better than that of milling hole edge, their contour are more clear and have less hard edges; grinding hole (e) and (f) edge are similar to that of boring hole, the outline is clear. The experiments show that the grinding process can effectively improve the hole edge and reduce machining hard edges.

3   CONCLUSION

By comparing a variety of experiments of small diameter of the blind hole processing method, it shows that quality parameters (surface roughness, surface residual stress, surface crack and hole edge) depend on the processing method. The surface quality of the grinding hole is the best, its surface roughness value is about Ra = 0.2 μm, drilling hole's surface roughness value is maximum, which shows that using grinding can get the best surface quality. Experimental results found that the surface residual stress of the grinding hole is larger, the edge of the hole is smooth, its crack depth is minimum because of extrusion between the role of grinding rod and the work-piece material, so it shows that the flexible automatic grinding method can effectively improve machining quality of the small diameter hole, and provide the research basis for improving the processing precision of micro parts.

ACKNOWLEDGEMENT

This research is supported by National Science and Technology Major Project 2012ZX04010061.

REFERENCES

Cao Zi-yang, He Ning, Li. Liang 2006, Micro machining technology. *Micro fabrication Technology*, (3), 1–5.

Kim YS, Kim YJ. Pariente F, Wang E. 1997, Geometric Reasoning for Mill-turn Machining Process Planning. Computers and Industrial Engineering, 33(3–4):501–504.

Kriangkrai Waiyagan, Erik LJ, Bohez. 2009, Intelligent feature based process planning for five-axis mill-turn parts [J]. *Computers in Industry*, 60(5):296–316.

Kriangkrai Waiyagan. 2002, Intelligent feature based process planning for five-axis lathe. Proceeding CAD-CG05 *Proceedings of the Ninth International Conference on Computer Aided Design and Computer Graphics*.

Li Sheng-yi., 2000, The development of ultra-precision machining and its key technologies. *China Mechanical Engineering*, 11(1–2):177–179.

Lee YS, Chiou CJ. Unfolded projection approach to machining non-coaxial parts on mill-turn machines [J]. *Computers in Industry*, 39(2), 147–173.

Yuan Zhejun, 2006, New Developments of Precision and Ultra-Precision Manufacturing Technology. *Tool Engineering*, 40, 3–9.

Manufacturing Engineering and Intelligent Materials – Lu & Abu Bakar (Eds)
© 2015 Taylor & Francis Group, London, ISBN 978-1-138-02832-6

Transient-micro thermal EHL analysis of power law water-based ferrofluid bearing

F. Ren
School of Mechanical Engineering, Qingdao Technological University, Linyi, Shandong, China

X.J. Shi
School of Mechatronics Engineering, Harbin Institute of Technology, Harbin, Heilongjiang, China

Y.Q. Wang
School of Mechanical Engineering, Qingdao Technological University, Qingdao, Shandong, China

ABSTRACT: In this article, considering the influencing factors of temperature, transient, roughness and the power law fluid, the Elastohydronamic Lubrication (EHL) simulation of water-based ferrofluid bearing were carried out. The stable, transient and the micro-transient solutions are compared. And the influence of wavelength and amplitude on the pressure and film thickness was discussed. The results show that while the load transient is considered, the lubrication film pressure and film thickness change apparently with the time, there are apparent bulges in the distribution of pressure and film thickness and they become more apparent with the decrease of the period and the increase of the amplitude. While the roughness was considered, the pressure and the thickness have obvious fluctuation, the number of the pressure and film thickness waves become less with the increase of the wavelength. The amplitudes of the waves become larger with the increase of the amplitude, the thinnest film thickness become thicker and the largest pressure become smaller.

Keywords: water-based ferrofluid; EHL; transient; roughness; wavelength; amplitude

1 INTRODUCTION

With the development of green chemistry, water-based ferrofluid get more and more international attention. The main characteristic of water-based ferrofluid is that the carried fluid is water, so it avoids the environmental pollution caused by organic solvents. In addition, it also has some other good characteristics, such as super paramagnetism, biological compatibility and good dispersion. So, water-based ferrofluid has a good application prospect in the aspects of self lubricating, grinding and polishing, dynamic seal and environmental protection.

In recent years, ferrofluid lubrication is mentioned in lots of literatures. Shah Rajesh C. and Bhat M.V. Shah study the effect of anisotropic permeability and slip velocity on load capacity and response time of the ferrofluid lubrication film of porous bearing. Chi C.Q. (2001) shows that the ferrofluid is a non-Newtonian fluid and the load-carrying capacity and friction force of plane slider lubricated with ferrofluid are given. Zhang Y. studies the static characteristics of ferrofluid journal bearing and found that the main feature of the magnetic force is the ability to change the size of the cavitation region. Mongkolwongrojn Mongkol presents the results of a transient analysis of elastohydrodynamic lubrication (EHL) of two parallel cylinders in line contact with a power law non-Newtonian lubricant under oscillatory motion. But the EHL analysis of ferrofluid lubricant has not been studied so far. In this paper, the micro-transient thermal EHL analysis of power law water-based ferrofluid will be carried out.

2 POWER LAW FLUID

The constitutive equation of power law fluid can be expressed as Liu & Yang (2007). The viscosity equation of the Newtonian fluid is replaced by m with η. The viscosity-magnetic field equation is given as (1)

$$m_0 = m_c\left[1 + \frac{5}{2}\phi + \frac{3}{4}\phi\frac{0.5\alpha L(\alpha)}{1 + 0.5\alpha L(\alpha)}\right] \tag{1}$$

where m is a physical quantity expression of viscosity, m_0 is the environmental viscosity in a magnetic field, m_c is the dynamic viscosity of water carrier fluid, p is the pressure, T is the temperature, z_0 is the viscosity-pressure coefficient, s_0 is viscosity-temperature coefficient, φ is the volume fraction of magnetic powder, $L(\alpha)$ expression is $L(\alpha) = \coth\alpha - 1/\alpha$, $\alpha = \mu_0 XD^3 B^2/6K_0T\mu^2$, X is magnetic susceptibility, μ_0 is the vacuum permeability, μ is the permeability and its expression is $\mu = 2\mu_0$, D is the diameter of magnetic powder particles, K_0 is the Boltzman coefficient.

3 EHL EQUATIONS

The journal-bearing thermal conductivity k_1 is 24.8 W/(m·K) and the specific heat capacity c_1 is 343 J/(kg·K); the bearing radius is 100.125 mm; the bearing elastic modulus E_1 is 113 GPa; the bearing Poisson's ratio is 0.3; the shaft thermal conductivity k_2 is 30 W/(m·K) and the specific heat capacity c_2 is 670 J/(kg·K); the shaft radius is 100 mm; the shaft elastic modulus E_2 is 206 GPa; the shaft Poisson's ratio is 0.28; the water-based ferrofluid thermal conductivity k is 0.586 W/(m·K) and the specific heat capacity c is 4200 J/(kg·K); the viscosity-pressure coefficient z_0 is 2.2×10^{-8} m²/N and the viscosity-temperature coefficient s_0 is 0.042 K⁻¹; the stable load F is 4000 kN and the journal bearing width L is 0.8 m, the particle diameter D is 10 nm, the environmental temperature T_0 is 313 K, the magnetic induction B is 20t, the magnetic susceptibility X is 1.

Governing equations. The governing equations include Reynolds equation considering transient effect, load equation changes with time as sine function, density equation, film thickness equation considering surface roughness, energy equation and their boundary conditions (Liu & Yang 2007). A_w is the load amplitude and t_p is the wavelength. w_0 is the load in per unit length. The magnetic force p_M and its equation is represented as (2)

$$p_M = 3\varphi\frac{\mu_0(\mu - \mu_0)}{\mu + 2\mu_0}\int_0^H HdH \tag{2}$$

where H is magnetic strength and its equation is $H = B/[\mu_0(1 + X)]$. ρ_0 is the environmental density of water-based ferrofluid and its calculation equation is given as (3).

$$\rho_0 = \frac{m_c + m_p}{V_c + V_p} = \frac{\rho_c V_c + \rho_p V_p}{V_c + V_p} \tag{3}$$

where V_C and V_p are the volume of carrier fluid and the magnetic particle, m_c and m_p are the quality of carrier fluid and the magnetic particle. In order to reduce parameters and improve the stability of the calculation, the governing equations were dealt with dimensionless method. An equivalent viscosity is introduced $\eta_0 = m_0^{1/n}|P_H|^{n-1/n}$. The other dimensionless parameters are the same as Liu & Yang (2007).

4 NUMERICAL METHOD

The dimensionless equations are discredited by finite difference method. The pressure is solved by multi-grid method, the temperature is solved by column scanning method and the film thickness is solved by multiple integration method. There are 6 layers in the process of solving pressure, and there are 961 grid nodes in the top layer. While the temperature field is calculated, uniform mesh is used within the lubricating film, and the number of the nodes is 9. Non-uniform mesh is used within the bearing and shaft, and the numbers of the nodes are both 5.

The time is divided into 30 instants, the pressure and the temperature will be solved alternately at every instant. The temperature is assumed that it is already known in the process of solving pressure. The pressure is solved with Reynolds equation, then the film thickness is solved with the pressure, the load equation will be satisfied by adjusting the center film thickness. The pressure and film thickness are assumed that they are already known in the process of solving temperature. The temperature field is solved by the energy equation and the heat conduction equation.

5 RESULT ANALYSIS

The influence of load periods and amplitudes. While the sine unimodal load transient is considered and the surfaces of the bearing and shaft are smooth, the load is 6000 KN, the velocity is 500 r/min and the magnetic induction is 20 mT. Figure 1 are the dimensionless pressure and thickness distributions of water-based ferrofluid under different load periods at the 15 instant, and the other conditions are the same.

It can be seen that there are apparent bulges in the distribution of pressure and film thickness with the change of the period and amplitude. They are caused by dynamic effect and they become more and more apparent with the decrease of the period and the increase of the amplitude.

The influence of roughness. Considering the surface roughness, the cosine function roughness is selected, the load is 6000 KN, the velocity is 5.2 m/s and the magnetic induction is 20 mT, the dimensionless pressure and thickness distributions of water-based ferrofluid under steady state condition and in different transients are compared. Figure 2 is the dimensionless pressure and thickness distributions of water-based ferrofluid under different roughness wavelengths and amplitudes at the 15 instant. It can be seen that the pressure and the thickness has obvious fluctuation than the smooth surface. The number of the pressure and film thickness waves become less with the increase of the wavelength, and the amplitudes of the waves become smaller. Because the number of the waves decreases, the inflow lubricants increase, the thinnest film thickness become thicker and the largest pressure become smaller. The amplitudes of the waves become larger with the increase of the amplitude. Because the amplitudes of the waves increase, the inflow lubricants decrease, the thinnest film thickness become thicker and the largest pressure become smaller.

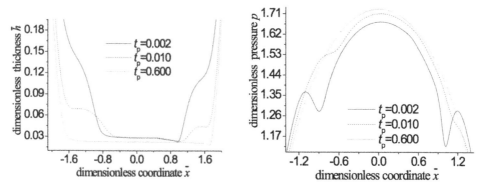

Figure 1.   The distribution of film thickness and pressure under different load periods.

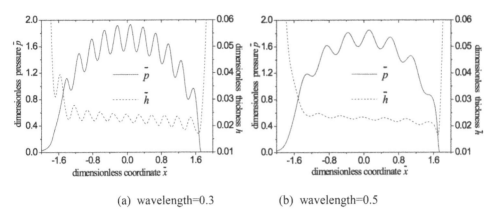

(a) wavelength=0.3   (b) wavelength=0.5

Figure 2. The distribution of pressure and film thickness under different wavelengths.

6 CONCLUSIONS

While the transient is considered, the lubrication film pressure and film thickness of the water-based ferrofluid change apparently with the change of the time, there are apparent bulges in the distribution of pressure and film thickness and they become more apparent with the decrease of the period and the increase of the amplitude. While the transient and the roughness are both considered, it reveals that the pressure and the thickness has obvious fluctuation than the smooth surface, the number of the pressure and film thickness waves become less and less with the increase of the wavelength, and the amplitudes of the waves become smaller and smaller, the thinnest film thickness become thicker and the largest pressure become smaller, the amplitudes of the waves become larger and larger with the increase of the amplitude, the thinnest film thickness become thicker and the largest pressure become smaller.

ACKNOWLEDGEMENTS

This work was financially supported by the National Science Foundation (51175275).

REFERENCES

Chi C.Q. (2001), "Non-Newtonian effects on ferrofluid lubrication," Beijing Hangkong Hangtian Daxue Xuebao," 27, 1, pp. 88–92.
Liu XiaoLing and Yang Peiran. (2007), "Analysis of Power Law Fluid Thermal Elastohydrodynamic Lubrication," Lubrication Engineering, 32, 8, pp. 19–23.
Mongkolwongrojn Mongkol, et al. (2008), "Elastohydrodynamic lubrication of rough surfaces under oscillatory line contact with non-newtonian lubricant," Tribol. Trans. 51, 5, pp. 552–561.
Shah Rajesh C. and Bhat M.V. (2004), "Ferrofluid lubrication of a porous slider bearing with a convex pad surface considering slip velocity," International Journal of Applied Electromagnetics and Mechanics, 20, 1, pp. 1–9.
Zhang Y. (1991), "Static characteristics of magnetized journal bearing lubricated with ferrofluid [J]. Journal of Tribology," 1991, 113, 3, pp. 533–538.

Manufacturing Engineering and Intelligent Materials – Lu & Abu Bakar (Eds)
© 2015 Taylor & Francis Group, London, ISBN 978-1-138-02832-6

Study on bifurcation and chaos in nonlinear vibration isolation system

X.R. Dai
Zhejiang Provincial Institute of Communications Planning, Design and Research, Hangzhou, China

S. Huan, W.J. Tao & X.Q. Tan
Civil Engineering Protection Research Center, Guangzhou University, Guangzhou, China

ABSTRACT: Bifurcation is the major route to chaos. In this paper, the mechanical parameters of nonlinear vibration isolation system are measured by tests, and the numerical model is established based on the tests. By analyzing harmonic excitation, global bifurcation diagram is achieved by using Poincare mapping method, and then the dynamics of the nonlinear vibration isolation system is revealed.

1 INTRODUCTION

Chaos exists widely in nature, since Lorenz discovered chaotic phenomena through simulation of a small climate using equation, understanding and concern of chaos have been more comprehensive and close. Chaos exhibits extremely complicated dynamic characteristics, but most of the time, the chaotic motion is harmful in engineering applications such as collision of machine. Therefore, in most projects cases, we hope to avoid the chaotic motions, or make the complex chaotic motions become more regular periodic motions.

However, more and more application of complex motion characteristics of chaos to solve practical engineering problems were discussed recently (Jiang & Zhu 2001, Ueda 1991 & Rega etal. 1991). Zhu Shijian proposed chaotic spectrum vibration isolation technique, and pointed out that the nonlinear vibration isolation system had good vibration isolation performance in the state of chaos and could greatly isolate the line spectrum part of structure noise in their paper, and also obtained explanation and application in the paper. Bifurcation is the major route to chaos. Bifurcation study of nonlinear vibration isolation system is important for the study of chaotic dynamics of nonlinear systems, thus wide attention is paid by scholars all over the world (Benedettini et al. 1992, Jiang et al. 2001, Lou et al. 2005, Tao et al. 2011 & Tao et al. 2011).

There are many methods about the research of nonlinear bifurcation, such as power series method, perturbation method, harmonic Melnikov function method, singularity method, Poincare-Bikhoff normal form method etc... Most of these methods are qualitative research, but qualitative research is not enough, because of the complexity of nonlinear systems, quantitative theory result is very difficult to acquire, and the research on the theory and method is often difficult. Therefore, it is very important to use numerical calculation to study the bifurcation of the system. It plays an irreplaceable role especially in the tracking of bifurcation solutions and determination of the location of bifurcation point (Huan et al. 2011, Tao et al. 2010 & He et al. 2002).

According to the experiments, the paper measures the mechanical parameters of nonlinear vibration isolation system, and numerical models of nonlinear vibration isolation system are created on the basis of the parameters. Global bifurcation diagrams which describe the variation of the solution of nonlinear vibration isolation system with the excitation force amplitude are obtained using the numerical calculation method, and the bifurcation behavior is studied.

2 VIBRATION ISOLATION SYSTEM

Devices of nonlinear vibration isolation system used in the experiment are shown in Figure 1. The devices are mainly composed of the following parts: nonlinear vibration isolator, hard spring, steel etc. The experimental devices are mainly used in ship machinery vibration isolation system, for marine machinery vibration isolation system, the base can no longer be regarded as rigid base, and should be expressed as elastic base with certain impedance. Therefore, the double hard spring at the bottom of the flexible foundation are used to simulate the ship vibration isolation system. Nonlinear vibration isolator is in the upper flexible foundation, the vibration isolation device is shown in Figure 2. The isolator is mainly composed of a concave plate, rubber, flat steel plate. Concave plate curve parabola, the gap is small at the two sides of the curve plate, and the biggest gap is at the middle part. Its working principle is: when the excitation force of mechanical equipment is put on the upper part of the concave plate, the lower rubber is squeezed. Because the stiffness of rubber is relatively small to steel plate rigidity, therefore, the main deformation is produced by rubber. With the increasing extrusion degree of steel plate and rubber, the contact area is increasing, the stiffness of isolator is also increased, resulting in nonlinear.

The nonlinear vibration isolator device designed in the experiment is shown in Figure 3. The two layer steel plates are precisely machined and grinded, the geometric dimensions are 800 mm × 400 mm × 20 mm, a part is cut at the lower surface of the upper plate forming a groove, the groove is composed of a parabola around two symmetrical composition. The parabolic equation of left side is $y = 0.01 \times \sqrt{x}$ (unit: m), The parabolic equation of right side is $y = 0.01 \times \sqrt{0.6 - x}$ (unit: m). Seeing the section from the direction of the icon, the groove length in the direction of X is 0.6 m, the maximum depth of groove of Y direction is 5.48 mm. There is a plane section which is 100 mm × 400 mm existing on each side of the upper plate, this part is mainly to generate linear stiffness. A rubber layer is placed between the upper and the lower plate, its size is 800 mm × 400 mm, the thickness is 8 mm, which provides the elastic restoring force as the elastic deformation body. The lower plate is a cuboid of no changes.

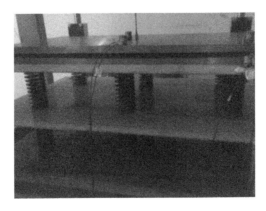

Figure 1.   Experimental isolation system device.

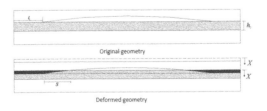

Figure 2.   Nonlinear vibration isolation device.

Figure 3. Nonlinear vibration isolation device on the upper and lower plate.

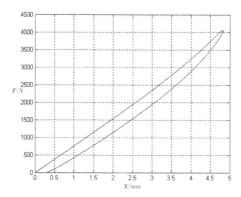

Figure 4. Force vs. displacement curve of rubber.

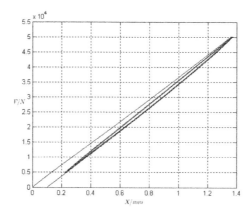

Figure 5. Force vs. displacement curve of vibration isolation device.

As shown in Figure 2, the distance between the two pieces of steel plate before deformation is h_0. Fix the lower plate, the downward motion displacement of the upper plate when it is under pressure is X, then the contact area between the two sides of rubber and the steel plate increases, and the contact area on both sides of the parabola is also gradually increasing, as shown by the black part in Figure 2. In the process of compression deformation, the rubber has always been in the elastic deformation range; therefore, the calculation formula of restoring force can be obtained according to the theory of elasticity.

$$F = 2 \times \left[\frac{l_1 DE}{h_0} X + \int_0^s \left(X - 0.01\sqrt{x} \right) dx \cdot \frac{DE}{h_0} \right] \qquad (1)$$

It is available through the integral changes of formula (1):

$$F = \frac{2ED}{h_0} \left(l_1 X + \frac{1}{3} \times \frac{X^3}{0.01^2} \right) \qquad (2)$$

213

Among them, $S=(X/0.01)^2$, D is the steel plate width, E is elastic modulus of rubber. According to the formula (2), the restoring force of the isolation device is a function of square party displacement variable X and the three party X, the form expressed by the equation (2) conforms to the need of hard spring characteristic of the Duffing equation which contains the items of stiffness linear and stiffness cubic. The material parameters involved in the model is primarily

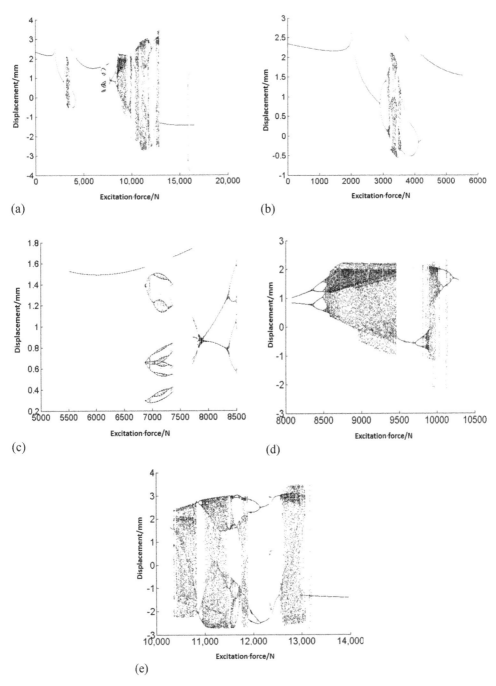

Figure 6. Bifurcation diagram varying with the excitation amplitude. (a) $0 \le F \le 1800$ N; (b) $0 \le F \le 5000$ N; (c) $5500 \le F \le 8500$ N; (d) $8500 \le F \le 10500$ N; (e) $10500 \le F \le 14000$ N.

elastic modulus of rubber, the elastic modulus of rubber obtained through the static experiment is 5.0 MPa, as shown in Figure 4. Figure 5 is the force-displacement curve of the whole device, the Monomial coefficient is 1.1×10^6 N · m^{-1}, the cubic coefficient is 1.2×10^{12} N · m^{-3}.

3 NUMERICAL ANALYSIS

In numerical analysis, taking the excitation amplitude as the control parameter of bifurcation analysis, using four orders Runge-Kutta method to solve the Poincare mapping of the system, the changes of the coordinate of the Poincare mapping points varying with the control parameter are plotted. The initial conditions for the calculation is (0.002, 0), the calculation integral step length is taken to be 0.01 times the incentive force cycle, the excitation force frequency is 30 Hz. The global bifurcation diagrams of the system varying with the excitation amplitude is obtained through numerical calculation of the model, as shown in Figure 6, the graph shows all the solution branches of the exciting force range from 0~18000 N. As can be seen from the graph, the system is a nonautonomous system, the bifurcation properties are quite complex, there are multiple solution branches, multiple attractors coexist in certain parameter, and even phenomenon of the coexistence of chaotic attractor and periodic attractor is found. In order to describe bifurcation characteristics of the system better, bifurcation condition will be detailed and sub regionally discussed here.

1. Bifurcation is found in range $0 \leq F \leq 5000$ N, as shown in Figure 6(a). In this range, when F = 2878.6 N, the system appears the period doubling bifurcation, period doubling bifurcation is one of the routes to chaos, therefore, by adjusting the excitation force amplitude, which makes the system reach chaos motion state through period doubling bifurcation. When F = 3825.1 N, when F = 3866.6 N, Hopf bifurcation is formed.
2. The bifurcation behavior of the range is quite complex, multiple branches exist, not only the periodic 1 motion is found, but also the periodic 6 motion. Hopf bifurcation and reverse bifurcation of the system happens, respectively, when F = 6840.3 N and F = 6845.7 N. At F = 7383.2 N a saddle node bifurcation happens.
3. In the range $8500 \leq F \leq 10500$ N, a period doubling bifurcation of system happens when F = 8311 N, then it enters a state of chaos. The system comes back to the period 1 attractors at F = 9000 N, when F = 10021 N it enters the period doubling bifurcation again, then the system enters chaotic state again.
4. Most of the range is the chaotic region, when F is about 13120 N, the chaotic motion returns to a periodic 1 attractors and produce a new branch.

4 CONCLUSION

Abundant bifurcation characteristic of nonlinear chaotic vibration isolation device are involved in the paper, including saddle node bifurcation, period doubling bifurcation and Hopf bifurcation. Both subcritical bifurcation and supercritical bifurcation are frequently found in the bifurcation path. Many cases of the nonlinear systems enter chaos through period doubling and intermittent route.

ACKNOWLEDGEMENT

The authors would like to thank the ministry of transport of science and technology project (grant number 2010-353-333-150).

REFERENCES

Benedetini F, Rega G, Salvtori A. 1992. Prediction of bifurcations and chaos for an asymmetric elastic oscillator. *Chaos, Solitons, and Fractals; 2(3): 303–321.*

He Q.W., Zhu S.J., Wong X.T. 2002. Study on the shock response character of nonlinear shock isolation systems under a type of exponential impulse. *Process in Safety Science and Technology, Taian, China, Oct 10–13 2002, 541–545.*

Huan S., Tao W.J., Zhu S.J., et al. 2011. Research on chaotic dynamics of characteristic of hardening nonlinear isolation device. *Journal of Vibration and Shock. 30(11): 245–248.*

Jiang R.J., Zhu S.J. 2001. Prospect of applying controlling chaotic vibration in the waterborne noise confrontation. *18th Binnial Conference on Mechanical Vibration and Noise, ASME, Pittsburgh, USA, 9. DETC 2001/VIB-21633.*

Jiang R.J., Zhu S.J., He L. 2001. Prospect of applying controlling chaotic vibration in waterborne-noise confrontation, *Proceeding of ASME 2001 Design Engineering Technical Conference and Computers and Information in Engineering Conference, Pittsburgh, September, DETC 2001/VIB-21663.*

Lou J.J., Zhu S.J., He L., et al. 2005. Application of chaos method to line spectra reduction. *Journal of Sound and vibration 286 645–652.*

Rega G., Benedetini F, Salvtori A. 1991. Periodic and chaotic motions of an unsymmetrical oscillator in nonlinear structural dynamics. *Chaos, Solitons, and Fractals; 1(1): 39–54.*

Tao W.J., Huan S., Zhu S.J., et al. 2011 Design and research on continuous linear chaotic vibration isolation system. *Journal of Vibration and Shock, 30(9): 111–114.*

Tao W.J., Jiang G.P., Huan S. 2011. Numerical study on multi-degree of-freedom chaotic vibration isolation system. *Water Resources and Power. 29(8): 90–92.*

Tao W.J., Huan S., LI X.Y., et al. 2010. Research on the chaotic vibration isolation of Duffing system. *Journal of Guangzhou University, 19(6): 69–71.*

Ueda Y. 1980. Explosions of strange attractors exhibited by Duffing's equation. *Annals New York Acad Sci, 357: 422–433.*

Manufacturing Engineering and Intelligent Materials – Lu & Abu Bakar (Eds)
© 2015 Taylor & Francis Group, London, ISBN 978-1-138-02832-6

Study on turning response characteristic of tracked vehicle based on hydro-mechanic differential

F.Y. Cao & Y.L. Lei
Henan University of Science and Technology, Luoyang, Henan, China

ABSTRACT: The turning response characteristic of tracked vehicle directly reflects its running maneuverability. The hydro-mechanic differential turning system is the double power flow turning system of tracked vehicle. Considering ground track skid (slippage), turning center offset etc., basing on the turning movement analysis of tracked vehicle, the calculation formula of turning radius and turning angular velocity is deduced in view of hydro-mechanic differential. By means of force analysis and calculation, and the turning dynamic model of tracked vehicle based on hydro-mechanic differential is established. Referencing the prototype, the response characteristic of turning radius, turning angular velocity and running velocity of tracked vehicle based on hydro-mechanic differential is simulated with the displacement ratio of hydraulic turning close loop system. The theory basis for hydro-mechanic differential turning system control of tracked vehicle is provided.

Keywords: tracked vehicle; hydro-mechanic transmission; turning; response characteristic

1   INTRODUCTION

The turning response characteristic of tracked vehicle is a characteristic that describes the transient changes of its turning radius, turning angular velocity and running velocity caused by the turning operation input when it transforms from the line running condition to the steady-state turning condition or from a steady-state turning condition to another steady-state turning condition. It reflects its running maneuverability which is influenced by many factors (Cao 2006 & Said 2010), not only the turning operation input, ground nature, but also the equipment of turning system. However, most of the study literatures on turning characteristic of tracked vehicle have nothing to do with the specific turning system (Hu 2009 & Li 2008). The hydro-mechanical differential turning system is a double power flow turning system of tracked vehicle which is compounded of hydraulic closed loop, fixed gear and planet row (Ali 2009 & Yuan 2008).

2   TURNING MOVEMENT ANALYSIS AND CALCULATION

2.1   *Analysis of turning movement relation*

Assuming that the tracked vehicle is turned on a horizontal ground, its turning kinematics relationship can be shown in Figure 1. The moving coordinate system x'o'y' is established, the vehicle centroid as its origin while the longitudinal centerline as its y' axis. The xoy is the static coordinate system, which coincides with moving coordinate system at the beginning of turning.

When the tracked vehicle turns at the angular speed of ω, the vehicle runs at a planar motion while the grounding track works at a composite motion. At any time, the planar motion of the vehicle can be considered as a rotation at speed v around a point, which is

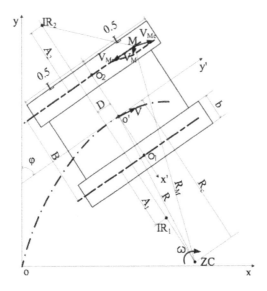

Figure 1. Turning movement of tracked vehicle.

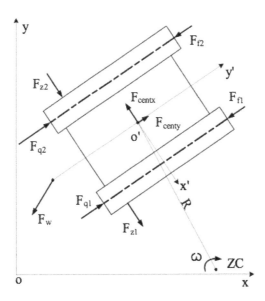

Figure 2. Turning forces of tracked vehicle.

the vehicle turning center named ZC. The distance between turning center ZC and vehicle centroid o′ is called the turning radius R. The point M of the grounding track is chosen as an analytical subject, its speed VM is compounded of two kinds of velocity, one is the bulk motion velocity, namely, the mean velocity V_{Me} of the coincident point of vehicle and grounding track. The other is the relative velocity, namely, the winding motion speed of the grounding track relative to the body of vehicle. At one point, the composite motion of any point on the grounding track can be regarded as a rotation around a point, which is called the instantaneous center of velocity of the grounding track. IR_1, IR_2 are, respectively, called the instantaneous center of velocity of the inner and outer track. As for the skid (slippage), the distance deflected from their respective geometric center O_1, O_2 are A_1, A_2. ZC, IR_1, IR_2 are situated on the same line, and the turning centerline of tracked vehicle is formed.

218

When tracked vehicle turns at a high speed or is equipped with working device, the turning resultant resistance is not equal to zero due to the effect of centrifugal force or working resistance. Thus, the forward (backward) offset D of the turning centerline relative to its horizontal centerline is engendered.

2.2 Turning force analysis and calculation

The horizontal forces that the vehicle bears are shown in Figure 2. They are driving force F_q (inner F_{q1}, outer F_{q2}), turning resistance F_z (inner F_{z1}, outer F_{z2}), working resistance F_w and running resistance F_f (inner F_{f1}, outer F_{f2}). As for high-speed turning vehicles, the effect of turning centrifugal force F_{cent} (F_{centx}, and F_{centy}) is taken into consideration as well.
In the coordinate system of Figure 2, the turning model is as follows:

$$F_{q1} + F_{q2} - F_{f1} - F_{f2} + F_{centy} - F_w \cos \beta = ma_x \tag{1}$$

$$F_{z1} + F_{z2} - F_{centx} + F_W \sin \beta = ma_y \tag{2}$$

$$\frac{B}{2}(F_{q2} - F_{q1}) - \iint x dF_{Z1} - \iint x dF_{Z2} - F_w l_T \sin \beta = J\dot{\omega} \tag{3}$$

The model is a nonlinear system of equations. In order to solve it, the Newton-Raphson method is adopted, the process is shown in Figure 3.

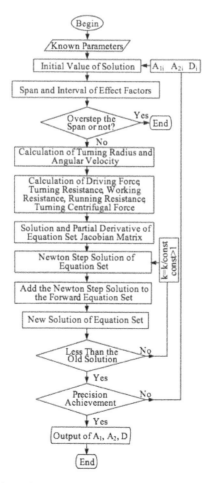

Figure 3. Solving flowing chart of equations.

3 SIMULATION OF TURNING RESPONSE CHARACTERISTIC

3.1 *Known Parameters*

According to application conditions, the track rolling resistance coefficient is set as 0.05, surface adhesion coefficient is set as 1. The structural parameters of tracked vehicle, the parameters of hydro-mechanical differential turning system, parameters of engine, and ratio of transmission case are shown in Table 1, Table 2 and Table 3.

3.2 *Simulation calculation*

When the ratio of transmission case is, respectively, first ($i_g = 3.5$), second ($i_g = 2.389$) and third ($i_g = 2.05$), the corresponding throttle opening will be 0.8, 0.9, 1, the responding curves of turning radius, turning angular velocity and running velocity are shown in Figure 4.

When the step signal on displacement ratio of hydraulic closed-loop is imposed, the inner track is slipped forward because of speed drop, and the driving force is decreased; however, the outer track is slipped due to speed up, and the driving force is increased. Thus, the difference between these two forces produces a turning torque which makes the tracked vehicle turn. The turning resistance on the grounding track is engendered by ground, forming the turning resistance torque. The turning torque is maximum at the beginning of turning and decreases thereafter. The turning resistance torque is zero at first and increases thereafter. When the turning torque and turning resistance torque are equal, the vehicle is gone into

Table 1. Parameters of tracked vehicle.

Parameter	m	B	L	b	h_g	i_m	r_q
Value	7000	1.435	1.615	0.7	0.6	5.5	0.346

Table 2. Turning system and engine parameters.

Parameter	i_f	i_y	i_z	α	p_H	n_e	P_e
Value	0.905	4.10	2.733	2.391	38	2300	95.6

Table 3. Ratio of transmission case.

Gears	F1	F2	F3	F4	F5	F6	R1	R2
Ratio	3.5	2.389	2.050	1.833	1.480	0.876	3.561	2.42

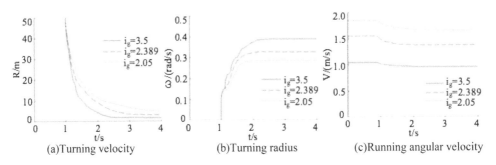

Figure 4. Turning response characteristic.

a steady-state turning condition. Meanwhile, the turning radius, turning angular velocity and running velocity is tended towards steady. The turning process which interprets that the model is reasonable is validated by the change of turning radius, turning angular velocity and running velocity in Figure 4.

From Figure 4, as for the same displacement ratio of hydraulic closed-loop, the larger the ratio of transmission case is, the smaller the turning radius is, the larger the turning angular velocity is, and the lower the running velocity will be, and the faster the vehicle is got into the steady-state turning. Besides, the increment of the turning angular velocity is larger than that of turning radius, so the vehicle is shown to the overturning characteristic. Because the skid (slippage) rate of track when turning is larger than that of straight driving, the turning speed is lower than straight driving speed at the same ratio of transmission case.

4 CONCLUSIONS

Considering grounding track skid (slippage), turning center offset etc., based on the turning movement analysis of tracked vehicle and force analysis, the turning dynamic model of tracked vehicle based on hydro-mechanical differential is established and the method is provided.

The turning response characteristic of tracked vehicle based on hydro-mechanical differential is studied using the simulation analysis method. The theory basis for hydro-mechanical differential turning mechanism design and driving control of tracked vehicle is provided.

REFERENCES

Ali, H.S. 2009. Hydrostatic Power Splitting Transmission Design and Application Example, *Journal of Terramechanics*, (3): 18–24.

Cao, F.Y. 2006. Design of Hydromechanical double power differential steering mechanism for tracked tractor, *Transactions of the CSAM*, 37(9): 5–8.

Hu, J.B. 2009. Characteristics on hydro-mechanical transmission in power shift process, *Chinese Journal of Mechanical Engineering*, (1): 50–56.

Li, W.Z. 2008. Bench test on steering of double flow driving tracked vehicle using steering wheel, *Transactions of the CSAE*, (8): 109–112.

Said, A.M. 2010. Track-terrain modeling and traversability prediction for tracked vehicles on soft terrain, *Journal of Terramechanics*, (4): 151–160.

Yuan, S.H. 2008. Design of two-range input split hydrostatic mechanical transmission, *Transactions of the CSAE*, (9): 109–113.

Manufacturing Engineering and Intelligent Materials – Lu & Abu Bakar (Eds)
© *2015 Taylor & Francis Group, London, ISBN 978-1-138-02832-6*

Effect of Cu doping on the structural and magnetic properties of $Ni_{0.5-x}Zn_{0.5}Cu_xFe_2O_4$ soft magnetic ferrites

P. Zhou, C.X. Ouyang & Y.H. Fan
Harbin Institute of Technology Shenzhen Graduate School, Shenzhen, China

X.Y. Xu, J.H. Zhu, Y.Y. Gao & Y.X. Fan
Zhen Hua Fu Electronics Co. Ltd., Shenzhen, China

ABSTRACT: Copper substituted soft magnetic ferrite samples with the chemical formula $Ni_{0.5-x}Zn_{0.5}Cu_xFe_2O_4$ ($0 < x < 0.5$) successfully fabricated by solid-state reaction method. Effect of Cu^{2+} content on the microstructure and magnetic properties of the samples have been analyzed using XRD, SEM, B-H analyzer and permeability studies, respectively. These studies showed that the density, grain size, saturation induction and initial permeability increased considerably with the optimum copper content of $x = 0.15$. The results indicated that the addition of CuO enhanced the sintering by the formation of liquid phase in the ferrites. A material with comprehensive magnetic performance of high Bs (405 mT), high initial permeability (467 at 100 KHz frequency), and low magnetic loss tanδ (0.009 at 100 KHz frequency), correlated with the microstructure characteristics of ferrite, was obtained, when Cu^{2+} doping content $x = 0.15$.

1 INTRODUCTION

In recent decades, power devices like switched-mode-power supplies and compact become smaller in size to meet demand. Ni-Zn ferrites material is essential to miniaturize magnetic components with high electrical resistivity (Shinde et al., 2013, Su et al., 2007). Ni-Zn ferrites have been one of the most popular materials with high resistivity, low dielectric losses, high Curie temperature, mechanical hardness and environmental stability (Lebourgeois et al., 1992). These properties are sensitive to the type of substitutes, so required ones can be selectively obtained by additions of various ions. Cu-substituted Ni-Zn ferrites have considerable influence on the magnetic and mechanical properties by improving densification and reducing porosity due to forming a liquid phase when sintering. There are many investigations (Yue et al., 2001, Murthy, 2001, Seetha Rama Raju et al., 2006) on Mg-Cu-Zn ferrites, especially the effects of Cu^{2+} on the sintering kinetics, atomic diffusivity and the magnetic properties. However, influence of Cu^{2+} in Ni-Zn ferrites on the structural and magnetic properties is seldom discussed. So in this study, we investigated the effects of $Ni_{0.5}Zn_{0.5}Fe_2O_4$ ferrite stoichiometry change invoked by partial substitution of Ni^{2+} with Cu^{2+} cations and relevant properties change and obtained a high-powered soft ferrite with relatively high permeability, low magnetic loss tanδ (1/Q) and good frequency characteristics.

2 EXPERIMENTAL

Conventional ceramic processing method was employed to fabricate $Ni_{0.5-x}Zn_{0.5}Cu_xFe_2O_4$ ($x = 0, 0.025, 0.05, 0.075, 0.1, 0.125$ and 0.15) ferrites using high-purity NiO, ZnO, Fe_2O_3, CuO in stoichiometric proportions. The mixed proportions were milled into homogeneous powders in the presence of distilled water, and then pre-sintered at 900 °C for 2.5 h in air.

After that, the calcined ferrite powders were grounded and mixed with 10% (mass fraction) polyvinyl alcohol. Powders were made into pellets and finally pressed into standard toroids. Toroid samples were sintered at 1100 °C for 3 h in muffle furnace, followed by furnace cooling.

The sintered samples were characterized by X-Ray Diffraction (XRD) with CuKα radiation with respect to phase purity and crystallinity. Microstructure was observed by low vacuum Scanning Electron Microscope (SEM). The inductance (L) and quality factor (Q) were measured by TH2828 impedance analyzer at 100 kHz frequency and 250 mV AC voltage. The initial permeability (μ_i) was calculated by the formula (1):

$$\mu_i = L \times Le/\mu_0 AeN^2,\tag{1}$$

where μ_0 is the vacuum permeability; Le is the effective magnetic circuit length; L is the inductance; Ae is the effective magnetic area; N is count of loops on a tested-toroid. The saturation magnetism (Bs) of the samples was measured by IWATSU SY-8218 B-H analyzer, at a frequency of 10 kHz. Permeability performance up to 10 MHz frequencies was performed by Agilent E4991 A impedance materials analyzer.

3 RESULTS AND DISCUSSION

3.1 *Structural properties*

Figure 1 shows XRD patterns of the general formula $Ni_{0.5-x}Zn_{0.5}Cu_xFe_2O_4$ ferrites (where x = 0, 0.025, 0.05, 0.075, 0.1, 0.125, 0.15). All samples were sintered at 1100 °C with heat

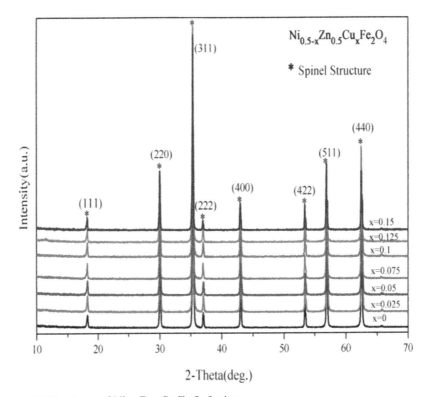

Figure 1.   XRD patterns of $Ni_{0.5-x}Zn_{0.5}Cu_xFe_2O_4$ ferrites.

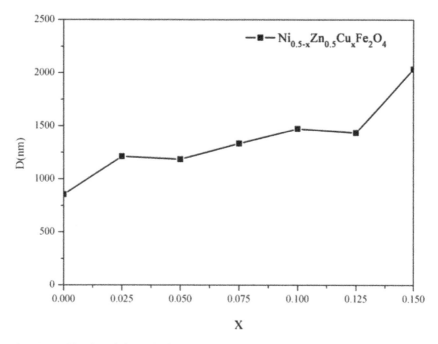

Figure 2.   Crystallite size of sintered $Ni_{0.5-x}Zn_{0.5}Cu_xFe_2O_4$ ferrites.

preservation for 3 hours. Note that a single-phase spinel structure peaks were obviously indexed in the XRD patterns, indicating that Cu^{2+} content x in ferrites had no effect on a new phase generating. The average crystallite size was calculated from the XRD line broadening using the Scherrer relationship (Cullity, 1978; Verma et al., 2011), as is shown in Figure 2. It can be seen that the average crystallite size grew larger, up to micron magnitude as Cu^{2+} substitute content x increases. The trend was proved as well as by electron micrographs in Figure 3.

The pictures under SEM illustrate ferrite particles distribution with Cu^{2+} content x increasing in Figure 3. The phenomena can be interpreted by the liquid phase mechanism due to CuO in the mixed powders with a low melting temperature point. During temperature rising when sintering, CuO, as a flux, was melting and packing other particles with a liquid phase, enlarging contact areas and promoting reaction between the flux and ferrites in mechanism. In result, grain size growth was accelerated, and partial air among particles became trapped intragranular pores. In fact, the whole process of grain growth is attributed to the competition between the driving force for grain boundary movement and the retarding force exerted by pores (Bellad et al., 1999). During the sintering, the thermal energy can generate a force and drive the grain boundaries to grow over pores, and densify the ferrite materials (Manjurul Haque et al., 2008), consequently decreasing the pore volume. The density variation of $Ni_{0.5-x}Zn_{0.5}Cu_xFe_2O_4$ ferrites was calculated and plotted into a curve in Figure 4. It is stated that when x = 0.15, the density reached 4.966 g/cm^3 much higher than 3.54 g/cm^3 when x = 0. Therefore, addition of Cu^{2+} definitely enhanced ferrite density.

3.2 Magnetic properties

The variation of saturation induction (Bs) measured at room temperature is shown in Figure 5. It is found that Bs increased with increasing Cu substituent content x up to a certain value. When x = 0.150, Bs achieved 405 mT, surpassing the value 145 mT as Cu substituent x = 0.

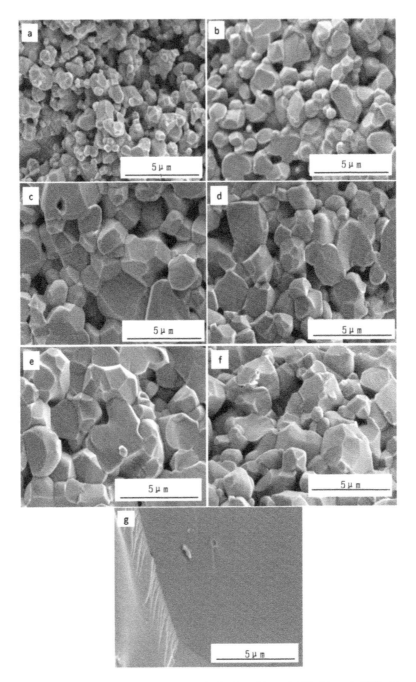

Figure 3.   SEM micrographs of $Ni_{0.5-x}Zn_{0.5}Cu_xFe_2O_4$ ferrites: (a) x = 0.000, (b) x = 0.0025, (c) x = 0.050, (d) x = 0.075, (e) x = 0.100, (f) x = 0.125, (g) x = 0.15.

The parameter is dependent on the microstructure and super-interchange effect caused by cations on the A- and B-sites of the spinel structure (Singh et al., 2001). On the one hand, the sintered ferrites are polycrystalline in microstructure, and the densification of ferrites with Cu^{2+}, as mentioned above, resulted in considerable enhancement of magnetic moment per unit volume, so a high Bs was obtained. On the other hand, Cu^{2+} changes to Cu^+ at a certain concentration (Rahman & Ahmed, 2005), and Cu^+ has a preference to occupy A-site,

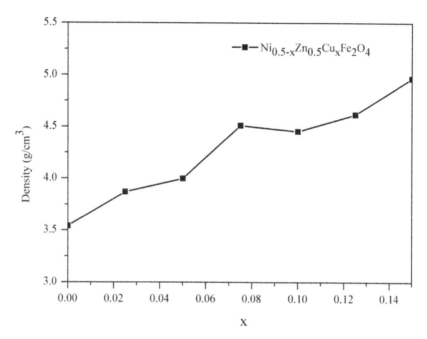

Figure 4.   Density of sintered $Ni_{0.5-x}Zn_{0.5}Cu_xFe_2O_4$ ferrites.

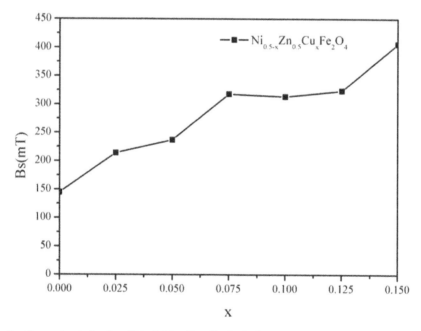

Figure 5.   Saturation induction (Bs) of $Ni_{0.5-x}Zn_{0.5}Cu_xFe_2O_4$ ferrites.

compelling Fe^{3+} in A-site diverted to B-site. And the magnetic moment of Cu^{2+} and Fe^{3+} have been regarded as 1 μ_B and 5 μ_B (Sun et al., 2001), respectively, so the net magnetic moment M ($M = M_B - M_A$) increased with Cu^{2+} substitutes content x increasing. Simultaneously, the saturation induction Bs increases. In theory, substitution of Cu^{2+} replacing Ni^{2+} at octahedral site would reduce the net magnetic moment at B site, so in the beginning, with Cu^{2+} content x in ferrites increasing, Bs decreases, as some research reported (Akther Hossain & Rahman, 2011).

However, there is not a declining line in Figure 5 in this study. The possible reason is that the enhancement of magnetic moment per unit volume attributed to densification exceeded the effect of the net magnetic moment resulted from the difference between Cu^{2+} (1 μ_B) and Ni^{2+} (2 μ_B).

The variation of initial permeability μ_i measured at the frequency ranging from 10^5 to 10^7 Hz is graphically represented in Figure 6. The lines of initial permeability μ_i versus frequency were stable almost at parallel levels, then subsequently increased to extreme high levels and abruptly dropped down as frequency increased. It is concluded that the permeability value increased with increasing Cu content x in $Ni_{0.5-x}Zn_{0.5}Cu_xFe_2O_4$, and the maximum value was about 467 at 100 kHz when x = 0.15. The contribution of permeability is owing to spin rotations as well as domain wall motion (Brooks et al., 1992; Tsutaoka et al., 1995). The mechanism can be described as follow formula (2) (Akther Hossain et al., 2007):

$$\mu_i = 1 + X_w + X_S \qquad (2)$$

where X_w is the domain wall susceptibility and X_s is the intrinsic rotational susceptibility. Meanwhile, $X_w = 3\pi M_s^2 D/4\gamma$, $X_s = 2\pi M_s^2/K$, and where K, Ms, γ, D are the total anisotropy, the saturation magnetization, the domain wall energy, and the average grain diameter, separately. From formula (2), we can see that the domain wall motion is determined by grain size and it is enhanced with the grain size increasing. Moreover, magnetization plays a significant role in permeability enhancement. In this study, permeability tendency was mainly due to contribution of increasing grain size and magnetization. Thus, the permeability of ferrites with copper content x increasing was demonstrated, and materials utility in frequency range was justified.

Figure 7 shows the frequency dependence of the magnetic loss tangent tanδ of the samples sintered at 1100 °C. And the magnetic loss tangent value tanδ was the reciprocal of quality value (1/Q) measured on the coil wound toroidal samples. The variation of loss tangent tanδ with frequency ranging from 0.1 MHz to10 MHz presented a similar trend for all the samples. It is seen that tanδ decreased a little with frequency, and sharply ascended beyond 3 MHz, except for x = 0.15. The loss is firmly associated with the pores, grain boundaries and

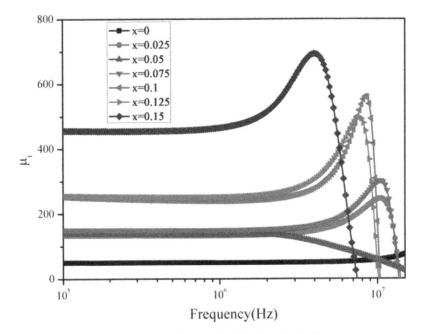

Figure 6.   The variation of initial permeability (μ_i) for $Ni_{0.5-x}Zn_{0.5}Cu_xFe_2O_4$ ferrites.

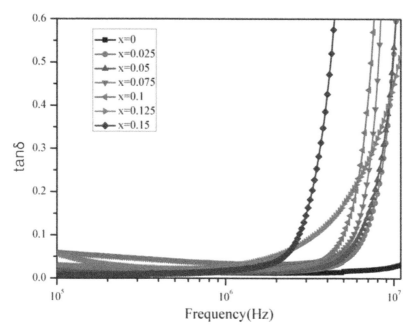

Figure 7. The influence of the Cu^{2+} content x for $Ni_{0.5-x}Zn_{0.5}Cu_xFe_2O_4$ ferrites on the tanδ.

grain size in polycrystal microstructure. Grain boundaries and pores prevent domain wall from motion, raising loss. Furthermore, lag of domain wall motion with the applied alternating magnetic field and various domain defects also increase the loss (Overshott, 1981). The graph is interpreted that ferrites with larger Cu^{2+} content x had larger average grain size and lower porosity; thereby the lowest loss tanδ in low frequency field was obtained. When energy in the applied magnetic field was transferred to the lattice at the resonance maximum, the magnetic loss tangent tanδ rapidly rose [10]. Moreover, that tanδ with Cu^{2+} content x = 0.15 increased at lower frequency compared with others was maybe related to the large number of grain boundaries, which is foundation for domain motion and energy transfer. Finally, the graph implied that the lowest loss tanδ was 0.009 in the frequency of 100 KHz, when Cu^{2+} content x = 0.15, but its applied frequency range was limited.

4 CONCLUSIONS

Structural and magnetic properties of $Ni_{0.5-x}Zn_{0.5}Cu_xFe_2O_4$ soft ferrites were studied. It is concluded that the variation composition ferrites sintered at 1100 °C formed a single phase (spinel structure), without redundance. CuO, as a flux, densified the microstructure and increased the grain size in $Ni_{0.5-x}Zn_{0.5}Cu_xFe_2O_4$ soft ferrites with Cu^{2+} content × increasing and while x = 0.15, $Ni_{0.5-x}Zn_{0.5}Cu_xFe_2O_4$ ferrite obtained excellent structural properties. Preferable ferrite material with magnetic properties of high Bs, high μ_i, and low magnetic loss was received, as well as Cu^{2+} content x = 0.15, meaning that $Ni_{0.35}Zn_{0.5}Cu_{0.15}Fe_2O_4$ is an ideal chemical formula.

ACKNOWLEDGEMENTS

This work was supported by the Innovation of Science and Technology Commission of Shenzhen Municipality (CXZZ20130515152233332).

REFERENCES

Akther Hossain, A.K.M., Mahmud S.T., Seki, M., Kawai, T., Tabata, H. (2007) Structural, electrical transport, and magnetic properties of $Ni_{1-x}Zn_xFe_2O_4$. *Journal of Magnetism and Magnetic Materials*, 312: 210–219.

Akther Hossain, A.K.M., Rahman, M.L. (2011) Enhancement of microstructure and initial permeability due to Cu substitution in $Ni_{0.50-x}Cu_xZn_{0.50}Fe_2O_4$ ferrites. *Journal of Magnetism and Magnetic Materials*, 323: 1954–1962.

Bellad, S.S., Watawe, S.C., Chougule, B.K. (1999) Microstructure and permeability studies of mixed Li-Cd ferrites. *Journal of Magnetism and Magnetic Materials*. 195: 57.

Brooks, K.G., Berta, Y., Amarakoon, V. (1992) Effect of Bi_2O_3 on impurity ion distribution and electrical resistivity of Li-Zn Ferrites. *Journal of American Ceramic Society*, 75(11): 3065–3069.

Cullity, B.D. (1978) Elements of X-Ray Diffraction, 2nd ed., Addison Wesley, Reading, MA, pp. 99–102.

Lebourgeois, R., Perriat, P., Labeyrie, M.(1992) High and low level frequency losses in NiZn and MnZn spinel ferrites, *ICF-6*, pp.1159–1164.

Manjurul Haque, M., Huq, M., Hakim, M.A. (2008) Influence of CuO and sintering temperature on the microstructure and magnetic properties of Mg-Cu-Zn ferrites. *Journal of Magnetism and Magnetic Materials*, 320(21): 2792–2799.

Murthy, S.R. (2001) Low temperature sintering of MgCuZn ferrite and its electrical and magnetic properties. *Bulletin of Materials Science*, 24(4): 379.

Overshott, K.J. The causes of the anomalous loss in amorphous ribbon materials, (1981) *IEEE Transactions on Magnetics*, 17(6): 2698–2700.

Rahman, I.Z., Ahmed, T.T. (2005) A study on Cu substituted chemically processed Ni–Zn–Cu ferrites. *Journal of Magnetism and Magnetic Materials*, 290–291: 1576–1579.

Seetha Rama Raju, V., Murthy, S.R., Gao, F., Lu, Q., Komarneni, S. (2006) Microwave hydrothermal synthesis of nanosize PbO added Mg-Cu-Zn ferrites. *Journal of Materials Science*, 41: 1475–1479.

Shinde, T.J., Gadkari A.B., Vasambekar, P.N. (2013) Magnetic properties and cation distribution study of nanocrystalline Ni–Zn ferrites. *Journal of Magnetism and Magnetic Materials*, 333: 152–155.

Singh, N., Agarwal A., Sanghi, S., Singh, P. (2001) Synthesis, microstructure, dielectric and magnetic properties of Cu substituted Ni-Li ferrites, *Journal of Magnetism and Magnetic Materials*, 323(5): 486–492.

Su, H., Zhang, H., Tang, X., Jing, Y., Liu, Y.(2007) Effects of composition and sintering temperature on properties of NiZn and NiCuZn ferrites. *Journal of Magnetism and Magnetic Materials*, 310: 17–21.

Sun, K., Lan, Z.W., Yu, Z., Jiang, X.N., Huang, J.M. (2001) Phase formation, grain growth and magnetic properties of NiCuZn ferrites. *Journal of Magnetism and Magnetic Materials*, 303: 927–932.

Tsutaoka, T., Ueshima, M., Tokunaga, T., Nakamura, T., Hatakeyama, K. (1995). Frequency dispersion and temperature variation of complex permeability of Ni-Zn ferrite composite materials, *Journal of Applied Physics*, 78(6): 3983.

Verma, S., Joy, P.A., Kurian, S. (2011) Structural, magnetic and Mossbauer spectral studies of nanocrystalline $Ni_{0.5}Zn_{0.5}Fe_2O_4$ ferrite powders, *Journal of Alloys and Compounds*, 509: 8999–9004.

Yue, Z.X., Zhou, J., Li, L.T., Wang, X.H., Gui, Z.L.(2001) Effect of copper on the electromagnetic properties of Mg-Zn-Cu ferrites prepared by sol-geo auto-combustion method. *Materials Science and Engineering*, B86: 64.

Manufacturing Engineering and Intelligent Materials – Lu & Abu Bakar (Eds)
© 2015 Taylor & Francis Group, London, ISBN 978-1-138-02832-6

Purification and characterization of an extracellular polysaccharide GS501 from *Ensifer* sp.

Z.H. Hu
Shandong Academy of Medical Sciences, Jinan, Shandong, China
School of Medicine and Life Sciences, University of Jinan-Shandong Academy of Medical Science,
Jinan, Shandong, China

Y.M. Li, Q. Li, Z.D. Lin, Q.S. Dong, W.W. Zhu & Y.Y. Zhao
School of Biological Science and Biotechnology, University of Jinan, Shandong, China

ABSTRACT: An extracellular polysaccharide from *Ensifer* sp, named GS501, was purified by a series of steps including ethanol precipitation, protein removal, ion exchange chromatography and gel filter chromatography. Then the characteristics of GS501 were analyzed. The monosaccharide composition analysis was performed by high performance liquid chromatography. The result showed that GS501 was mainly composed of glucose, which was in accord with infrared spectroscopy analysis. Scanning electron microscope analysis showed that GS501 had a gully and loose surface structure, which may be in favor of adsorption and flocculation.

Keywords: *Enfiser* sp; exopolysaccharide; purification; characteristics

1 INTRODUCTION

Microbial Extracellular Polysaccharides (EPS) are mucus compounds that were secreted by some microbes during their growth process. Their main physiological function is to protect the strains against the external hazards (Philippe, Beat. 2001). The activities of EPS depended on its structure and composition that could be analyzed after EPS purified (Andaloussi et al. 1995). Based on some reports, EPS are composed of one or many kinds of monosaccharide or their derivatives, which was analyzed by means of high performance liquid chromatography with a multi angle laser detector (Mata, et al. 2006). The structure of EPS was very complex, which gave rise some considerable difficulty to analyze its structure (Pan, Wang & Chen, et al. 2009).

In this study, an extracellular polysaccharide named GS501 was extracted from *Ensifer* sp, and was characterized by a series of analytical techniques including High Performance Liquid Chromatography (HPLC), infrared spectroscopy analysis and scanning electron microscope analysis. These analyses indicated that the GS501 would be a good material of flocculant.

2 MATERIALS AND METHODS

2.1 *Preparation of crude polysaccharides*

The fermentation broth of *Ensifer sp.* was centrifuged at 8000 r/min for 15 min. The resulting supernatant was collected and precipitated by cooled ethanol at 4 °C overnight. Then the precipitation was dissolved in deionized water, and the resulting solution was added into chloroform and n-butylalcohol (v/v 4:1) for removing protein moiety. The dialysis was performed using a dialysis bag with a molecular weight of 3.5 kDa for 24h. The contents of EPS and protein were determined using the methods of phenol sulfuric acid method (Dubois et al. 1956) and Coomassie brilliant blue method (Liu & Wang 2007), respectively.

2.2 *Purification*

GS501 was dissolved in 10 mL of distilled water and applied to DEAE-Sepharose anion exchange column chromatography using 0–1 mol/L sodium chloride gradient elution at a flow rate of 0.4 mL/min. Then the elute was further applied to Sepharose-4B gel filter chromatography using deionized water elution at a flow rate of 0.25 mL/min. The elute containing polysaccharide was lyophilized using a vacuum freeze-drying method.

3 MONOSACCHARIDE COMPOSITION ANALYSIS

The polysaccharide and monosaccharides as a standard were hydrolyzed in 2 mol/L trifluoroacetic acid at 105 °C for 6 h. The derivation of 1-phenyl-3-methyl-5-pyrazolone (PMP) was carried using a reaction volume of 120 uL at 70°C for 1 h (Li et al. 2014). The mobile phase was 0.1 mol/L phosphate buffer-acetonitrile (83:17) at a flow rate of 1.0 mL/min. The column temperature was 35°C and the detection wavelength was 250 nm.

4 RESULTS

GS501 was purified by DEAE-Sepharose anion exchange column chromatography (Fig. 1a) and Sepharose-4B gel filter chromatography (Fig. 1b). As shown in Figure 1, only a single

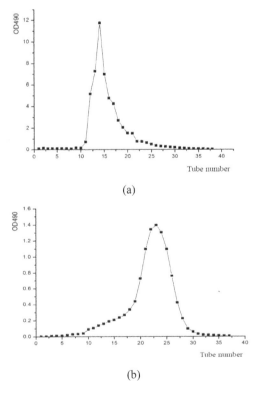

(a)

(b)

Figure 1. The elute profile of DEAE-Sepharose anion exchange column chromatography and Sepharose-4B gel filter chromatography.

232

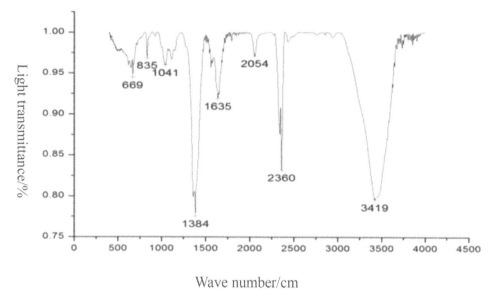

Figure 2.　Infrared spectroscopy of GS501.

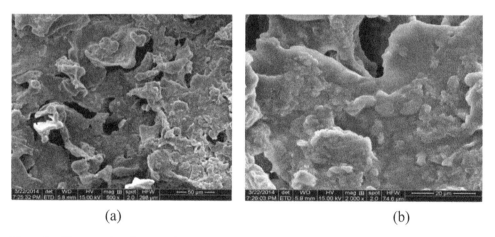

(a)　　　　　　　　　　　　　　　　　(b)

Figure 3.　The scanning electron microscope image of GS501.

peak occurred after column chromatography, which means that GS501 may be a homogeneous polysaccharide with one kind of monosaccharide.

　Figure 3a showed that GS501 was relatively loose and more void at the magnification of 500 times and it maybe have a larger surface area according to the magnification of 2000 times Figure 3b. This means GS501 may be a good material of flocculant.

5　CONCLUSIONS

An extracellular polysaccharide, named GS501, was isolated and purified from a strain of *Ensifer* sp. GS501 was purified by DEAE-Sepharose anion exchange column chromatography and Sepharose-4B gel filter chromatography, which indicated that GS501 may be a homogeneous polysaccharide with one kind of monosaccharide. And it was mainly composed of glucose. FT-IR analysis indicated that GS501 exhibited typical major broad stretching peak appeared at 3419 cm^{-1} for the–OH group, and the strong absorption at 1384 cm^{-1} was due to the–CO stretching, the absorption at 1041 cm^{-1}–1669 cm^{-1} range was the pyran ring. Element analysis showed that GS501 containing 44.38% C and 46.60% O. Finally, GS501 was determined as relatively loose, void and with a large surface area according to the magnification of 500 times and 2000 times implying it may be a good material of floccula.

ACKNOWLEDGMENTS

This work was supported by the National Natural Science Foundation of China (Project No. 31300045, 31100088), Foundation of University of Jinan (XKY1324), Open fund of Xinjiang Production & Construction Corps Key Laboratory of Protection and Utilization of Biological Resources in Tarim Basinand (BYBR1405), Shandong province science and technology development plan (Grant No. 2013GSF12006).

REFERENCES

Andaloussi, S.A., Talbaoui, H., Marczak, R., et al. 1995. Isolation and characterization of exocellular polysaccharides produced by bifidobacterium longum. *Applied Microbiology and Biotechnology*, 43: 995–1000.

Dubois, M., Gillers K.A., Hamilton J.K. 1956. Colorimetric method for determination of sugars and related substances. *Anal Chem*, 2: 350–352.

Liu, Y.H., Wang, F.S. 2007. Structural characterization of an active polysaccharide from phellinus ribis. *Carbohydrate Polymers*, 70: 386–392.

Li Yumei, Li Qiang, Hao Dakui, Hu Zhiheng, Song Dongxue, Yang Min. 2014. Characterization and flocculation meclhanism of an alkali-activated polysaccharide flocculant from Arthrobacter sp. B4. *Bioresource Technology*, 170: 574–577.

Mata, T.A., Victoria, B., Inmaculada, L., et al. 2006. Expolysaccharides produced by the recently described halophilic bacteria Halomonas ventosaeand Halomonas anticariensis. *Microbiology*, 157: 827–835.

Pan, D., Wang, L.Q., Chen, C.H., et al. 2009. Structural characterisation of a heteropolysaccharide by NMR spectra. *Food Chemistry*, 112: 962–966.

Philippe, D., Beat, M. 2001. Applications of exopolysaccharides in the dairy industry. *International Dairy Journal*, 11: 759–768.

Manufacturing Engineering and Intelligent Materials – Lu & Abu Bakar (Eds)
© *2015 Taylor & Francis Group, London, ISBN 978-1-138-02832-6*

Design of hydraulic manifold block based on UG software

H.M. Zhang, J.B. Xie & X.H. Li
Guangxi Hydraulic and Electric Polytechnic, Nanning, Guangxi, China

ABSTRACT: This article introduces the process and methods for 3D parametric design of hydraulic manifold block based on UG software. An example was given to prove that the application of UG software has practical significance for hydraulic integrated block design. It can improve the accuracy and efficiency of the design, as well as the cost of the design.

Keywords: UG software; hydraulic manifold block; the 3D parameterization design

1 INTRODUCTION

Integrated hydraulic system is the most common structure of hydraulic system. The most critical and difficult to control link in this system is the design of hydraulic manifold blocks. The internal part of manifold blocks has many interlaced channel which connected the bottom channel of the hydraulic components. The hydraulic system circuit is constituted by the channel. Of the two important parts for the design of hydraulic manifold block, the first is to ensure that the internal channel of hydraulic manifold blocks are connected and meet the requirement of the action of the hydraulic system, and the second is to ensure that the element installed on the integrated block, non-interference in each other, assembly is reasonable.

Hydraulic manifold block is the center of the integrated hydraulic system, due to the oil circuit between the hydraulic components connected by hundreds of channels inside manifold block; if the design is unreasonable, mutual interference between the channels will occur. In addition, the design must also meet the design quality requirements such as the security wall thickness between disconnect channel should be appropriate; the flow area should be enough. These problems lead to the traditional manual layout, the channel connecting and checking is very complex. Therefore, it requires that the designers have very good space imagination ability, and the ability to maintain highly concentrated attention for long, design errors will be easy to occur and afterwards also difficult to find.

With the rapid development of three-dimensional CAD software, at present has been widely used in the area of design. Scholars at home and abroad extensively studied the hydraulic integrated block design and checking based on 3D software. Now, the three-dimensional design software such as UG, CATIA, Pro/E, Solid Works, Solid Edge, etc. has been commonly used in the area of design. 3D software mostly used the concept of parametric design, design changes flexible and has relevance, greatly improve the design efficiency and accuracy. This article introduces the process and methods for 3D parametric design of hydraulic manifold block based on UG software. An example was given to prove that the application of UG software has practical significance for hydraulic integrated block design. It can improve the accuracy and efficiency of the design, as well as the cost of the design.

2 UG SOFTWARE INTRODUCTIONS

UG software is launched by Unigraphics Solutions Company integrating CAD/CAM/CAE 3D parametric design software and represents the most advanced mainstream trend of computer aided design, analysis, and manufacturing software. It's widely used in aerospace, shipbuilding, automobile, machinery and electronics industries in the international and domestic companies. UG has strong abilities of modeling, virtual assembly and flexible engineering drawing design ability; the new version-NX has more powerful function, which makes design more convenient and quick. UGNX 3D parametric software has many application modules, when establishing or opening a parts file, use the start menu to switch between the application modules.

The main characteristic of UG software is as follows:

1. Providing a product design environment based on process. Integrated the entire product development process from the concept design to product modeling, analysis and manufacture.
2. With unique Knowledge Driven Automation function. Make the product and process knowledge integrated in a system, can greatly improve the productivity of enterprises.
3. Product design technology based on assembly. By applying the master model method, can start from the product's overall design, detailed design every part step by step. It also can realize the correlation between application modules from design to manufacture and realization to work together.
4. The application of high precision boundary entity modeling module, characteristic, parameterized and variation design, with functions of parametric modeling.
5. With good secondary development interface and tools, support a variety of general and popular data interchange standard.

3 THE DESIGN PROCESS OF HYDRAULIC MANIFOLD BLOCK

As shown in Figure 1, the process of using UG software to design hydraulic manifold block according to three steps for the preliminary design, 3D modeling and assembly to the output 2D engineering drawings. In recent years, with the intensification of market competition

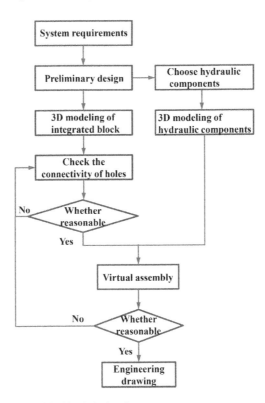

Figure 1.   The hydraulic manifold block design flow.

236

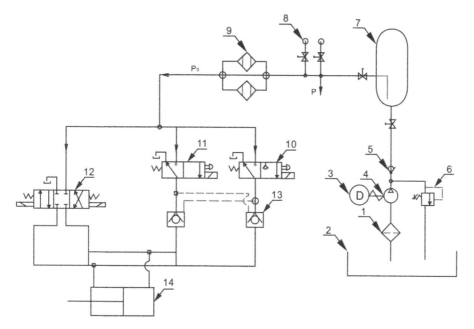

Figure 2. The principle diagram of the hydraulic control system. 1 Oil filter 2 Fuel tank 3 motor 4 Oil pump 5 Check valve 6 Relief valve 7 Accumulator tank 8 Pressure gauge 9 Binocular oil filter 10–11 Electromagnetic ball valve 12 Electromagnetic directional valve 13 Hydraulic controlled check valve 14 Oil cylinder.

and the rapid development of information technology, the mainstream companies are using concurrent and collaborative design; greatly improving the speed of product design, save the design cost, at the same time also can further enhance the quality of the products and services, to meet changing market needs.

The following take the design of hydraulic manifold block in hydraulic mechanical electrical hydraulic governor as an example, brief description of the process of using UG software design hydraulic manifold block. The hydraulic control system of hydraulic mechanical electrical hydraulic governor is shown in Figure 2.

3.1 *Understanding system requirements, selecting hydraulic components and modeling*

According to the principle of the hydraulic system diagram and the actual working parameter to select the matching check valve, throttle valve, relief valve, electromagnetic directional valve, accumulator tank and other hydraulic components. Then set up 3D model according to the size of the selected hydraulic components, deposited in the hydraulic components model library, ready for the call.

3.2 *The design of hydraulic manifold block*

Hydraulic manifold block is the center of the hydraulic system oil intersection; all hydraulic components are installed on the surface of the manifold block make up the hydraulic system control core. The chief problem of hydraulic manifold block design is to ensure that the oil channel of various hydraulic components are connected. Secondly, to ensure that the flow area is enough. In addition, in order to reduce the loss of oil, as far as possible to reduce the corner of important hydraulic channel.

Hydraulic manifold block is to create the corresponding hole feature in the basic model to complete oil channel connected, UG software uses the parametric design, the specific modeling process is: First, set up a slightly bigger parameterized cuboid in modeling module of

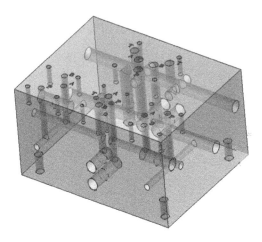

Figure 3.   The 3D model of hydraulic manifold block.

UG software; and then, create the corresponding holes at each plane of cuboid to connect the hydraulic components. The purpose of the oil channel connected can be implemented by adjusting the position and depth of the holes.

After completion of the preliminary modeling, the 3D model need to be further modified and perfected. In the hydraulic integrated block with different types of holes features such as countersunk holes, ladder holes, threaded holes and auxiliary holes. As shown in Figure 3, the connectivity between all kinds of holes can be visually seen through the 3D model perspective and different projection view in modeling module of UG software. Ensure that each oil channel within manifold block conform to the requirements of the hydraulic components installed, there won't be interference between each other. For hydraulic system with relatively complicated functions, can carry out finite element analysis to choose the optimal solution, to avoid repeated prototype experiments and analysis in the laboratory, so as to improve the speed of hydraulic manifold block design, ensure the quality of products. After model was preliminary established, in order to save materials, reduce cost, under the premise of ensuring manifold block has enough intensity and the hydraulic components have enough installation space, and the assembly workers have enough operation space to minimize the overall dimension of hydraulic manifold blocks to reduce weight.

3.3   The whole assembly of hydraulic manifold block

In the assembly module of UG software, establish a parametric assembly; call in the hydraulic components 3D model established before, through the assembly constraints such as alignment, matching, distance to take the various hydraulic components assembly to the corresponding position on the hydraulic manifold block. Different parts of the assembly can be seen by transforming perspective, but also to check whether the design process is reasonable, parts size is appropriate, whether there is interference between the hydraulic components and whether the installation position of manifold block on the machine is appropriate, and whether there is interference between manifold block and other parts etc. by the explosion diagram and assembly process video replay; if design problem is discovered, it can be modified in a timely manner.

3.4   Generate engineering drawing of hydraulic manifold block

After the completion of whole assembly for the hydraulic manifold block, validated that there is no design errors, can be generated by engineering drawings automatically by the drawing module of UG software. UG 2D engineering drawings are based on 3D model created in modeling and assembly module. It is obtained by projection in the engineering drawing module, the 2D engineering drawings and 3D models are interrelated. When there is any change in the 3D model, corresponding 2D engineering drawings will be updated, so the designer just focus on the 3D models, don't have to worry about discrepancy of 2D engineering drawings and 3D models, so as to reduce duplication of work, improve the efficiency of the drawing. UG software can generate not only the basic view, but also can automatically generate the section view, local amplification view, axonometric drawing and other auxiliary view and so on. For assembly, the drawing module can generate parts diagram, and can automatically tag part numbers, automatically generate parts subsidiary bar and material consumption sheet, etc.

4 CONCLUSIONS

Hydraulic manifold block is the core of the integrated hydraulic system, the application of UG software to design, can carry on the omni-directional observation of hydraulic manifold block entity model, can be more reliable to check the correctness of the whole system in hydraulic manifold block and the rationality of the layout for hydraulic components, assembly interference is more intuitive, more convenient to generate and modify engineering drawings. So as to shorten the design time and reduce the design cost, and better ensure the accuracy of the design, improve the design efficiency and quality, can meet the design requirements as much as possible.

ACKNOWLEDGEMENT

Fund project: Guangxi university scientific research project *Research of hydraulic machinery governor based on PWM* (YB2014510).

REFERENCES

Chen Jiu. 2003. Manifold Design Based on Solid works Software. *Journal of Hydraulic and Pneumatic* 7: 5–6.
Fan Jianliang. et al. 2008. Application of Pro/E software in Design of Hydraulic Integrated Circuit. *Journal of Coal Mine Machinery* 29 (1): 174–176.
Gao Weiguo. et al. 2007. Study on the Method of Hydraulic Integrated Block 3D Design. *Journal of Hydraulic and Pneumatic* 2: 23–26.
Wang Chunmei. et al. 2008. The Design of Hydraulic Integrated Valve Block. *Journal of Coal Mine Machinery* 29 (9): 40–41.
Wei Huanhuan. et al. 2010. 3D-design and Outlook to Hydraulic Manifold Block. *Journal of Coal Mine Machinery* 31 (3): 12–14.

Manufacturing Engineering and Intelligent Materials – Lu & Abu Bakar (Eds)
© *2015 Taylor & Francis Group, London, ISBN 978-1-138-02832-6*

A new type of automatic induction parking lock

Z.R. Li, Y. Hou & X.B. Wang
North China Institute of Aerospace Engineering, Langfang, Hebei, China

ABSTRACT: This paper introduces the automatic induction parking lock based on single chip processor, wireless communication module and corresponding control circuit. The system using STC89C52 single-chip processor as the core, comprises a transmitting part and a receiving part for information transmission, and according to the judgment result control the state of the parking lock switch. Because of the system hardware overhead is small, high intelligence, and low error rate, the system has a high application value.

1 INTRODUCTION

With the progress of the times, the automobile has become part of ordinary families. Increasingly strained parking resources have become the major challenge that the mankind must face, especially in the residential areas, hospitals and other public places. Therefore, the parking lock becomes an effective way to solve this problem. However, the traditional parking lock used have many inconveniences; this makes it necessary to design an auto-sensing parking lock.

2 SYSTEM DESIGN

The composition of this system is divided into the transmitting part and the receiving part. The transmitting part generate data coding, send data to a certain range by the wireless transmitter module, the receiving part, installed on the vehicle's interior, receive data through a wireless receiver module, the single-chip processor drive the motor and control the lock movements.

The transmitting part is installed in the car, the keys of the car control the work of the system. When the car starts, it will auto-start the transmitter, the information generated by the data generation chip and send the code to the modulation chip, and then transmit the code by the wireless transmitter module to the receiving end located in the parking lockin order to achieve the purpose of opening or closing the parking lock. The transmitting part diagram is shown in Figure 1.

The receiving part receives the information code through the wireless receiver module, decode by the decoding module and send to the micro controller of the receiving part to compare. After comparison, control the subsequent circuit lock or unlock, and the motor forward or reverse. In this process, the position sensors installed on the lock detect the lock's height, and judge the motor rotating direction. The receiving part diagram is shown in Figure 2.

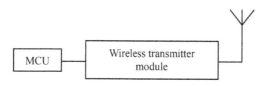

Figure 1. The transmitting part diagram.

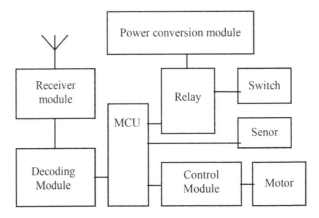

Figure 2. The receiving part diagram.

3 HARDWARE CIRCUIT DESIGN

3.1 *The transmitting part*

The transmitting part is installed in the vehicle, the power supplied by the vehicle battery. This part consists of a single-chip smallest system and a wireless transmitter module. The control module using the STC89C52 as the core, the instruction systems, the hardware structure and the resources on-chip are fully compatible with the MCS-51 micro controller.

The transmitter module use the DF transmitter module, the operating frequency is 31.5 MHz, using surface acoustic resonator SAW frequency stabilization, has very high frequency stability, frequency drift of only 3 ppm/kWh when the ambient temperature changes between −25 and +85 degrees, which is particularly suitable for multiple received wireless remote control and data transmission system.

3.2 *The receiving part*

The receiving part of the parking lock installed in the ground, power supplied by the AC mains embedded underground, the receive module consist of a single-chip processor, a Dual Full-Bridge motor driver module, a wireless receiver module, a 220V AC transfer power conversion module and other components. The MCU module is used to receive the information, contrast, judgment and control, the Dual Full-Bridge module complete the motor drive control tasks for the parking lock. The system uses the L298N as the motor drive chip. The Dual Full-bridge L298 Ncontains 4 channels logic drive circuit, can drive two 2-phase or one 4-phase stepper motor, containing two H-Bridge high voltage and large current full bridge driver, receiving standard TTL logic level signal, it can drive the following 46V, 2 A stepper motor, and can directly through the power to regulate the output voltage. This chip can be controlled by the microcontroller I/O port to provide analog signal.

Above the circuit design, it can complete the wireless communication for the information, as well as the automatic control tasks for the parking lock.

4 SOFTWARE DESIGN

The software design is also divided into the transmitting part and the receiver part, the major role of the transmitter is to send the information. Each system has a unique information code composed by four group8-bits data. The first and second bits are address codes, avoid four group information code in conflict and influence, the impact of the encoding scheme is shown in Figure 3.

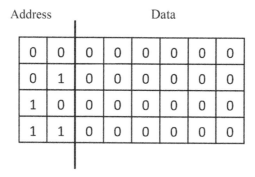

Address		Data					
0	0	0	0	0	0	0	0
0	1	0	0	0	0	0	0
1	0	0	0	0	0	0	0
1	1	0	0	0	0	0	0

Figure 3. The code with the address bits.

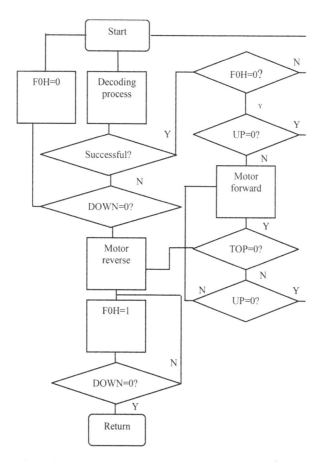

Figure 4. The flow chart of receiving part.

The receiving part receive the information code and complete the task of decoding judgment, but also control the motor in the correct information code. To lock up and down, there are many situation to raise or down the lock, I will have the car into the exit process may encounter to be analyzed.

When the vehicle approaches, the parking lock and the parking lock decoding success in the setting time, then detect the lowest point of the switch (**DOWN**) is closed or not. If it is closed, the lock has been reduced to the lowest point, then the program returns to the start and receive re-coding; if the lock does not reach the lowest point, then start the motor decline program to make lock decline, in the process to determine the lowest point switch is closed, if the lock not closed, then continue decline, otherwise, the program returns to the start.

At this time, the parking lock down to the lowest point, the car slowly pulls into the parking spaces; turn off the engine and power, and stop the transmitting part send the encoded data. At this time, receiver module cannot be received within the specified time and complete the data decoding, the lock automatically detects the highest point of the switch (UP) is closed. If the switch is closed, so the lock had risen to the highest point, and return to the start. If it does not get to the highest point, then drive the lock rise. At the same time, the system will detect the switch at the top of the lock is closed. If the switch touches the bottom of the vehicle, then the top switch will be closed, the system calls the motor down program, in order to avoid destroying the vehicle. If the switch is not closed and direct the up switch is closed, so the parking lock is not touching the car chassis is going directly to the top.

When the owner restarts the car, the transmitting part send data, the receiving system receive and decode the data, then the lock drops to the lowest point once again and the car can be pulled out. The flow chart is shown in Figure 4.

5 CONCLUSIONS

The automatic induction parking lock can be realized above the options of the text, the system uses the micro controller as the core, through the wireless communication, complete the active judgment and automatic processing. The system is more intelligent, sensitive and can be operated without driver. It solves the parking problem very well. The system has high accuracy, sensitivity, and reliability, while the price is low, so the system has a high popularization value.

ACKNOWLEDGEMENTS

The work was supported by the Youth Fund Project of North China Institute of Aerospace Engineering. Project ID is KY-2014-18, and the work was supported by the Major Project of North China Institute of Aerospace Engineering. Project ID is ZD-2014-02.

REFERENCES

Cherrill M. Spencer, Seung J. Rhee. 2004. Comparison Study of Electromagnet and Permanent Magnet Systems for an Accelerator Using Cost-Based Failure Modes and Effects Analysis. *SLAC-PUB* 2004:1–4.

Gao Qiang, Zhang Nan. 2009. Intelligent parking lock control system based on microcontroller technology research. *Innovative technology*, (2):6–7. (In Chinese).

Minoru Sakair, Daisuke Suzuk, Ayako Nishimura, Yukiko Ichige and Masashi Kiguchi, Simultaneous detection of breath and alcohol using breath-alcohol sensor for prevention of drunk driving, *IEICE Electronics Express*, Vol. 7, No. 6, 467–472.

Rhee S.J., Ishii K. 2002. Life cost-based FMEA incorporating data uncertainty, Proc. ASME Design Engineering Technical Conf, Montreal.

Zhang Hui, 2009. Intelligent electronic remote control parking lock design based microcontroller. *Journal of Nantong shipping Vocational and Technical College*, 8(3):68–70. (In Chinese).

Manufacturing Engineering and Intelligent Materials – Lu & Abu Bakar (Eds)
© *2015 Taylor & Francis Group, London, ISBN 978-1-138-02832-6*

Finite element analysis of rolling shearing process of medium plate based on upper blade geometric parameters

Y.G. Li, J.L. Nie & Y.R. Xu
Taiyuan University of Science and Technology, Taiyuan, Shanxi, China

ABSTRACT: The upper blade geometric parameters are the important technical parameters of rolling shear. The process of rolling shear with upper blade geometric parameters was simulated by using FEM software DEFORM-3D, and the influence on the equivalent stress distribution, the cross-section quality of plate and the maximum shear force by the upper blade geometric parameters was analyzed. The simulated best upper blade geometric parameters are given as follows: the rack angle is 10°, the back angle is 5°, the corner radius is 0.2 mm. Comparing the numerical simulation with the experiment, the deviations of the maximum shear force are 6.8%–11.6% in cutting and 8.1%–12.6% in rolling shearing. After the upper blade geometric parameters are optimized, the simulated maximum shear force is reduced compared with theoretical values, the reduced rates are about 4.3% in cutting and 11.8% in rolling shearing, the energy saving shear is realized.

Keywords: upper blade geometric parameters; rolling shear; FEM software DEFORM-3D; equivalent stress distribution; cross-section quality of plate; maximum shear force

1 INTRODUCTION

The blade is the "cutting tool" of the medium plate shearing process. The reasonable blade geometric parameters not only affect the shear quality and the shear force, but also affect the service life of the blade on rolling shearing process (Wang & Wu 2013).

Aiming at solving the frequent problems such as the damaged shear blade, the large shear force and the low quality of cross section, which happened to the rolling-cut dividing shear, an upper blade model with blade geometric parameters was designed. By defining the rake angle, the back angle and the corner radius of upper blade, numerical simulation of rolling shearing process was done, and the influence of the upper blade geometric parameters to the shear force and the cross-section quality was analyzed (Li 2013, Yang 2002). On this basis, the best upper blade geometric parameters were determined, the maximum shear force was reduced on plate shearing process and the theoretical method was provided to seek for the energy saving shear (Yang et al. 2008).

2 THE MECHANISM OF ROLLING SHEAR

This paper takes the 4300 mm hydraulic rolling-cut dividing shear as the research object. Figure 1 is the schematic diagram of hydraulic rolling-cut dividing shear.

The hydraulic rolling-cut dividing shear is the hinge point movement which drives linkage by two hydraulic cylinders, and the rolling shear movement which drives the upper blade by composite link mechanism (Han et al. 2011). The movement process of rolling shearing is composed of cutting stage, rolling shearing stage and cutting off stage. The role of cutting stage is to make the upper blade reach a predetermined position and cut into the steel plate at the right angle, thereby preparing for the rolling shearing stage. The role of rolling shearing

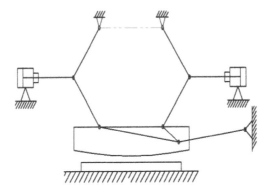

Figure 1. The schematic diagram of hydraulic rolling-cut dividing shear.

stage is to make the upper blade return to the original position, so that the steel plate can be smoothened through the upper and lower blade. In addition, the way of hydraulic transmission improves the force situation of key parts and the reliability of equipment (Murakawa & Lu 1995).

3 FINITE ELEMENT SIMULATION OF ROLLING SHEAR

3.1 The fracture criterion of rolling shear

The fracture criterion plays an important role in accurate simulation on the process of fracture. The simulation can use the Cockroft & Lathem criterion (Cockcroft & Latham 1986). Its function is described as follows:

$$C = \int_0^{\bar{\varepsilon}_f} \left(\frac{\sigma^*}{\bar{\sigma}} \right) d\bar{\varepsilon}$$

where $\bar{\sigma}$ is the equivalent stress; σ^* is the main maximum stress; ε is equivalent strain; $\bar{\varepsilon}_f$ is the strain when the material failures; C is the critical damage value of materials; From the searchable literature, $C = 0.45$ (Hu & Li 2011).

3.2 The description of upper blade geometric parameters

The upper blade can be used as "cutting tool" on medium plate rolling shearing process, therefore, there should be the upper blade geometric parameters, such as the rake angle γ, the back angle α, the corner radius r. Definition is shown in Figure 2.

3.3 The establishment of finite element model

For the rolling shearing process and the description of the upper blade geometric parameters, the finite element model of the rolling shear is established by using the software of Pro/E and Deform-3D. In order to reflect the real shear deformation, the local mesh subdivision can be applied to the contact area between the blade and the steel plate as shown in Figure 3. This material model is a mixture model of isotropic and kinematic hardening, which is associated with strain rate. Detailed material properties are shown in Table 1.

The material of medium plate adopts Q235A steel in the cold state, the number of elements is 80000–200000, the width of medium plate is 1100 mm, the shearing velocity is 450 mm/s. The shear process in simulation is a large deformation, and the plastic deformation of the material is far larger than the elastic deformation, so the steel plate can use plastic kinematic model.

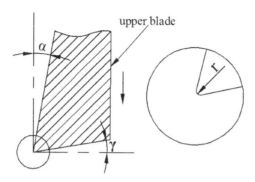

Figure 2. The upper blade geometric parameters of rolling-cut dividing shear.

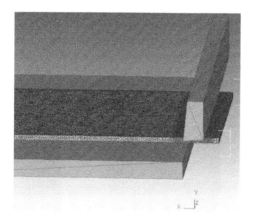

Figure 3. Finite element simulation of rolling shear.

Table 1. Material properties in finite element simulation.

Material properties	Q235A
Elastic modulus/N·m^{-2}	2.12×10^{11}
Poisson ratio	0.288
Mass density/kg·m^{-3}	7.86×10^3
Shearing modulus/MPa	8.23×10^{11}
Yield strength/MPa	2.35×10^8
Fracture criterion	0.45

4 RESULTS AND DISCUSSION

Under similar other conditions of shear parameters, the process of rolling shear was simulated by setting different upper blade geometric parameters. By analyzing the emergence, fracture, extension of crack in steel plate, the influence of upper blade geometric parameters to the shear stress distribution, the cross-section quality of plate and the maximum shear force was obtained.

4.1 *Influence of upper blade geometric parameters to the shear quality and the shear stress*

The cross-section quality of steel plate and the distribution of equivalent shear stress were mainly studied under different upper blade geometric parameters.

Figure 4(a) shows that when the rake angle and the back angle were all negative, the contact area between the upper blade and the steel plate was increased, thereby increasing the shear force on plate shearing process. In addition, the cross sections of steel plate were prone to squeeze high burr. Figure 4(b) shows that when the blade geometric parameters were not given, the distribution of equivalent shear stress was not uniform on the shearing area, the cross section of steel plate was also not smooth. Figure 4(c) shows that when the blade geometric parameters were appropriate, the distribution of equivalent shear stress was relatively uniform on the shearing area, the cross-section quality of steel plate was also very high.

From the distribution of equivalent shear stress and the cross-section quality of steel plate, the best upper blade geometric parameters were obtained. The rake angle $\gamma = 10°$, the back angle $\alpha = 5°$, the corner radius r = 0.2 mm.

4.2 *Impact on the shear force of the best blade geometric parameters and experiment*

Figure 5 and Figure 6 are the change curves of shear force and the cross section of steel plate, respectively, at the best blade geometric parameters. In Figure 5, the shear force of steel plate increased sharply in cutting, and it reached the maximum until the fracture of steel plate occurred instantaneously. The steel plate began to enter the rolling shearing stage after cutting, and the shear force of steel plate fluctuated with time. When the contact area of the steel plate and the upper shear blade reached the maximum, the shear force of steel

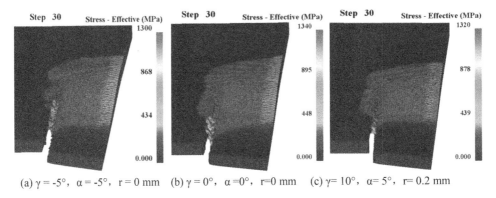

(a) $\gamma = -5°$, $\alpha = -5°$, r = 0 mm   (b) $\gamma = 0°$, $\alpha = 0°$, r = 0 mm   (c) $\gamma = 10°$, $\alpha = 5°$, r = 0.2 mm

Figure 4.   Simulation of 18 mm plate at different blade geometric parameters.

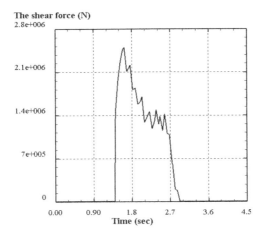

Figure 5.   The change curve of shear force at the best blade geometric parameters.

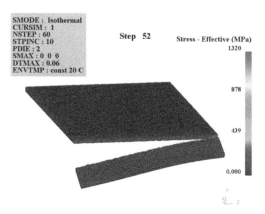

Figure 6.　The cross section of steel plate at the best blade geometric parameters.

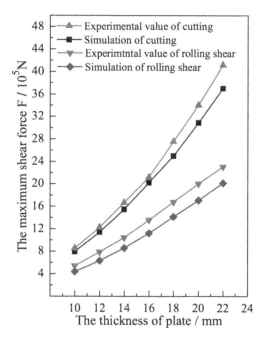

Figure 7.　The maximum shear force comparison of simulated results and experimental results.

plate also reached the maximum, and then quickly reduced to 0 N. Figure 6 shows that the cross section of steel plate was smooth and the quality of shearing was high at the best upper blade geometric parameters.

In order to verify the reliability of numerical simulation, the maximum shear force data which was collected from the hydraulic rolling-cut dividing shear at Linfen Iron and Steel Company were compared with the simulation results as shown in Figure 7.

When the range of the thickness of plate is from 10 mm to 22 mm, the deviation of the maximum shear force in cutting between the experimental results and simulated results is from 6.8% to 11.6%, and the deviation of the maximum shear force in rolling shear between the experimental results and simulation results is from 8.1% to 12.6%. Considering blade clearance and other factors, the results of numerical simulation are very reliable.

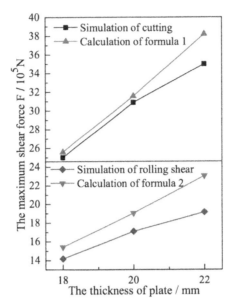

Figure 8. The maximum shear force comparison of simulated results and calculated results.

4.3 *Comparison and analysis of simulated values and calculations of theoretical formulas*

Theoretical equation of the maximum shear force in cutting and rolling shearing (Zhang et al. 2010) is given as follows:

$$P = k_1 k_2 k_3 k_4 k_5 \sigma_b \delta_s \frac{Bh^2}{r} \tag{1}$$

where k_1 = comprehensive stress transformation coefficient; k_2 = blade gap effect coefficient; k_3 = blunt knife effect coefficient; k_4 = entrance angle arc curvature effect coefficient; k_5 = entrance angle correction factor; σ_b = material strength limit; δ_s = material elongation; B = distance between the connecting rod and the upper blade articulation point; r = crank radius.

$$P = k_2 k_3 k_\phi (1 + \mu)(0.5 - \mu) \beta \sigma_s \frac{h^2}{\tan \alpha} \tag{2}$$

where k_2 and k_3 is same as above; k_ϕ = shear equivalent angle effect coefficient; ϕ = cutting angle of upper blade; μ = Poisson's ratio; β = fracture coefficient. σ_s = material yield limit.

The shearing process of Q235A steel plates with the thickness of 18 mm, 20 mm, 22 mm were simulated, and the simulated values of the maximum shear force were obtained in cutting and rolling shearing. The theoretical values of the maximum shear force were calculated by the theoretically-calculated formula. Comparative analysis results are shown in Figure 8.

It could be obtained that the maximum shear force was decreased in cutting and rolling shearing under the optimization of upper blade geometric parameters. The reduced rates were, respectively, about 4.3% and 11.8%, and it had achieved the goal of reducing the maximum shear force. Therefore, it was confirmed that the reasonable blade geometric parameters played a good role in promoting the highly efficient energy-saving shear.

5 CONCLUSIONS

Under the different upper blade geometric parameters, the finite element model of rolling-cut dividing shear is built, and the rolling shear process is simulated by using FEM. For the analysis and simulation of the equivalent stress distribution and the cross-section quality of steel plate, the best upper blade geometric parameters are obtained: the rake angle $\gamma = 10°$, the back angle $\alpha = 5°$, the corner radius r = 0.2 mm. Comparing the maximum shear force between the numerical simulation and the experiments, the deviations are in the range of 6.8%–11.6% in cutting and 8.1%–12.6% in rolling shearing, the results of numerical simulation are close to facts. Comparing simulated values with theoretically-calculated results, it can be concluded that the maximum shear force is decreased after the optimization of upper blade geometric parameters. The reduced rates are about 4.3% in cutting and 11.8% in rolling shearing, and it has achieved the goal of reducing the maximum shear force.

ACKNOWLEDGEMENTS

The authors would like to express their appreciation to the support of the project of Shanxi Scholarship Council of China (2014-065), Shanxi Province Natural Science Fund Project (2014011024-5), Shanxi provincial Key Tackle Projects on Science and technology (20090321028) and Taiyuan Science and technology Stars Projects (20111076).

REFERENCES

Cockcroft M.G. & Latham. D.J. 1986. Ductility and the Workability of Metals [J]. *Journal Institute of Metals*, 96: 33–39.

Han Heyong, Huang Qingxue, Ma Lifeng, Wang Jing. 2011. Research on the Hydraulic System of Hydraulic Rolling Shear [J]. *Journal of Sichuan University*, 43(3): 239–243.

Hu Jianjun, Li Xiaoping. 2011. DEFORM-3D Plastic Forming CAE Application Tutorial [M]. Peking University press, 234–239.

Li Ying. 2007. Elementary introduction the application of rolling cut shear in heavy plate mill [J]. *Metallurgical Standardization & Quality*, 45(5): 53–55.

Murakawa M & Lu Y. 1995, Study on precision shearing of sheet II Burr-free shearing by means of rolling cut shear [J]. *Journal of the Japan Society for Techno logy of Plasticity*, 410: 242–246.

Wang Dazhong, WuShujing 2013. Metal Cutting Control and Tool [M]. Beijing, Tsinghua University press, 91–95.

Yang Guchuan. 2002. Analysis of the influence factors on the efficiency and quality of shear to length for plate [J]. *Steel Rolling*, 19(3): 58–59.

Yang Gangjun. 2008. The Study for Reducing and Eliminating the Shearing Force Peak Value of the Rolling Shear Machine [J]. *Heavy Machinery Science and Technology* (2): 25–26.

Zhang Jidong, ChuZhibin, Chang Yu, Huang Qingxue. 2010. Formula build and industrial research for maximal shear stress of rolling shears [J]. *Heavy Machinery* (5): 59–62.

Manufacturing Engineering and Intelligent Materials – Lu & Abu Bakar (Eds)
© 2015 Taylor & Francis Group, London, ISBN 978-1-138-02832-6

Design of automatic door control system based on PLC for multimedia classroom

Q.L. Yang, W.G. Li, Y.C. Chen & L.L. Huang
Faculty of Mechanical and Electrical Engineering, Kunming University of Science and Technology, Kunming, Yunnan, P.R. China

ABSTRACT: This paper presents a design approach of automatic door control system based on PLC for multimedia classroom. The designs of control interface of front door and back door are introduced, and touch screen soft is used to simulate the proposed approach. The detailed system design is given, and control ladder diagrams of front door are detailed. The simulated experiment results testify the correctness of the proposed approach. This design will reduce the workload in inspections of the classroom for workers, and is convenient for management and use of the multimedia classroom.

1 INTRODUCTION

With the enrolment expansion of colleges and universities, the scale of application for multimedia classroom in colleges and universities is all-time high, which results in many problems such as the use and management of multimedia classroom.

With the development of computer networks, many automatic control systems enter the lives of people, which make the lives easier. The control system based on PLC has the merits such as the strong capacity of resisting disturbance, high reliability, small volume, convenient design, use and maintenance, and have been applied in many marketplaces, public buildings, banks, hospitals etc. The automatic door control systems based on PLC have been applied in many fields (Liu 2009).

Automatic door system control systems based on PLC detect the infrared ray with 8–13 micrometer that is coming from human body to control the opening and the closing of the door. Infrared sensor detects the moving object that enter into induction zone; if the detected object enter into induction zone, the infrared sensor sends signal to PLC, and PLC controls the motor rotation, and then the door will open. If the moving object is outside the induction zone, the door will remain closed (Liu 2009).

The traditional automatic door control systems commonly adopt relay logical control, which have the disadvantages such as high failure rate, low reliability and inconvenient maintenance will be replaced by control system based on PLC (Yao 2002).

In this paper, the detailed design approach of automatic door control system based on PLC is presented. This paper mainly aims to reduce the required time consumed in inspections of the classroom for audio-visual workers, and audio-visual workers have to open and close doors every day, and the design of automatic door control system for multimedia classroom is introduced. According to current application status of multimedia classroom, and with the increase of investment in education, this design of multimedia classroom control system will have a good prospect of applications.

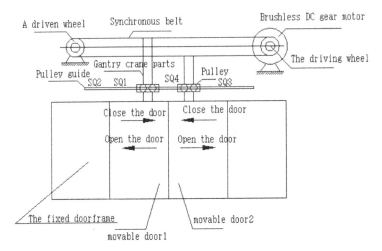

Figure 1. Automatic door drive system.

2 DESIGN OF AUTOMATIC DOOR CONTROL SYSTEM OF MULTIMEDIA CLASSROOM

2.1 *Transmission mechanism*

The transmission system of the automatic induction door of the classroom is shown in Figure 1-door hanging piece is provided with a pulley sliding on the rail, the upper part is connected with a synchronous belt by a connecting piece, the lower part is connected with the movable door leaf, when brushless DC gear motor rotates, the synchronous belt is driven, at the same time, the door hanging pieces is driven to slide on the rail, and then the door leaf is driven to open and close. In Figure 1, four limit switches are used, both sides are placed two limit switches to control the door, respectively. While opening the door, if the door touch the limit switch SQ1, then the door speed will slow down, and if it touches SQ2, then the door will stop. While closing the door, if the door touches SQ3, then the speed will slow down, and if the door touches SQ4, then the door will stop (Li 2009).

2.2 *The sensor selection*

The microwave radar sensor is selected as the classroom door sensor. At the same time, in order to avoid the emergence of clamping phenomenon, the ejection type photoelectric switch is chosen. In order to realize the closed-loop speed control, the speed measuring encoder has been chosen to detect speed.

2.3 *Touch screen selection*

Touch screen is an induction type liquid crystal display device that is capable of receiving contact input signal, which is contacted with the graphics on the screen button. The screen tactile feedback system can drive a variety of connecting devices according to the prepro-grammed program, and can be used to replace the button panel.

Touch screen is regarded as man-machine interactive equipment of control system, and not only has the advantages mentioned above, but also can greatly reduce the input points of PLC, and greatly extend the control function.

3 CALCULATION OF POSITION FOR DOOR LEAF

The principle of door leaf position calculation is as follows: chooses eight timers, front door four timers: T252, T253, T254 and T255; back door four timers: T246, T247, T248 and T249.

The work process is as follows: (1) calculate timing value $t1$ while opening the door with high speed $v1$, (2) calculate timing value $t2$ while opening the door with low speed $v2$, (3) calculate timing value $t3$ while closing the door with high speed $v3$, (4) calculate timing value $t4$ while closing the door with low speed $v4$, (5) the absolute position of the door leaf can be calculated by following equation:

$$S = (v1 \cdot t1 + v2 \cdot t2) - (v3 \cdot t3 + v4 \cdot t4) \tag{1}$$

4 DESIGN OF CONTROL INTERFACE OF FRONT-DOOR AND BACK-DOOR FOR MULTIMEDIA CLASSROOM

Control interface functions of front-door and back-door of multimedia classroom include: automatic control and manual control of front-door and back-door, visual monitoring of the classroom doors, fault alarm etc.

Control interface of front-door and back-door of multimedia classroom are introduced as follows: Twelve position switches, those are M131, M416, M419, M130, M121, M122, M133, M417, M418, M132, M144, M146; One switch buttons, while click the control interface, will turn back to the main interface; Six indicating lamps, those are M203, M202, M200, M205, M204, M201; One alarm lamp (Y026), the display of front-door and back-door (Y011) (Jiang, Ouyang & Zeng 2008).

5 DESIGN OF LADDER DIAGRAM FOR AUTOMATIC DOOR CONTROL SYSTEM

Here shows only one ladder diagram and state diagram for sequence control function of front door that is shown in Figure 2, the left is state diagram of sequence control for front-door,

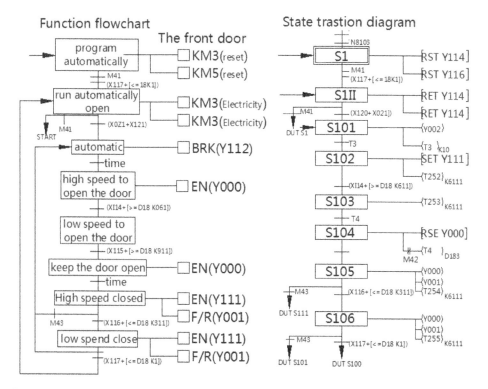

Figure 2. Ladder diagram and state diagram of sequence control for front-door.

Figure 3.   Figure of opening front-door by clicking touch screen.

and the right is the control ladder diagram of sequence control for front-door. The running result for this ladder diagram while opening the front-door by clicking touch screen is shown in Figure 3 (Sun, Yang & Zhang 2010, Li & Li 2008).

6   EXPERIMENT

The proposed approach is testified by our simulated experiment. After the ladder diagram is designed, we firstly download the ladder diagram into the PLC and run the ladder diagram, and then the control interface soft is implemented. After setting up the connection between the PLC and touch screen soft, we click the "open front-door" button by touch screen. Figure 3 shows the figure of the front-door has been opened, which validate that the proposed design approach is correct.

7   CONCLUSIONS

This paper presents the approach of automatic door control system of multimedia classroom applying the PLC sequence control technique. The detailed automatic door control ladder diagram is designed; the door leaf position is calculated by two high-speed timers. The control of front-door and back-door of multimedia classroom is realized. The proposed design approach can reduce the workload of management personnel, through the touch screen, can automatically/manually switch interface; the operation is simple, practical, convenient, stable and reliable. Through the control of the alarm, it can run the fault alarm system and teachers' quick help. The simulated experiment result and the result of touch screen online simulation validate the correctness of the proposed design approach.

REFERENCES

Jiang, H.S., Ouyang, Y.X. & Zeng, Y.B. 2008. Design of the Control System of Step Motor Based on PLC and Touch Screen. *Machinery & Electronics in Chinese.* (10): 75–78.
Li, W.H. & Li, G.Q. 2008. Design of Step Motor Control System Based on PLC. *Instrumentation Technology in Chinese.* (9): 23–25.
Liu, S.M. 2009. Application of PLC in Automatic Door control System. *Electrotechnology in Chinese*, (11): 31–33.
Li, H.X. 2009. *Automatic door engineering*, Beijing: China Water Power Press.
Sun, K.L., Yang, Z.W. & Zhang, Y. 2010. Control system design of automatic door based on PLC. *Journal of Mechanical & Electrical Engineering in Chinese.* 27(11): 123–126.
Yao, W.F. 2002. Automatic Door Control System based on PLC. *Building automation in Chinese.* (6): 64–65.

Manufacturing Engineering and Intelligent Materials – Lu & Abu Bakar (Eds)
© *2015 Taylor & Francis Group, London, ISBN 978-1-138-02832-6*

Design of projection control system based on PLC for multimedia classroom

Q.L. Yang, W.G. Li, Y.C. Chen & L.L. Huang
Faculty of Mechanical and Electrical Engineering, Kunming University of Science and Technology, Kunming, Yunnan, P.R. China

ABSTRACT: This paper presents the design of projection control system based on PLC for multimedia classroom. The proposed approach can control the rising and the dropping of projection screen, as well as the opening or closing of projector. The detailed control ladder diagram is designed. In addition, touch screen is used for control panel, which makes the operation and the maintenance more convenient. The simulated experiment results testify the correctness of the proposed design approach.

1 INTRODUCTION

With a lot of advantages, multimedia teaching is a new modern way for teaching, which can make teaching more easy, cultivate and stimulate the interests of students in learning, make the classroom teaching effective, focus on the importance and difficulties in teaching, and ensure to apply some effective ways to improve the teaching efficiency.

Projector is a high-tech digital display instrument. The projector can display the image information that is coming from computer, television and video, which makes up the deficiency of small screen of computer or television. Its application is more and more widely used such as in classroom, video reproduction, academic hall, commodity sales exhibition etc., and it is especially important in computer-assisted instruction. However, the projector is suspended in the air while in use, and is beyond the reach of the operator, which is very inconvenient since the operator cannot touch the projector directly. If operated by RCP (remote control panel), RCP may be insensitive because of low battery or old remote control board key. If remote control board is operated by many persons, parameter setting of projector may be changed by disoperation. If above situations take place while in class, the teacher will have to take some time to adjust the parameters of projector. It is inevitable that there are some faults, such as automatic shutdown because of projector's high temperature; projection screen is locked with some problems. In traditional multimedia classroom management, teachers only can call for help from management office, which is waste of time, and affects the normal teaching order (Zhang, Zhu & Luo 2009).

This design approach mainly seeks to solve these above mentioned problems. According to the current application status of multimedia classroom, with the increase of investment in education, the application of multimedia classroom control system will produce good prospects for development, and its market will be wide (Qin 2009).

2 HARDWARE DESIGN OF PROJECTION CONTROL AND SCREEN CONTROL

2.1 Principle of projection control

The opening and closing of the projector is achieved by the control of coil of AC contactor, which is also controlled by the output of PLC. The projector cannot be directly closed, and the projector needs to be kept cool for a few minutes, and then shut off. Therefore, to ensure the life of projector, we need a controller to realize the transition. That is, the output of PLC is used to control the coil of AC contactor, and the coil of AC contactor is used to control the controller of projector, and the controller is used to control the opening and closing of the projector.

2.2 *Necessity of selection of controller of projector*

As a sophisticated equipment, the projector has specific operating requirements that are the remote shutdown. Since the machine temperature is very high, we have to rely on the cooling fan to discharge heat outwards, the power will be turned off after the heat decline. Because of oversight, the projector is directly turned off while the projector is still working, which is likely to cause the explosion of bulb, or even burn out the motherboard, and cause many unnecessary economic losses to the user (Liu, Wang, Xu & Wang 2005).

The controller is designed according to the specification of the projector operation, you just turn on the power switch of controller, the power supply is switched on, and then the projector immediately works. When the projector finishes the task, then directly shuts off the power of controller, the controller will send "shut down" signals to the projector immediately, and then the projector will enter the state of delayed heat. If the delay is over, the power supply of the controller and the projector will be automatically turned off.

Therefore, the controller makes it is very convenient for the use of projector, and does not need to wait for heat dissipation of the projector. Also, there is no need to worry about the operation error.

2.3 *Hardware design of projection screen*

Projection screen is one of most commonly used and indispensable projector accessories for multimedia classroom.

2.3.1 *Selection of projection screen*

On market, the size of projection screen is about 50–300 inches, but considering the space constraints and the practicality, 100 and 120 inches are generally the best sizes for the projector. The distance between the projector and the screen is only 4 meters, if the picture is very large, then the distance between the projector and the screen also requires to be greater, in that case, the brightness of the projector is a challenge.

The current most used projection screen aspect ratio is 4:3 in the business and education fields, and the 4:3 screen can achieve better results. According to the above content, select the projection screen aspect ratio as 4:3.

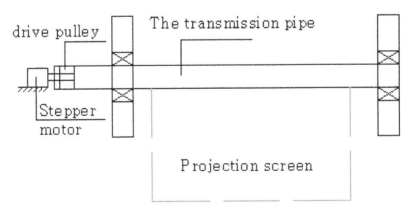

Figure 1. Screen transmission diagram.

2.3.2 *Design of transmission mechanism of projection screen*

The screen roll up or down mainly relies on the rotation of rotating wheel that is mounted on the shaft of the tubular motor that drives the transmission tube (outer tube) to rotation. As shown in Figure 1, a drive wheel is installed in stepping motor shaft, a transmission tube is sleeved outside the driving wheel, and a transmission tube is fixed by two bearings. When the stepper motor rotates, a transmission tube is driven by the drive wheel, and then the screen will roll up or down. Tubular motor has adjustable stroke by mechanical limit, which can properly arrange tubular motor stroke according to the projection screen height (Zhang, Ni & Zhang 2006).

3 DESIGN OF LADDER DIAGRAM FOR SCREEN CONTROL

Here we do not use the limit switch, but according to the travel time to realize the program limit. If the screen height $H = 1850$ mm, tubular motor speed $n = 28$r/min, the drive pipe diameter $D = 47$ mm, we have:

$$t_{Travel\ time} = \frac{H.60}{\pi D n} = \frac{1850}{3.14 \times 47 \times 28} \times 60s = 27s \qquad (1)$$

We can use two power-off-keep type timers (see T250 and T251 in PLC ladder diagram) to calculate fall time T1, and rise time T2. Since the rise and fall of the screen are uniform, we can obtain the position value only by calculating the time. The operation time of screen is equal to T1-T2, which denotes the real time corresponding to screen position, and the minimum value is 0 s, the maximum value is 27 s. Figure 2 shows the ladder diagram of screen control (Jiang & Ouyang & Zeng 2008).

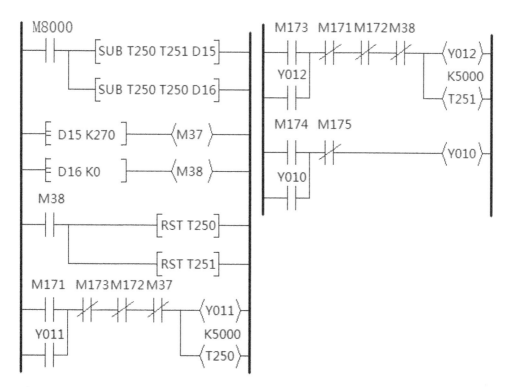

Figure 2.   Ladder diagram of projector screen control.

259

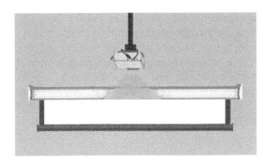

Figure 3. Projector has been opened, and screen is falling down but not reaching the lowest position.

4 INTERFACE DESIGN OF SCREEN CONTROL FOR PROJECTOR

Project control interface includes: projector switch on, projector switch off, the screen rise, the screen fall, the screen stop, limit switch, indicating lamp of projector, indicating lamp of screen rising, indicating lamp of screen falling, parts movements, parts display, interface change switch.

5 EXPERIMENT

The proposed approach is testified by our simulated experiment. After the ladder diagram is designed, we firstly download the ladder diagram into the PLC and run the ladder diagram, and then the control interface soft is implemented, after setting up the connection between the PLC and touch screen soft, we click the "open projector" button by touch screen. Figure 3 shows the figure of the projector has been opened, and screen is falling down but do not reach the lowest position. Figure 3 validates that the ladder diagram is correct, which can be correctly implemented. Therefore, the simulated experiment validate that our proposed design approach is completely correct.

6 CONCLUSIONS

This design applies sequence control based on PLC to realize the control of the projector and screen, the intelligent fault detection and alarm. The teachers can call for help quickly applying the proposed approach. By software simulation and touch screen online simulation, the basic functions of this design have been achieved; the simulated experiment result validates the correctness of the proposed approach. I believe that the proposed design approach will be widely applied in the future.

REFERENCES

Jiang, H.S., Ouyang, Y.X. & Zeng, Y.B. 2008. Design of the Control System of Step Motor Based on PLC and Touch Screen. *Machinery & Electronics in Chinese*. (10): 75–78.
Liu. H.H., Wang, Y., Xu, W.D. & Wang, Q.Q. 2005. The Control System for Projector Based on USB Bus. *XianDai DianZi Jishu in Chinese*. (22): 37–40.
Qin, Z.H. 2009. *Electrotechnics*. Beijing: China Higher Education Press.
Zhang, L.F., Ni, B.X. & Zhang, P. 2006. Design of the Control System of Stepping Motor Based on PLC. *Micromotors in Chinese*. 39(1): 86–88.
Zhang, Y.Y., Zhu, Y. & Luo, H.W. 2009. Design and Realization of Automatic Control System for Projector Screen. *Computer Knowledge and Technology in Chinese*. 5(11): 2886–3043.

Manufacturing Engineering and Intelligent Materials – Lu & Abu Bakar (Eds)
© *2015 Taylor & Francis Group, London, ISBN 978-1-138-02832-6*

Implementation of green and intelligent building based on information fusion technology: A new perspective

L.N. Ma

Shenyang Urban Construction University, Shenyang, China

ABSTRACT: In this paper, a novel information fusion based intelligent building methodology is proposed. Intelligent lighting system is the core component of the smart home. In order to achieve user-centric intelligent control, improve indoor visual comfort level and saving energy at the same time, we have designed a multi-agent intelligent lighting system based on multi-source information fusion. Multi-agent system can improve the system of interactive, so distribution subsystem parallel processing of data and information decision to realize closed loop control. Intelligent data fusion model proposes the fusion ambient light, occupancy rates, energy restriction, user preferences for decision making, and other information. ANFIS is adopted to control the environmental condition. Through the FIS algorithm, we eventually finalize the implementation steps.

1 INTRODUCTION AND PREVIOUS WORK

1.1 *General introduction*

Intelligent system is the core component of the smart home. A well-designed, precision intelligent lighting control system not only meet the demand of people's daily lighting, it also can save energy and time, achieve the goal of all kinds of control. The intelligent lighting system we propose will not only artificially control family equipment and intelligent equipment, but also through the intelligent decision-making system manage all kinds of automatic devices using multi-source information fusion. With this ultimate goal, the novel information fusion technique based intelligent system is proposed. The control goal is to maintain the high-level comfort with the minimum power consumption in different operating conditions.

1.2 *Background analysis*

There are plenty of researches aiming at dealing with the problems mentioned. Some of the studies focus on automatic control with intelligent decision-making based on occupancy estimation and daylight-linked lighting (Wang, Wang & Dounis et al 2012, Pilecki, Marek & Wojciech 2015), but with the popularity of smart mobile devices, we tend to control lighting model combined with people's decisions and smart mobile devices. Multi-agent technology has been in many studies, using the development of the intelligent building energy management control system. In (Wen et al 2014), Analysis of Farmer Household Decision-Making and Its Effects on Environment Change by Multi-Agent Systems are discussed. In (Budyal & Manvi 2014, Yu & Chen 2013) Modeling of Intelligent Command and Control System based on the Technology of Multi-Agent are analyzed. This paper developed a novel multi-agent system model in intelligent building energy and comfort management according to the behavior of the residents, but it is not a public information fusion model. To integrate information environment and the human behavior from the outside world combined with automatic control and manual control. In this paper, four types of agents can be arranged at different platforms including android system and personal computer.

ANFIS algorithm based information fusion model is designed and discussed with simulation in the end.

2 PRINCIPLES OF MULTI-AGENT INTELLIGENT SYSTEM

2.1 The overview

Multi-agent control system is composed of several agents. Although functionally each agent has a different function, each agent still has common characteristics. For example, each agency must have a certain degree of autonomy, each agent can communicate with each other by using their own language, and the most important that each agent can determine all of their own behavior in order to achieve the main goal. Figure 1 shows the prior discussion.

2.2 Four individual agents

In a multi-agent lighting suggested system, there are four types of agents: central agent, local lighting agent, agent sensor and personal agent. (1) Personal agent: In order to finalize the user-centered intelligent control. The private sector is responsible for promoting response to the needs of the owner. It is an agent's personal assistant that is responsible for the management of information from the host, and then provides the orders to the host, and provides feedback. At the same time, personal agent terminal have diversity of control mode. (2) Local lighting agent: In addition to mediate and information providers of responsibility, there is also an important role in the decision-making and execution control in multi-agent systems. Local lighting agent server is located in the control center, receives the sensor data sensor agent, and then transmit data to the data fusion center and carries on the analysis and processing. (3) Sensor agent: multiple sensor sensing the environment the existence of a large number of redundant data, a direct impact on the performance of network communication. (4) Central agent: The central agent has two main functions: agent system configuration and promote cooperation between the local agent. Intelligent lighting system is a subsystem of smart home, there are many other subsystems managed by the local agent directly. Central agent allows administrators to enable or stop a local agent.

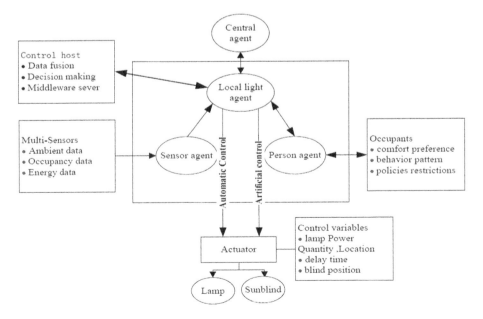

Figure 1. Principles of proposed system.

3 OUR PROPOSED INFORMATION FUSION MODEL

3.1 System design and implementation

Ambient visual data, occupancy condition data and energy data, these three types of data from the sensor network. Visual data refers to the construction of environment parameters, such as solar radiation, light intensity, etc. Light intensity directly reflect the degree of indoor light and shadow. This is the most important factor in the lighting control. The main goal of this work is the decision-making level, based on environmental vision occupancy conditions, the preference of inhabitant, energy and information. We can estimate the global environment vision and light intensity and solar radiation data based on adaptive fuzzy inference system.

3.2 ANFIS based information fusion model

ANFIS is a system that integrates neural networks and fuzzy inference system, which combines the learning ability of neural network and the ability to make decisions on fuzzy logic. Figure 2 indicates the basic architecture of ANFIS. We will discuss the proposed five layers.

First layer: Fuzzification layer. Each node in this layer is to generate output in the form of member based on fuzzy set value. Node function can be a common bell or Gaussian membership function. This layer of output different node consists of the following (equation one):

$$O_{1,j} = \mu_{AI}^k(x_i) \tag{1}$$

In the equation, x_i denotes the input value, $\mu_{AI}^k(x_i)$ represents the membership function, usually depicted bell-shaped function with a maximum value 1 and minimum value 0, in the formula, $\{a_i^k, b_i^k, c_i^k\}$ are the parameter sets need to be adjusted.

$$\mu_{AI}^k(x_i) = \cfrac{1}{1 + \left(\cfrac{x - c_i^k}{a_i^k}\right)^{2b_i^k}} \tag{2}$$

Layer two: Rule layer. Layer two executes the fuzzy "AND" of the antecedent part of the fuzzy rules, the output of a node in this layer represents the firing strength of a fuzzy rule:

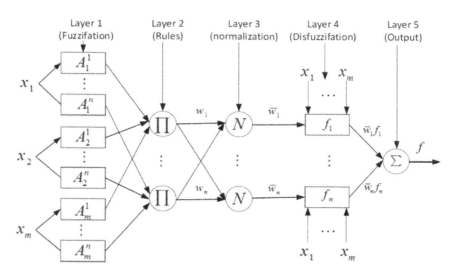

Figure 2.   The basic structure of ANFIS model.

$$O_{2,k} = w_k \prod_{i=1}^{m} \mu_{Al}^{k}(x_i) \tag{3}$$

Layer three: Normalization layer. In this crucial layer, process and calculation are conducted at each single nodes, normalization of the fire strength is also conducted.

$$O_{3,k} = \overline{w_k} = \frac{w_k}{\sum_{i=1}^{n} w_i} \tag{4}$$

Layer four: Defuzzification layer. We can define the output of the layer in the formula 5, in which the conclusion parameters are defined as: $\left\{ d_0^k, d_1^k, d_2^k, ..., d_m^k \right\}$.

$$O_{4,k} = \overline{w} f_k \left(d_0^k + d_1^k x_1 + d_2^k x_2 + ... + d_m^k x_m \right) \tag{5}$$

Layer five: Output Layer. The formula 6 illustrates the output result.

$$O_{4,k} = \sum_{i=1}^{n} \overline{w_k} f_k = \frac{\sum_{k=1}^{n} w_k f_k}{\sum_{k=1}^{n} w_k} \tag{6}$$

4   EXPERIMENTAL ANALYSIS AND CONCLUSION

Multiple sensors to collect solar radiation of the actual environment, outdoor light intensity and indoor light intensity. We control window curtain open rate and using ANFIS, respectively, control the brightness of the LED lights. In the first ANFIS, we use outdoor light intensity values between the differences and indoor light intensity, solar radiation as input variables. Scope of the proportion of open (0, 1, 0 means fully closed, 1 is completely open) output variables. In the second ANFIS, we use the percentage of the two input variables than

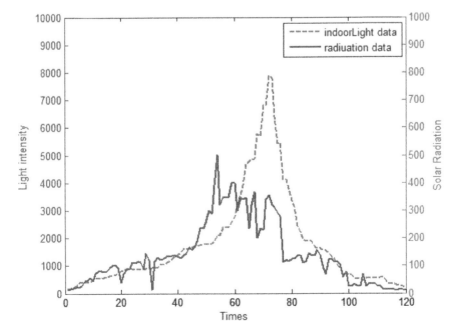

Figure 3.   The experimental simulation.

open, indoor light intensity difference between values and user preferences, and the output of the brightness of LED lights the second ANFIS. From 7.00 am to 7.00 pm, we do data collection every 5 minutes. There are 120 groups of indoor light intensity and solar radiation data completely. Outdoor light intensity is usually greater than 10000 according to our monitor, so we assume that all of the 11000; Time sequence diagram is shown in Figure 3.

In this work, a novel information fusion-based intelligent building methodology is proposed. Sensor agent specifically designed for multi-sensor data, and individual agent's purpose is to make the interaction between the residents and the surrounding environment by learning the behavior of the residents. According to the case study and simulation results, it puts forward that the multi-agent system can control the light effectively and meet the needs of residents, and based on ANFIS model algorithm is very suitable for lighting control. In addition, it provides an open structure. In this kind of structure, the agent can be easily configured, and also can adapt to different family environment and different functions. Information fusion model can also be extended to other smart home system. In future research, we plan to modify the proposed framework and use more neural network based method to conduct more theoretical and experimental analysis to achieve better result.

REFERENCES

Budyal V.R. & Manvi S.S. 2014. ANFIS and agent based bandwidth and delay aware anycast routing in mobile ad hoc networks. ELSEVIER, *Journal of Network and Computer Applications*:140–151.

Pilecki, Marek A. Bednarczyk, and Wojciech Jamroga. 2015. "Model Checking Properties of Multi-agent Systems with Imperfect Information and Imperfect Recall." *Intelligent Systems' 2014*. Springer International Publishing. 415–426.

Wang Z, Wang L, Dounis A.I, et al. 2012. Multi-agent control system with information fusion based comfort model for smart buildings [J]. *Applied Energy*, 99: 247–254.

Wen Guanghui, et al. 2014. "Node-to-node consensus of multi-agent systems with switched pinning links." Control and Decision Conference (2014 CCDC), The 26th Chinese. IEEE.

Yu Weihong, Chen Yan. 2013. Implementation of Lightweight Embedded Agent on Android Platform [J]. *Computer Engineering*. 39(7):298–301.

Manufacturing Engineering and Intelligent Materials – Lu & Abu Bakar (Eds)
© 2015 Taylor & Francis Group, London, ISBN 978-1-138-02832-6

The implementation of Fast Fourier Transform based on FPGA

Y. Dai
Department of Information Engineering, College of Tangshan, Hebei, China

ABSTRACT: This paper introduces the arithmetic of FFT processor-Radix-2, and introduces the method and process of design and realizes a FFT processor, which is integrated in FPGA chip, regarding FPGA as design carrier. The results of hardware test of FFT processor show that the processor works well and has high speed. The design has the advantages of simple, agility and small bulk.

1 INTRODUCTION

Fast Fourier Transform (FFT) algorithm is the highest algorithm to calculate the Discrete Fourier Transform (DFT) (Antonio, Eduardo, Adoracion & Jose 2000). Because of FFT supercomputing capacity, it is widely used in radar processing, observation, time orientation processing, high speed image processing, secure wireless communications, digital signal processing and so on (Szadkowsk 2006). As a basic operation of digital signal processing, there are different requirements of FFT processor in different areas. The traditional method of FFT processor design is DSP by software programming. Limit the performance of DSP itself, it is difficult to achieve high-speed and large-scale FFT operations. Researching high-speed and high-precision FFT processor becomes a point of digital processing (Ayan, Anindya & Swapna 2001).

This paper describes the architectural design of FFT processor which is implemented using Field Programmable Gate Arrays (FPGA). FPGA realization is very much useful, particularly in the developing countries, due to low investment required for the same compared to the prototyping cost of an ASIC. Among other benefits are the short design cycle and the scope of reprogram ability for improvement in the design without any additional cost along with the facility of desktop testing (Satake, Hiroi, Suzuki, Masuda & Ito 2008).

2 FFT ALGORITHM

FFT algorithm can be divided into two categories. Firstly, points—N of FFT is equal to 2^m, (m is any natural number); secondly, N is not equal to 2^m. The second situation can be transformed into the first case, so we used FFT which belongs to the first category. FFT algorithm in the first category included the radix-2, radix-4 and so on. In theory, the radix number is larger, the number of operation is fewer, but the process will become complicated. The relatively widespread use of the algorithm is radix-2.

In radix-2 algorithm, $N = 2^M$. The discrete Fourier transform of N complex samples $x(k)$, $k = 0, 1, \ldots\ldots, N-1$ is defined as Eq. (1):

$$X(k) = \sum_{n=0}^{n-1} x(n) W_N^{kn} \tag{1}$$

$x(n)$ is divided into groups, $n = 2r$ and $n = 2r + 1$. The N point DFT of $x(k)$ as depicted in Eq. (1) may be rewritten as

$$X(k) = \sum_{r=0}^{\frac{N}{2}-1} x(2r)W_N^{2rk} + \sum_{r=0}^{\frac{N}{2}-1} x(2r+1)W_N^{(2r+1)k}$$

$$= \sum_{r=0}^{\frac{N}{2}-1} x(2r)W_N^{2rk} + W_N^k \sum_{r=0}^{\frac{N}{2}-1} x(2r+1)W_N^{2rk} \quad \left(r = 0, 1 \ldots\ldots, \frac{N}{2} - 1 \right) \qquad (2)$$

where $W_N^{2rk} = e^{-j\frac{2\pi}{N}2rk} = e^{-j\frac{2\pi}{N/2}rk} = W_{N/2}^{rk}$. The Eq. (2) can be written as Eq. (3):

$$X(k) = \sum_{r=0}^{\frac{N}{2}-1} x(2r)W_{\frac{N}{2}}^{rk} + W_N^k \sum_{r=0}^{\frac{N}{2}-1} x(2r+1)W_{N/2}^{rk} = G(k) + W_N^k H(k) \qquad (3)$$

where $G(k) = \sum_{r=0}^{\frac{N}{2}-1} x(2r)W_{N/2}^{rk}$, $H(k) = \sum_{r=0}^{\frac{N}{2}-1} x(2r+1)W_{N/2}^{rk}$.

$G(k)$ and $H(k)$ are DFT of $N/2$. $G(k)$ includes even sequence from original sequence. $H(k)$ includes only the odd sequence from original sequence. In addition, their period is $N/2$. Because of $W_N^{N/2} = -1$, $W_N^{k+N/2} = -W_N^k$, we can get the Eq. (4) and Eq. (5):

$$X(k) = G(k) + W_N^k H(k), k = 0, 1, 2, ⬚, \frac{N}{2} - 1 \qquad (4)$$

$$X(k + \frac{N}{2}) = G(k) - W_N^k H(k), k = 0, 1, 2, ⬚, \frac{N}{2} - 1 \qquad (5)$$

From the above, DFT of N point sequence is calculated from two DFT of $N/2$ sequence. And so on, $G(k)$ and $H(k)$ can continue to be decomposed and they synthesize DFT of N sequence finally.

The FFT algorithm in which the input data samples are split into odd- and even-numbered ones, is called decimation in time.

3 THE IMPLEMENTATION OF FFT

3.1 *Butterfly processing module*

The butterfly is the basic operator of the FFT. It computes a two-point FFT. It takes two data words from memory and computes the FFT. The butterfly processing module computes one butterfly every four cycles. It consists of one multiplier and two adders. The RTL for it is shown in Figure 1.

Because of 4 cycles required to calculate one butterfly, note that during data input and data output the butterfly is incremented by the clock while during FFT computation mode

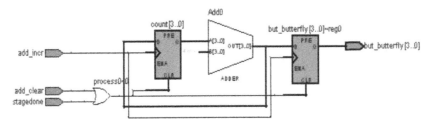

Figure 1. RTL of butterfly processing module.

whenever "clear" or "stage done" signal goes high, the butterfly generator is reset. The simulation waveform of the butterfly generator is shown in Figure 2.

3.2 *The design of FFT processor*

The FFT processor architecture consists of a single radix-2 butterfly, a dual-port FIFO RAM, a coefficient ROM, a controller and an address generation unit.

The address generation unit mainly generated ROM, RAM access address which is used in the FFT processor.

The controller of FFT processor is the core of FFT processor. It enabled or disabled the work of FFT processor, memory modules, address generation and data distribution. The controller is modeled as a finite machine. It has states ranging from rst1 and rst7. The actions performed in each state are clearly commented in the code.

3.3 *Simulation and result*

Validation data is x = [4000, 3000, 2000, 6000, 7000, 8000, 9000, 0], The simulation waveform of FFT processor based FPGA is shown in Figure 3. The simulation results in MatLab is shown in Figure 4. The comparison between them is shown in Table 1.

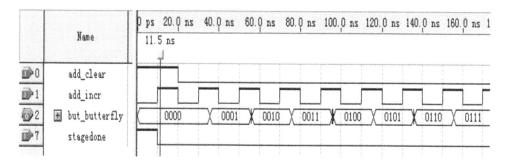

Figure 2. Simulation waveform of the butterfly generator.

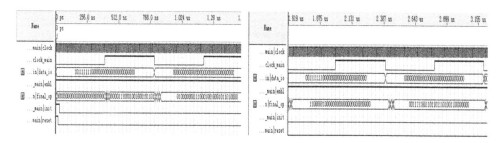

Figure 3. Simulation of FFT processor based on FPGA.

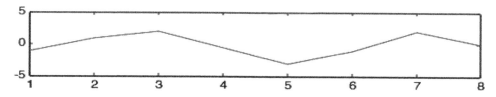

Figure 4. Simulation of FFT Processor in MATLAB.

269

Table 1. The comparison between theoretical and calculated value.

	Theoretical real	Theoretical imaginary	Calculated real	Calculated imaginary	Error real	Error imaginary
X(0)	0	0	0	0	0.00%	0.00%
X(1)	3.768	−1.061	3.76775	−1.06065	0.01%	0.03%
X(2)	−8	−0.5	−8	−0.5	0.00%	0.00%
X(3)	0.2322	−1.061	0.23225	−1.06065	0.02%	0.03%
X(4)	0.5	0	0.5	0	0.00%	0.00%
X(5)	0.2322	1.061	0.23225	1.06065	0.02%	0.03%
X(6)	−8	0.5	−8	0.5	0.00%	0.00%
X(7)	3.768	1.061	3.76775	1.06065	0.01%	0.03%

4 CONCLUSIONS

The design of FFT processor algorithm using FPGA is presented. The ASIC implementation would be faster in terms of the maximum operating clock frequency but the design with FPGA offers a more cost-effective solution having short design cycle and the desktop development capability with the speed which is adequate for our specific application[5]. This paper introduced butterfly element, address generation unit and controller. Finally, it shows that the FFT processor operational results can meet requirements.

REFERENCES

Antonio J. Acosta, Eduardo J. Peralias, Adoracion Rueda, Jose L. Huertas, 2000. VHDL behavioural modeling of pipeline analog to digital converters, Measurement, vol. 31, no. 2, pp. 47–60, Decemeber.

Ayan. Banerjee, Anindya Sundar Dhar, Swapna Banerjee, 2001. FPGA realization of a Cordic based FFT processor for biomedical signal processing, *Microprocessors and Microsystem*, vol. 25, no. 1, pp. 131–142, February.

Satake Shin-ichi, Hiroi Yoshiaki, Suzuki Yuya, Masuda Nobuyuki, I to Tomoyoshi. 2008. Special-purpose computer for two-dimensional FFT, *Computer Physics Communications*, vol. 179, no. 6, pp. 404–408, June.

Szadkowsk Z., 2006. 16-point discrete Fourier transform based on the radix-2 FFT algorithm implemented into cyclone FPGA as the UHECR trigger for horizontal air showers in the Pierre Auger Observatory, *Nuclear Instruments and Methods in Physics Research A*, vol. 560, no. 2, pp. 309–316, February.

Manufacturing Engineering and Intelligent Materials – Lu & Abu Bakar (Eds)
© 2015 Taylor & Francis Group, London, ISBN 978-1-138-02832-6

The reactor temperature regulation of PWR nuclear power plant based on fuzzy theory

H.R. Li & X.H. Yang
Shanghai Key Laboratory of Power Station Automation Technology, Automatic Engineering of Shanghai University of Electric Power, Shanghai, China

H. Wang
Guangdong Power Grid, Yangjiang Power Supply Bureau, Yangjiang, China

ABSTRACT: By assuring the fuzzy set, membership function and rules of fuzzy control reasonably, the fuzzy controller of reactor temperature is researched and designed. And the simulation result shows that the fuzzy controller has obvious advantages on improving the dynamic and steady-state performance of reactor temperature compared with traditional proportional control, integral control and differential control.

1 INTRODUCTION

In recent years, with the continuing growth of global oil demand, the soaring price of fuel and steam energy, a large number of research institutions consider that the nuclear power will be an important way to solve energy crisis in the situation of energy supply increasingly reduced. In the long term, nuclear power is indispensable energy which can reduce the environmental pollution gradually and improve the energy structure of the world. The basic characteristics of nuclear power determine its irreplaceable status, which are: nuclear power is clear energy which does not generate pollutant like 502 and COZ; the safety and reliability of nuclear power are being enhanced constantly and all countries are studying control systems such as AP1000 to improve and enhance its safety. So, it is very important to ensure the reactor core safety and prevent bad accident like reactor core melting. Therefore, there is actual significance to study the temperature control of reactor core.

2 REACTOR TEMPERATURE CONTROL

The core temperature control of nuclear reactor is one of the most important parameters controls in nuclear power plant, for the safety, reliability and economic run of units are determined heavily by the performance of temperature control. The changing range of temperature in reactor core internal can be limited by temperature control and bad accidents like overpower of reactor and reactor core melting, can be prevented.

The fundamental mission of reactor control is: in normal running conditions, it controls the upgrade and transform of power and usual shutdown, and maintains steady-state operation and can always keep in the specified operating range with given load disturbances; at any cases, it can ensure a safe shutdown and the safety of staff and equipment. In order to fit for the requirements of loads changing in nuclear power plant, the reactor power control is implemented by shifting the control rod.

In the year 2001, basing on the optimal control theory, Khajavi designed the control structure of SFAC which can regulate parameters automatically. Then, a robust and optimal controller of self-tuning parameters was gained to control operating range of large spectrum (Khajavi et al. (2001)). In 2003, H. Arab-Alibeik proposed the LOG/LTR controller based on improved structure of SFAC to control the coolant temperature of nuclear reactor. The controller has a good control performance on the power and temperature of nuclear power plant (Arab-Alibeik & Setayeshi (2003)).

In 2009, ZheDong proposed a new nonlinear controller, which has sufficient conditions for asymptotic stability of the closed-loop. The simulation result shows that the temperature of transient response value of power and coolant can be controlled rapidly by controlling the speed of control rod (Dong et al. 2009).

In this paper, the reactor has a better running in steady-state condition through the control of temperature in PWR nuclear power plant. Through the simulation platform, the temperature curve of reactor can be obtained and then two orders mathematical model of reactor temperature is built. After that, PID controller is designed as well as its parameters to get a excellent control strategy. On the basis of the strategy, fuzzy controller is designed to obtain a better control performance.

3 THE DESIGN OF FUZZY CONTROL

In the process of experiment, the real-time curve of temperature control system is recorded when control rod R is in different rod positions. The real-time curve of −5% step signal is selected to model. Through the two points method, the transfer function of controller object is:

$$G(s) = \frac{1.25}{1000s^2 + 25s + 1} \tag{1}$$

And the maximum error of fitting is within 0.1%. The fitting result is shown in Figure 1. The errors are recorded in Table 1.

By comparing the fitting transfer function curve and experimental data curve, the maximum absolute value of error is 0.00091, namely 0.09%.

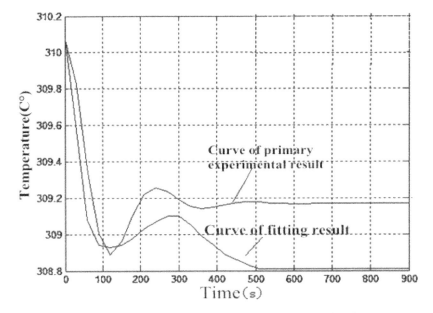

Figure 1.   The comparison of experimental data curve and fitting transfer function curve.

Table 1. The error analysis of fitting curve and actual curve.

Time(s)	0	30	60	90	120	150
Experimental data	310.07	309.57	309.08	308.94	308.93	308.94
Absolute value of error	0	0.00082	0.00091	0.00021	0.00013	0.00007
Time(s)	180	210	240	270	300	330
Experimental data	308.98	309.03	309.07	309.1	309.1	309.05
Absolute value of error	0.00040	0.00061	0.00060	0.00044	0.00029	0.00034
Time(s)	360	390	420	450	......	330
Experimental data	308.99	308.94	308.86	308.83	......	
Absolute value of error	0.000049	0.00067	0.00088	0.00091	......	

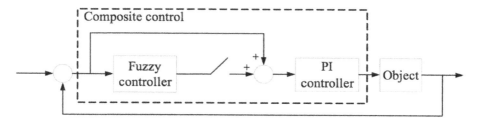

Figure 2. Fuzzy controller cascaded with PID controller.

The design of PID parameter self-tuning fuzzy controller: the general two-dimensional fuzzy controller takes linguistic variables of error E and error rate EC as inputs, besides, the actual control system of transition process requires different emphasis of the performance index in different periods. The system expects to have a fast response in the initial state and have high steady-state accuracy in the steady state. And namely, the parameters of K_p, K_i and K_d are different in different periods. Therefore, the correcting values ΔK_p, ΔK_i, ΔK_d of K_p, K_i and K_d are not only related to E, $\sum E$ and EC, but also connected to time factor t. Consequently, the time factor are taken as an input variable of fuzzy self-tuning parameters controller.

1. The output variable is ΔK_p: during the initial control, K_p is designed as a small value to reduce the rush of all system physical properties. During the middle control process, K_p should be increased to enhance the response of system and in the end control, the K_p' should be reduced sightly to avoid big overshoot.
2. The output variable is ΔK_i: in the initial control, in order to prevent the process of system response appearing integral saturation phenomenon which would cause a large overshoot, the K_i is designed as a small value to avoid influence on stability of system. In the middle of control process, the K_i should be selected suitably and in the end of control process, K_i should be a big value to reduce the steady error and enhance the steady-state accuracy.
3. The output variable is ΔK_d: during the initial control, in order to avoid overshoot, the differential effect can be enhanced appropriately and K_d can be designed as a larger value. During the middle control process, K_d should be a smaller value for the response process is sensitive to the change of K_d. During the end, K_d should be a smaller value than before to improve the ability of restraining disturbances.

The structure of fuzzy controller cascaded with PID is shown in Figure 2.

Compared with control performance of fuzzy PID controller and PID controller, the result is shown in Figure 3.

Through the comparison, the system adjusting time of fuzzy controller is shorter than the system controlled by PID obviously. The difference of adjusting time is 130 s. Therefore, the control performance of adding fuzzy control is better than simple PID control.

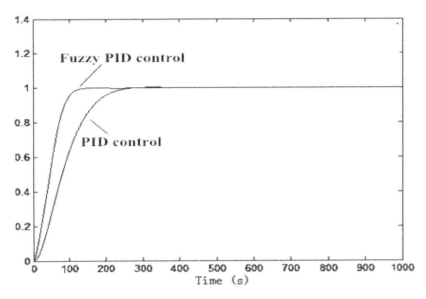

Figure 3. The comparison of response curve between fuzzy PID controller and PID controller.

4 CONCLUSION

The paper builds the mathematical model of reactor temperature through the reactor temperature curve from simulation platform. PID controller is designed as well as its parameters, which achieve the temperature control of PWR nuclear power plant.

On the basis of that, fuzzy controller is designed and better control performance has been achieved.

ACKNOWLEDGEMENTS

This paper is supported by National Natural Science Foundation of China (Project Number: 61203224), Shanghai Science and Technology Commission Key Program (No. 14511101200), Shanghai Natural Science Foundation (13ZR1417800), the Foundation of Shanghai Key Laboratory of Power Station Automation Technology (No. 13DZ2273800), project of Science and Technology Commission of Shanghai (project number: 12510500800).

REFERENCES

Arab-Alibeik H., Setayeshi S. 2003. Improved temperature control of a PWR nuclear reactor using LQG/LTR based controller [J]. *IEEE Trans. Nuel.* Sci. 50(1); 211–218.

Dong Zhe, Feng Junting, Huang Xiaojin. 2009. Nonlinear Observer-Based Feedback Dissipation Load-Following Control for Nuclear Reactors [J]. *Nuclear Science.* 56(l):272.

Khajavi M.N., Menhaj M.B., Ghofrani M.B. 2001. Robust optimal self-tuning Regulator of nuclear reactors [J]. *in Proc. 1st Conf. Applications of Physics and Nuclear Science in Medical and Industry.* 1235–1238.

Manufacturing Engineering and Intelligent Materials – Lu & Abu Bakar (Eds)
© 2015 Taylor & Francis Group, London, ISBN 978-1-138-02832-6

Analysis on combined flanges heat-structure coupling sealing reliability

S.D. Wen & H.H. Wang
School of Mechanical Power Engineering, East China University of Science and Technology, Shanghai, China

ABSTRACT: Concerning the randomness of design parameters, the sealing reliability analysis of combined flanges of a high temperature tower inlet was down, and the parameter sensitivities were analyzed. By taking the random variable according to the practical situation, the reliability analysis model of combined flanges was built and solved with ANSYS Probabilistic Design Module (PDS), and the sealing reliability of combined flanges was obtained. Based on the parameter sensitivity analysis, the most influencing parameters were specified.

1 INTRODUCTION

Bolted flange joints, often used to connect the pressure vessel and piping system is the most widely used removable sealing device, its role is to guarantee sealing of flanges connected by bolt joints (Zheng et al., 2008). The bolted flange joints, concentrated in the high-risk industries, once the device leaks, it may cause property damage, energy waste, even irreversible damage (Cai et al., 2012). Traditional design method usually assume material parameters, structure size, work load and other parameters as constant. This design method usually introduces considerable safety factor, thus causing unnecessary material waste. Different from traditional design method, reliability design method uses random variables that truly reflects the real situations, and gives failure probability and sensitive analysis (Peng et al., 2009). With the ANSYS Probabilistic Design Module (PDS) and the Monte Carlo method, the sealing reliability analysis and sensitivity analysis of the combined flanges of a high temperature tower inlet were done.

2 COMBINED FLANGES MODEL

Bolted flange joints, consist of 24-inch flange, 8-inch flange, pipes, container receivership, gaskets, bolts and other components, also known as combined flanges. The material of flange are forged 316 L (00Cr17 Ni14Mo2) austenitic stainless steel, bolts are 25Cr2MoVA, nuts are 35CrMoA, and gaskets are PTFE spiral wound gasket with inner and outer ring (Liu, 2013, Wang & Guan, 2012).

The main indicators of combined flanges sealing include installation bolt load, flange deflection angle and gasket stress. In the non-standard flange design, the installation bolt load is calculated from gasket coefficient m and preload pressure ratio y, both of them are given based on the design specifications (ASME, 2007). According to Waters Law, the required gasket stress under preload conditions must be greater than the pressure ratio y; and under operating conditions it must be greater than mp (p is the working pressure) (Cai et al., 2012). Take $y = 20$ MPa, $m = 4.5$, $p = 1$ MPa. According to ASME-PCC specification (ASME, 2010), the allowable flange deflection angle generally provided by the factory test data or gasket manufacturer (here takes 0.3°).

According to GB150 Steel Pressure Vessel or ASME VIII-1, under preload condition, the bolt preload should be greater than the required minimum gasket pressing force W_a.

$$W_a = \pi b D_G y$$

where b = gasket effective width; D_G = gasket average diameter; and y = gasket sealing pressure.

Under operating conditions, bolt preload consists of two parts:

$$W = P + F$$

where medium pressure: $P = 1/4\ \pi D_G^2 p$; gasket required compression load: $F = 2b\pi D_G mp$.

The required minimum cross-sectional area of bolt A_m is determined by $Max\{W_a, W_p\}/[\sigma_b]$, final selected cross-sectional area is A_b, the actual design bolt preload is:

$$W = 0.5(A_m + A_b)[\sigma_b]$$

where $[\sigma_b]$ = Allowable stress of bolt material.

The single bolt preload of 8-inch flange is 14.4 KN; the 24-inch flange is 57.5 KN. Actual value for FEM model take 16 KN and 60 KN, respectively (Wang et al., 2012).

3 FINITE ELEMENT ANALYSIS

Considering the symmetry of the flange periodic structure, the 1/4 model was established, and set symmetric constraints on the side surfaces, axially fixed constraints imposed on the end of the barrel to eliminate the rigid displacement. Adopt Indirect Coupling Method, thermal analysis first, then structural analysis.

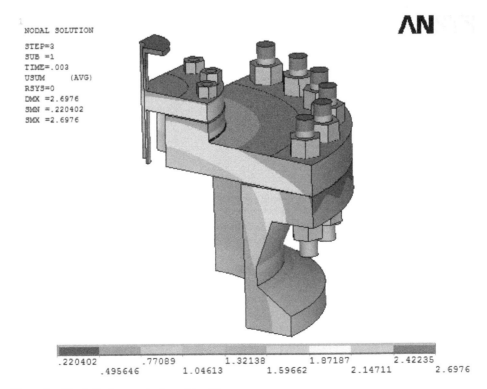

Figure 1. Total displacement of combined flanges.

Ambient temperature set $20°C$, the inner wall surface temperature load set $500°C$. Set corresponding convective heat transfer coefficient for every surface. Conduct thermal analysis.

When conducting structural analysis, three load steps are used: in the first step bolt preloads are applied; in the second step, pressure loads are applied; and in the third step, temperature loads calculated from thermal analysis are inputted.

Figure 1 is the Y direction displacement contours of the combined flanges, the closer to the container receivership, the greater the displacement of the flange. Thus, the displacement difference of inner and lateral of flanges generated, namely, flange deflection. Too large flange deflection angle will increase the leak rate, so one must strictly limit the flange deflection angle.

4 RELIABILITY ANALYSIS

In reliability analysis, take operating temperature, operating pressure, bolt preload, material parameters as the input variables, and follow normal distributions; and according to the sealing indicators, take deflection angle of 8-inch flange and 24-inch flange, gasket minimum stress of 8-inch flange and 24-inch flange as the output variables.

4.1 *Reliability calculation*

Limit state function: $\begin{cases} Z_1(X) = \sigma_{min} - [\sigma] \\ Z_2(X) = \alpha - [\alpha] \end{cases}$

where σ_{min} = the minimum stress flange gasket combination; $[\sigma]$ = allowable stress gasket paper; α = flange deflection angle; and $[\alpha]$ = allowable deflection angle.

Approximate the distribution of 8-inch flange gasket minimum stress to a normal distribution, the mean value $u = 22.15$ MPa, the variance value $\sigma = 2.84$ MPa, and the allowable value $u_0 = 4.5$ MPa, reliability index: $\beta = u - u_0/\sigma = -6.1$.

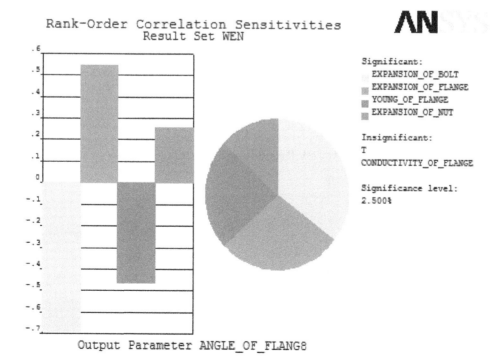

Figure 2. Sensitivity map of 8-inch flange deflection angle.

Reliability: $R(t) = 1 - \Phi(\beta) \approx 1.0$

Other reliabilities about the rest output variables calculated similarly, also close to 1.0, the sealing of combined flanges meet the requirements.

4.2 *Sensitivity analysis*

The sensitivity analysis results of about 8-inch flange deflection angle is shown in Figure 2. The main affecting factors include Young's modulus of flange and thermal expansion coefficient of flange, bolt and nut; for 24-inch flange, it also includes temperature. The main factors that have significant effect on 8-inch flange gasket minimum stress include thermal expansion coefficient of flange and thermal expansion coefficient of bolt; in addition to these, the 24-inch flange gasket also includes bolt Young's modulus.

5 CONCLUSION

The sealing reliability analysis of combined flanges is shown by ANSYS probabilistic design module, the probability distribution of gasket stress and flange deflection angle were obtained, the parameter sensitivities were analyzed, visually displayed the influence level of design parameters on the flange sealing performance, and provides a good reference for the combined flanges design.

The reliability of the combined flanges close to 1.0, reflected that the sealing performance of combined flanges designed by traditional method is higher than the actual required value, which is actually a waste of resources. Compared to traditional design method, the reliability design method has a huge advantage.

REFERENCES

ASME (2007) VIII-1-2007. Boiler and Pressure Vessel Code. The first volume of VIII book: pressure vessel construction rules.
ASME (2010) PCC-1-2010. Guidelines for Pressure Boundary Bolted Flange Joint Assembly.
Cai, R.L., Cai, A.M. You, Y.H., Zhang, L.Z., Fan, S.L. & Ying, D.Y. 2012. Safe Sealing Technology for Bolted Flange Connection (1)—Assembly Bolt Load. *Process Equipment & Piping* 49 (3): 12–17.
Liu, C. 2013. Integrity and Sealing Anslysis of Assembly Flanged Joints under High Temperature. Shanghai: East China University Of Science and Technology.
Peng, C.L., Ai, H.N., You, Liu, Q.S. & Xiang, W.Y. 2009. Reliable Analysis for Pressure Vessel Based on ANSYS. Nuclear Power Engineering 30(1):109–111.
Wang, H.H, Lu, J.C., Guan, K.S. & Wang, Z.W. 2012. Strength and Sealing FEM Analysis of Combined Flanges with a Pipe. *Pressure Vessel Technology* 29(2):22–29.
Zheng, J.Y., Dong, Q.W. & Sang, Z.F. 2008. *Process Equipment Design.* Beijing: Chemical Industry Press.

Manufacturing Engineering and Intelligent Materials – Lu & Abu Bakar (Eds)

On the hole cleaning optimization of drilling engineering

H.C. Gu

Offshore Drilling Company, Shengli Petroleum Engineering Company, SINOPEC, Dongying City, Shandong Province, China

ABSTRACT: In petroleum drilling engineering, wellbore cleaning problem has been an important issue. If the borehole is polluted, torque and drag will be abnormal in the construction process, or even lead to accidents. Now researchers have developed a number of models, and summed up a lot of methods to calculate the minimum drilling fluid flow out in cuttings, and processing of wellbore cleaning problems. In this paper, by comparing different models, analysis obtained the application range of different models, and pointed out the direction for guiding engineering practice.

Keywords: petroleum engineering; offshore drilling; cutting removal

1 INTRODUCTION

The bad hole cleaning may result in lost circulation, hinder the casing or liner running jobs, excessive overpull on trips, high rotary torque, excessive equivalent circulation density, formation breaking down, slow ROP, excessive bit wearing and pipe sticking problems. The main causes that influence cutting removal are drill pipe eccentricity, wellbore size and inclination, drilling fluid density, cuttings size, cuttings density, drill pipe rotation, drilling rate, drilling fluid rheology and flow rate. The relationship of these factors is shown in Figure 1.

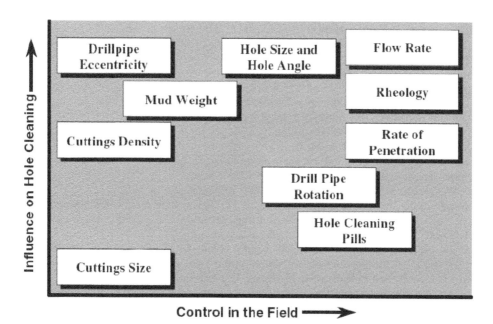

Figure 1. Controllability of drilling parameters in field.

Drilling fluid rheology and flow rate are the two main parameters which would influence cuttings transport strongly while their control in the field is relatively easy (Patrick & EgdSunde 1998). Based on these facts, several methods and models have been introduced that can be used to find the minimum flow rate required to remove the cuttings from the well for a specific drilling condition and drilling fluid rheological properties (Rishi 2000). Larsen's model can be used to find the minimum flow rate for cuttings removal from 55 to 90 degrees of inclination. (Rudi Rubiandini 1999). Another one is Moore's model that is used to find the slip velocity of cuttings in vertical wells. In this study, a computer programming in VB was developed that combines these two methods and predicts the minimum flow rate for cuttings transport from 0 to 90 degrees of inclination. Another computer program was written that calculates the optimum flow rate for different drilling fluid rheological properties using both hydraulic horsepower and jet impact force criteria. Then, these two programs are combined to find that rheological properties of the drilling fluid that gives the optimized flow rate higher than the flow rate to remove the cuttings from a horizontal well called Cao129-H29, drilled in Shengli oil field in China (Larsen 1997).

2 FLOW RATE OPTIMIZATION

In order to achieve a good hole cleaning in Cao129-H29 well, which is located in Shengli oil field in China, Larsen's model and Moore's correlation were combined to calculate the optimum hole cleaning parameters in the mentioned well. To present the procedure and results, three points in well trajectory are chosen, first one in 300 meters TVD, located in vertical part of well, the other one is in 690 meters TVD in deviated section and the last point in horizontal section, 990 meters TVD. This will illustrate the usability of this approach in all ranges of inclinations.

1: Depth of 300 m TVD (300 m MD).

This point is located in vertical part of well (0 degree inclination). The following data set belongs to this point while drilling.

Drill pipe: length = 133 m, OD = 127 mm, ID = 108.6 mm;

Drill collar: length = 166 m, OD = 203.2 mm, ID = 76.2 mm;

Drilling fluid density = 1.15 g/cm^3; Hole size = 444.5 mm; inclination = 0°; ROP = 9.14 m/hr; Maximum allowable surface pressure = 20.69 MPa; cuttings density = 2.6 g/cm^3; average cuttings size = 4.45 mm; YP = 2*PV.

The result of optimization is shown in Figure 2. As the drilling fluid Plastic Viscosity (PV) and Yield Point (YP) increases, the flow rate required for transporting the cuttings decreases. As a result, in the case of vertical wells or vertical portions of deviated wells, to achieve a good hole cleaning condition, PV and YP of drilling mud should be increased. As is shown, for values of PV less that 44 cp, the optimized flow rate is quite suitable for hole cleaning, but for PV greater than 44 cp, the optimized flow rate is not sufficient for cutting removal; thus, higher values of flow rate should be selected.

2: Depth of 690 m TVD (750 m MD).

This point is located in deviated section of well (46-degree inclination)

Drillpipe: length = 386 m, OD = 127 mm, ID = 108.6 mm;

Drillcollar: length = 200 m, OD = 127 mm, ID = 76.2 mm;

Drilling fluid density = 1.32 g/cm^3; Hole size = 311.2 mm; ROP = 9.14 m/hr; Maximum allowable surface pressure = 20.69 MPa; cuttings density = 2.6 g/cm^3; average cuttings size = 4.45 mm; YP = PV.

From the Figure 3, it can be concluded that for the angle of inclination of 46 degrees (within 35°–65°), as drilling fluid plastic viscosity and yield point increases, the flow rate required to achieve good hole cleaning increases. The inclination range between 35° to 65° has the most difficulty in transporting the cuttings. Because in this section gravity force causes settlement of the cuttings, additionally, cuttings have a tendency to roll down to the bottom of the hole and rapidly accumulate. In order to prevent this process to occur, more turbulent drilling fluid regime is desirable. This can be done either by increasing flow rate or decreasing

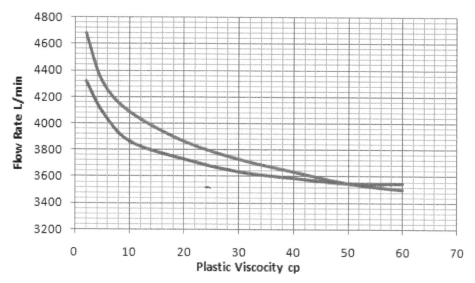

Figure 2. Optimum flow rate and the flow rate for hole cleaning of the depth of 300 m TVD.

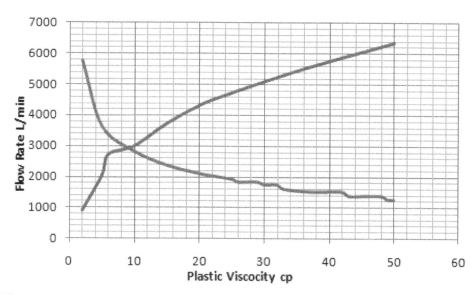

Figure 3. Optimum flow rate and the flow rate for hole cleaning of the depth of 690 m TVD.

mud viscosity. As in Figure 3, for the values of PV less than 9 cp, the optimized flow rate can be used, but for values more than 9 cp, the flow rate should be higher to achieve good hole cleaning. To select the higher flow rates (GPM), one should carefully consider the pumping pressure and surface facility pressure rating aspects.

3: Depth of 990 m TVD (1050 m MD).

This point is located in horizontal section of well (90-degree inclination)

Drillpipe: length = 586 m, OD = 127 mm, ID = 108.6 mm;

HWDP: length = 200 m, OD = 127 mm, ID = 76.2 mm;

Drillcollar: length = 176 m, OD = 152.4 mm, ID = 71.5 mm;

Drilling fluid density = 1.44 g/cm³; Hole size = 311.2 mm; ROP = 6.1 m/hr; Maximum allowable surface pressure = 20.69 MPa; cuttings density = 2.6 g/cm³; average cuttings size = 4.45 mm; YP = 0.5*PV.

281

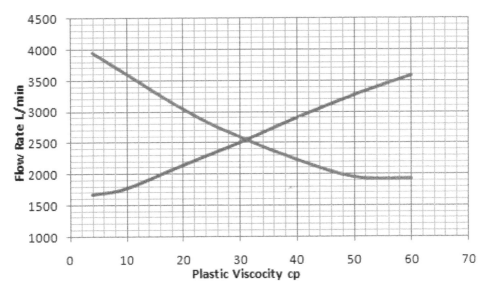

Figure 4. Optimum flow rate and the flow rate for hole cleaning of the depth of 990 m TVD.

As it can be seen in Figure 4, in the horizontal section of the well, reducing drilling fluid plastic viscosity and yield point reduces the required flow rate for hole cleaning. In horizontal well, it has been approved that the lower the yield point, the better the hole cleaning. So yield point of the drilling fluid is considered to be one-third of its plastic viscosity. At the lower values of YP, both good hole cleaning and optimized drilling hydraulics can be achieved.

3 CONCLUSIONS

The new method to find the minimum flow rate to achieve good cuttings transport can be used to find the minimum flow rate for all range of inclination from 0° to 90°. As the results show, behavior of flow rate selection varies by well inclination. Within the range of 0 to 55 degrees of inclination, as the rheological properties of drilling mud goes higher, the value for required flow rate decreases. The major force acting on cutting settlements is gravity force and can be overcome by increasing fluid-carrying capacity. Within high inclination intervals, because cuttings tend to settle on low side of well to make a bed, lower rheological properties of drilling mud is more desirable and produce better hole cleaning effects. To adjust and choose the best values for flow rate, bit horse power or impact force, surface and pumping limitations should be considered as well.

REFERENCES

Larsen T.I., SPE. 1997. Unocal Cap. A.A. Pilehvari SPE. Texas A&M U. and J.J. Azar, SPE. U. of Tulsa "Development of a New Cuttings-Transport Model for High-Angle Wellbores Including Horizontal Wells" SPE 25872.

Patrick Kenny and EgdSunde.1998. Staloil A/S. and Terry Hernphill. Baroid Drilling Fluids "Hole Cleaning Modelling: What's 'n' Got To Do With It?" SPE 35099.

Rishi B. 2000. Adari, SPE, Stefan Miska, SPE. and Ergun Kuru, SPE, University of Tulsa, Peter Bern, SPE, BP-Amoco, and ArildSaasen, SPE, Statoil. "Selecting Drilling Fluid Properties and Flow Rates For Effective Hole Cleaning in High Angle and Horizontal Wells" SPE 63050.

Rudi Rubiandini R S. 1999. SPE-1172519, Bandung Institute of Technology, Indonesia "Equation for Estimating Mud Minimum Rate for Cuttings Transport in an Inclined-Until Horizontal Well" SPE/IADC 57541.

Manufacturing Engineering and Intelligent Materials – Lu & Abu Bakar (Eds)
© *2015 Taylor & Francis Group, London, ISBN 978-1-138-02832-6*

On the experimental investigation of rock abrasiveness

S.L. Guo

Drilling Technology Research Institute, Shengli Petroleum Administration Bureau, SINOPEC,
Dongying City, Shandong Province, China

ABSTRACT: Rock abrasiveness is a key parameter of rock mechanical character, which can be used in selected bit, improving the efficiency of breaking rock, prolonging the working time of the bit and optimizing the structure of the rock breaking tool. Based on the technology of the bit designation and manufacturing, the experimental investigation on rock abrasiveness that simulates the actual drilling surroundings in petroleum engineering had been conducted by means of developing new rock abrasiveness device and using quenching steel spiral simulation micro-bit. Based on the rock abrasive tests on more than 150 pieces of rock which came from the main seven geological profiles in Shengli oil field, the rock abrasiveness was developed with the wear weight of microbit which broke volume rock.

Keywords: rock; abrasiveness; test; bit; classification

1 INTRODUCTION

In the drilling operation of petroleum engineering, there are continuous or discontinuous contacts and friction between drilling tools and rock breaking surface. The drilling tools would be worn by the rock, and gradually became dull, even fail. The rock abrasiveness is defined by the wear between the rock-breaking tools and the rock, which is the ability of the rock damaging the rock-breaking tools. The rock abrasiveness is one of the fundamental parameters of the rock's mechanical and physical properties, so the investigation on rock abrasiveness can guide the appropriate selection of the bit, the improvement of rock-breaking efficiency, the prolongation of the working time, the reduction of the cost and the improvement of the rock-breaking tools design (Zou & Wang 2003). By now, the experimental investigation on the rock abrasiveness (Xing 1992, Zhao, Deng & Xie 2012, Wu, Zhang & Li 1985, Kong 1985 Yin & Yan 1992, Kong 1985, Okubos, Fukuik & Nishimatsuy 2010, Mohamad 2012, Tomazl, Sling 2010, Thuro 2009) is rare, and the existing research methods could not simulate the real drilling environment in petroleum engineering very well.

The experimental method of the rock abrasiveness based on the real drilling environment in petroleum engineering conducted by designing and manufacturing new rock abrasiveness device and quenching steel spiral simulation micro-bit. The evaluation index, the classification method and criterion of rock abrasiveness is proposed. Finally, the relationship between the rock abrasiveness index and rock mechanics parameters is discussed. The research of rock abrasiveness in the future could be defined by the result of this paper, and the basic data could be provided to bit manufacturers for bit selection.

2 IN-HOUSE LABORATORY INVESTIGATION

2.1 *Test device and simulation bit*

The rock abrasiveness test device is developed by Drilling Technology Research Institute of Shengli Petroleum Administration, as shown in Figure 1 (SINOPEC Corporation). The

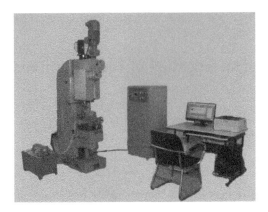

Figure 1. Test device of rock abrasiveness.

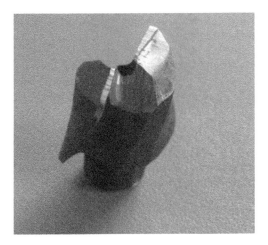

Figure 2. Quenching steel spiral simulation micro-bit.

diameter of the quenching steel spiral simulation micro-bit is 22 mm, and the length of the micro-bit is 20 mm, and the material hardness is 45~50 HRC, as shown in Figure 2. The hard metal bit drilled the pre-hole where the diameter is 22 mm, as shown in Figure 3.

2.2 *The method of rock abrasiveness test*

2.2.1 *Drill core in house*
There were more than 850 rock samples, and they were cored from seven main geological sections of Shengli oil field. The formations of the samples were from Guantao formation to Simian system, and the depth is from 850 m to 4,507 m. Nearly all of the common kinds of lithology such as sandstone, mudstone, conglomerate, and esite, basalt, and limestone and granite gneiss during drilling were included. The rock drillability, hardness and plasticity coefficient were tested before the conduction of the rock abrasiveness experiment. The drillability grade value of the selected core was from 2.01 to 8.78.

2.2.2 *Test method*
On the basis of analyzing the experimental data, it was found that the drill depth was shallow when the drillability was more than 5. Because only the bottom of the bit was worn by the rock, cuttings was produced too little to reflect the cuttings wearing the lateral of the bit,

Figure 3. Pre-hole bit.

Table 1. The experimental results of rock abrasiveness of different kinds of rock.

Technical parameters: Rate of revolution [200 r/min], pressure [500 N], time [15 min]

Lithology	Drillability	Hardness [Mpa]	Hole depth [mm]	Wearing loss [mg]
Mudstone	3.85	796	18.6	218
Sandy mudstone	3.76	889	6.2	279
Calcareous sandstone	5.53	1562	0.5	89

Calcareous Sandstone   Sandy Mudstone   Mudstone

Figure 4. The experimental results of rock abrasiveness of different kinds of rock.

and this situation did not conform to the real drilling environment in petroleum engineering. The experimental results of rock abrasiveness of different kinds of rock are shown in Table 1 and Figure 4.

Isodiametric carbide wedge bit was used to drill a pre-hole with the depth of 20 mm to approach the real drill behavior of bit wear, and the cuttings produced was retained in the pre-hole, then the test of rock abrasiveness was conducted with quenching steel spiral simulation micro-bit in the pre-hole.

According to the result of the experiment, the wearing loss of using pre-hole method was more than the wearing loss without pre-hole, and the effect was well. The method of the experiment had been determined. Step one: a pre-hole which was 20 mm is drilled with isodiametric carbide wedge bit, and the cuttings is retained in the pre-hole. Step two: the test of rock abrasiveness is experimented with quenching steel spiral simulation micro-bit in the pre-hole. Both before and after the rock abrasiveness experiment, the simulation bit should be cleaned, dried and weighted to determine wearing loss as the index of rock abrasiveness.

2.2.3 *The technical parameters of rock abrasiveness*

The hardness of the rock, the size and shape of the particle, the nature of the cement, the quartz content and the percentage of the other solid minerals are the key factors of rock abrasiveness. In the experiment of rock abrasiveness, the critical technical parameters of rock abrasiveness are pressure, grinding speed, grinding time and cooling medium. The pressure is the positive pressure added on the rock. The grinding speed is the relative velocity between the cutting tool and the rock. The grinding time is the contact time between the cutting tool and the rock. The characteristic between the cooling tool and the rock could be changed by the cooling medium.

Therefore, the reasonable determination of the pressure, the grinding time and speed is the key of this study. Combined with technical parameters, collected through debugging the device, the orthogonal experimental method with three factors and four levels is used to determine the drilling parameters of rock abrasive experiment.

According to the analysis result of the orthogonal experiment, the drilling depth of the simulation bit reaches the maximum value, when the rate of revolution is 200 r/m, the pressure is 500 N, the grinding time is 20 min, and the wearing loss is noticeable. Finally, 200 r/m, 500 N and 20 min is, respectively, determined as the rate of revolution, the pressure and the grinding time of the experiment of rock abrasiveness. In this study, the cooling medium is water without any additive.

2.2.4 *Test method of rock abrasiveness*

The procedure of testing method of rock abrasiveness is as follows:

1. Preparation of rock samples
   With automatic double rock cutting machine, the chosen core was prepared into standard specimen of which the two faces was parallel, and then put in the oven for drying.
2. Physical treatment and weighing to simulation bit before the experiment
   With KQ-200 KDE ultrasonic bath, at 70°C, the rock sample was cleaned for 20 min, and dried for 30 min at 100°C, and weighed with FA1104 high precision electronic balance. Then, its weight was recorded.
3. Drilling the pre-hole
   A 20 mm pre-hole which was drilled with isodiametric carbide wedge bit.
4. Test of rock abrasiveness
   In the speed of 200 r/min and pressure of 500 N, the sample is drilled with quenching steel spiral simulation micro-bit in the pre-hole for 20 min, and the hole depth should be recorded.
5. Physical treatment and weighing to simulation bit after the experiment
   With KQ-200 KDE ultrasonic bath, at 70°C, the rock sample is cleaned for 20 min, and dried for 30 min at 100°C, and weighed with FA1104 high precision electronic balance. Last, its weight is recorded.

   In order to obtain the accurate rock abrasiveness, the experiment of rock abrasiveness should be repeated three times, and then the results were averaged.

3 EVOLUTION INDEX AND CLASSIFICATION METHOD OF ROCK ABRASIVENESS

3.1 *Evolution index of rock abrasiveness*

After having determined test method of rock abrasiveness, the experiments of rock abrasiveness to the chosen rock samples had been completed.

In the experiment of rock abrasiveness, the rupture degree, how much the rock was affected by the simulation bit in the condition could be reflected by the rock volume which was broken, and the wearing degree how much the cutting tool is affected by the rock in the experimental condition could be reflected by the wearing loss of the simulation bit. For different kinds of rock, if the wearing loss of the simulation bits is the same, the more rock volume was broken, the lower rock abrasiveness is. Conversely, if the rock volume which is broken by the simulation bit is the same, the more wearing loss of the bit, the higher rock abrasiveness is.

In this study, the rock abrasiveness index is expressed by the wearing loss of the simulation against the rock volume which was broken.

$$R_a = \frac{W}{V} \times \theta \qquad (1)$$

The rock volume which is broken by the simulation bit could be calculated with the depth of the drilled hole and the diameter of the simulation bit.

$$V = \frac{\pi \times d^2 \times h}{4} \qquad (2)$$

The wearing loss of the simulation bit could be calculated by the weight loss of the simulation bit.

$$W = W_1 - W_2 \qquad (3)$$

3.2 Classification method of rock abrasiveness

The experimental data had been processed with formula 1, 2 and 3, and then the indexes of rock abrasiveness for different samples are obtained. The lithology and main mechanical parameters of some rock samples are shown in Table 2.

Table 2. The lithology and main mechanical parameters of some rock samples.

No	Sample number	Lithology	Drillability	Hardness [MPa]	Quartz content [%]	Rock abrasiveness index [mg/cm³]
1	Z83	Packsand	6.32	2061	45	206
2	G79	Packsand	5.08	1327	42	196
3	YI67	Sandstone	5.72	1167	48	143
4	Z177	Sandy mudstone	5.89	646	41	133
5	Z179	Pebbly sandstone	4.20	544	47	108
6	D8	Conglomerate	3.50	760	28	82
7	Z82	Sandy mudstone	4.36	676	25	78
8	H74	Siltstone	5.66	1200	47	77
9	H24	Argillaceous siltstone	5.68	628	47	73
10	G31	Pebbly sandstone	2.01	1641	26	68
11	Z155	Siltstone	2.81	309	60	46
12	Z168	Argillaceous dolomite	5.98	1112	20	55
13	G82	Sandstone	4.34	1212	42	41
14	K22	Mudstone	4.22	539	12	28
15	C34	Basalt	5.66	1517	10	26
16	D6	Granitic gneiss	8.10	1875	21	18
17	YI103	Siltstone	6.68	694	30	18
18	Z42	Mudstone	6.35	1070	47	17
19	Z181	Dolomitic sandstone	6.46	1189	35	7
20	K41	Sandy mudstone	4.98	641	25	5

Table 3. Recommended classification standard.

Rock abrasiveness level	Rock abrasiveness index [mg/cm³]	Rock abrasiveness strength
I	<10	Weak
II	10~35	
III	35~60	Medium
IV	60~85	
V	85~110	High
VI	110~135	
VII	135~160	Extremely high
VIII	>160	

Based on the above study, the rock abrasiveness is divided into eight grades with Sturges empirical formula, and the recommended classification standard is shown in Table 3.

4 THE RELATIONSHIP BETWEEN ROCK ABRASIVENESS INDEX AND MINERAL COMPOSITION AND ROCK MECHANICAL PARAMETERS

4.1 *The relationship between rock abrasiveness index and mineral composition*

The finding is that the rock abrasiveness index of clasolite was high, because the quartz content is high in the clasolite. The rock abrasiveness indexes of dolomite and limestone were low because calcite, feldspar and biodetritus content was high in this carbonate rock. The rock abrasiveness indexes of basalt and andesite were medium, because the content of feldspar was high and the content of quartz was low in the igneous rock. The analysis result showed that the size of rock abrasiveness index was related with the petrogenetic mineral. Normally, the quartz content was higher, the rock abrasiveness index was higher.

4.2 *The relationship between rock abrasiveness index and hardness and drillability*

The relationship between rock abrasiveness index and harness is shown in Figure 5 and the relationship between rock abrasiveness index and drillability is shown in Figure 6.

1. From the result of the experiment, the rock abrasiveness index is high for some high harness rock, and the formations of high rock abrasiveness and high harness are often drilled, because the detrital intergranular cement types and mineral compositions are different. For the clasolite, the common types of cement are iron, siliceous, argillaceous, calcium and so on, and the siliceous is the most hard, so rock of high rock abrasiveness and high harness could appear. For dolomite, limestone and some carbonate rocks, the quartz content is low and the main components are calcite, feldspar and biodetritus, the rock abrasiveness index is less than 20. For basalt, and esite and igneous rocks, the hardness is high, because the quartz content is low, the rock abrasiveness is medium, and less than 50. For F verification, the value is less than $F_{0.05,32}$ ($F_{0.05,32} = 4.08$). The finding is that the relation between rock abrasiveness index and hardness is poor.
2. The drillability is related to mineral type and content, mineral particle size and shape, rock structure, cement property and cementing form. According to the fitting result, the value is less than $F_{0.05,32}$ ($F_{0.05,32} = 4.08$). The relation between rock abrasiveness index and drillability is poor.

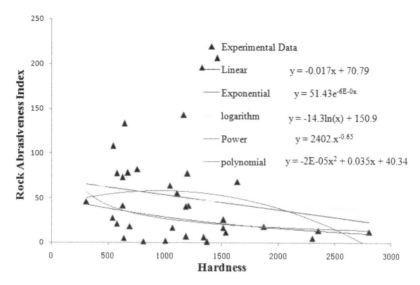

Figure 5.   The relationship between rock abrasiveness index and hardness.

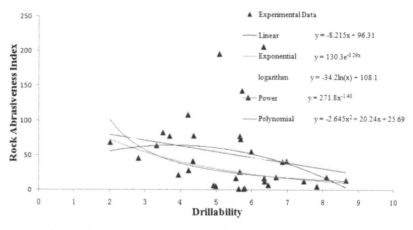

Figure 6.   The relationship between rock abrasiveness index and drivability.

5   CONCLUSIONS AND SUGGESTIONS

1. The simulation micro-bit has satisfied the simulation in which the rock is broken by bit and the bit is worn by the rock because of the spiral structure and quenching steel material, and it is beneficial to analyze the experimental data.
2. Test method of rock abrasiveness which had been developed on the basis of the micro-bit drilling that simulates the actual drilling surroundings to get the rock volume broken by the micro-bit and the wearing loss of the simulation bit is practical and scientific.
3. According to the rock abrasiveness index tested in the experiment, rock abrasiveness is divided into eight grades with Sturges empirical formula, and the bit type selection and the prediction of bit life could be conducted with the rock abrasiveness index.
4. The suggestion is to research the relationship between rock abrasiveness and physical mechanics property in the direction of rock mineral component and cement property to deduce the corresponding relation.

ACKNOWLEDGEMENTS

This work was financially supported by the Science and Technology Research Project of Shengli Petroleum Administration Sinopec Corporation (GS1003).

NOMENCLATURE

R_a: Rock Abrasiveness Index, mg/cm^3.
W: Wearing Loss of the Simulation bit, mg.
V: Rock volume which was broken by the simulation bit, mg.
d: Diameter of the simulation bit, mm.
h: Depth of the drilled hole, mm.
W_1: Weight of the simulation bit before the experiment, mg.
W_2: Weight of the simulation bit after the experiment, mg.

REFERENCES

Kong Jian. 1985. Experimental study on rock abrasiveness with artificial diamond [J]. *Earth science*: 10(3):53–63.

Mohamad. Assessment on abrasiveness of rock material on the wear and tear of drilling too l[J]. *Electronic Journal of Geotechnical Engineering*, v 17 A, p 91–100, 2012; E-ISSN: 10893032.

Okubos, Fukuik, Nishimatsuy. 2010. Estimating abrasivity of rock by laboratory and in situ tests [J]. *Rock Mech Rock Eng*.

Sling, H.L. 2010. Determining rock abrasivity in the laboratory [J]. Rock Mechanics in Civil and Environmental Engineering—*Proceedings of the European Rock Mechanics Symposium*, EUROCK 2010, p 425–428.

Thuro, Kuroschl. 2009. Assessment of abrasivity by physico-mechanical properties of rocks [J]. *Journal of Mining Science*, v 45, n 3, p 240–249, May.

Tomazl. Application of an instrumented tracer in an abrasion mill for rock abrasion studies [J]. *Strojniski Vestnik/Journal of Mechanical Engineering*, v 58, n 4, p 263–270.

Wu Guanglin, Zhang Meinan, Li Lin. 1985. Determination of the abrasiveness of rock by using the wearing capacity of a copper needle [J]. *Journal of Chendu College of Geology*: 04:65–76.

Xing Jiguo. 1992. Test study for rock abrasiveness [J]. *West-china Exploration Engineering*: 2003,15(3):25–26. [5] Jinhong Yin, Yihua Yan. Rock abrasiveness measurement with ultrahard material[J]. *Journal of the University of Petroleum, China*: 05:25–29.

Yin Jinhong, Yan Yihua. 1992. Rock abrasiveness measurement with ultrahard material [J]. *Journal of the University of Petroleum, China*: 05:25–29.

Zhao Qianying, Deng Jingen, Xie Yuhong. 2011. Establishment and application of a universal prediction model of formation abrasivity [J]. *China Offshore Oil and Gas*: 23(5):329–334.

Zou Deyong, Wang Ruihe. 2003. Experimental study on rock abrasiveness with PDC bit [J]. *Journal of the University of Petroleum, China*: 27(2):41–43.

Manufacturing Engineering and Intelligent Materials – Lu & Abu Bakar (Eds)
© 2015 Taylor & Francis Group, London, ISBN 978-1-138-02832-6

Dynamic characteristics measurement of a torsion damper

R. Zeng & Lei Chen
School of Mechanical and Electronic Engineering, Wuhan University of Technology, Wuhan, China

X.J. Ren
Qingdao Sifang Rolling Stock Research Institute Co. Ltd., Qingdao, China

ABSTRACT: As a new type of torsion damper for automobile, Dual Mass Flywheel (DMF) characterizes nonlinear hysteresis, due to the coupling of elastic force and damping force. The research on hysteresis, generally based on experimental data, helps to obtain the nonlinear dynamic characteristics of DMF. In order to measure the dynamic parameters of DMF, experiments with and without excitations are conducted, according to the working principle of DMF. The hysteretic and rotational speed-dependent characteristics of DMF are observed in the experiments. Further, the nonlinear dynamical regulation of stiffness and damping of DMF are investigated, providing a reliable experimental verification method to study the nonlinear dynamical characteristics of DMF.

1 INTRODUCTION

Reciprocating motion of engine generates torsional vibrations. If the frequency of reciprocating torque of the engine is the same with the natural frequency of the power train, the resonance will happen. Resonance can bring about serious transmission gear noises and result in mechanical damage, which influences the service life of the components of the power train and the ride comfort. Dual Mass Flywheel (DMF), a new type of torsion damper, can effectively eliminate torsional vibrations caused by engine, which has been widely used in vehicles. DMF changes torsional vibration characteristics of automotive transmission system by adjusting rotational inertias of flywheels and torsional stiffness of springs, making the first-order natural frequency of the transmission system lower than the frequency corresponding to the engine idle speed. It can effectively avoid resonances under driving conditions. When the engine starts or stops, it is inevitable to go through the resonant region (Lei, Ming-ran & Zhengfeng 2012). In this case, the damping element of DMF works, which can reduce the resonant amplitude.

As a part of the power transmission system, DMF transfers torque from engine to clutch and attenuates the torsional vibration from the engine. The dynamical characteristics of DMF reflect the relations between the transferred torque and torsion angle when it works (Yuan & Yuxue 2009). Because the transferred torque is composed of the elastic torque and the damping torque, of which the features will further illustrate the dynamic regulation of the torsional stiffness and damping of DMF. What's more, the coupling of the elastic torque and damping torque results in hysteretic characteristics of DMF (Walter, Kiencke, Jones & Winkler 2008, Kukhyun et al. 2004). Experimental data will contribute to study the hysteretic transferred torque of DMF.

Previous studies on DMF are focused on the theoretical and experimental static elastic characteristics (Tao, Zhenhua & Chengqian 2006, Shengtian 2004) and the theoretical dynamic elastic characteristics (Zhe & Jaspal 2013, Schaper, Sawodny, Mahl & Blessing 2009). However, the relations between the dynamic transferred torque, the rotational speed of the engine, the torsional stiffness and damping of DMF are less involved. Thus, the experimental method of dynamical characteristics is investigated in this paper. The hysteresis and rotational speed-dependent features of DMF are observed from the experimental results. Furthermore, the dynamic regulations of torsional stiffness and damping of DMF are found out, which provides a reliable experimental method to verify the theoretical research.

2   WORKING PRINCIPLE OF DMF

A basic Dual Mass Flywheel (DMF) consists of two separated flywheels, connected by a spring-damping damper, as shown in Figure 1. The primary flywheel is connected with crankshaft, and the secondary flywheel with clutch (Lei, Ming-ran & Zhengfeng 2012). The arc springs are compressed or decompressed along the shell, transmitting the engine torque to the transmission system. To reduce wear, lubricating grease is filled in spring channels, generating damping as well. A simplified torsional vibration model of the power train is established to research the damping performance of DMF, which is shown in Figure 2.

Where, T denotes the engine torque, here $T = T_0\sin(\omega t)$, and T_0 and ω represent the amplitude and frequency, respectively. J_1 and J_2 denote the moment of inertia of the crankshaft combining with the first mass of DMF and the moment of inertia of the second mass of DMF combining with the rear components, respectively. In addition, K_{td} and C are the torsional stiffness and viscous damping constant of DMF, and θ_1 and θ_2 represent the torsional displacements of the first mass and the second mass of the system.

The equation of motion of the two degree-of-freedom system is given by

$$\begin{bmatrix} J_1 & 0 \\ 0 & J_2 \end{bmatrix}\begin{Bmatrix} \ddot{\theta}_1 \\ \ddot{\theta}_2 \end{Bmatrix} + \begin{bmatrix} C & -C \\ -C & C \end{bmatrix}\begin{Bmatrix} \dot{\theta}_1 \\ \dot{\theta}_2 \end{Bmatrix} + \begin{bmatrix} K_{td} & -K_{td} \\ -K_{td} & K_{td} \end{bmatrix}\begin{Bmatrix} \theta_1 \\ \theta_2 \end{Bmatrix} = \begin{Bmatrix} T_0\sin(\omega t) \\ 0 \end{Bmatrix} \tag{1}$$

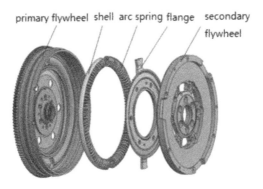

Figure 1.   A schematic diagram of basic DMF.

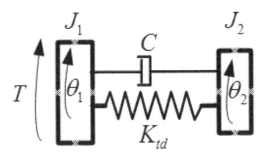

Figure 2.   Dynamic model of power train with DMF.

The natural frequency of the system can be deduced as

$$\omega_n^2 = K_{td}\frac{J_1+J_2}{J_1 J_2} \tag{2}$$

Hence, selecting appropriate torsional stiffness and moment of inertia, the natural frequency of the transmission system can be adjusted to be lower than the idle frequency (about 13 Hz). Solving the equation of motion, the relative torsional angle between the primary and secondary masses is expressed as

$$\Delta\theta = \frac{T_0}{(1+\lambda)\sqrt{\left(K_{td}-\frac{\lambda(J_1+J_2)}{(\lambda+1)^2}\cdot\omega^2\right)^2+(C\omega)^2}}\cdot\sin\left(\omega t-\arctan\frac{C\omega}{K_{td}-\frac{\lambda(J_1+J_2)}{(\lambda+1)^2}\cdot\omega^2}\right) \tag{3}$$

where λ denotes the moment of inertia ratio of the primary and secondary flywheels, which is J_1/J_2. The result shows that damping and stiffness of DMF play important roles in the vibration reduction performance of DMF, when the inertia ratio is determined. In addition, friction exists between the arc springs and the shells, resulting in the coupling of the torsional stiffness and damping and further the hysteretic transferred torque. Owing to the rotational motion of the transmission system, the arc springs are subjected to centrifugal forces. Thus, DMF will characterize the rotational speed-dependent. Accurately grasping the above characteristics of DMF is of greatest importance to study the damping performance of DMF.

3 DYNAMICAL EXPERIMENTS OF DMF

3.1 Experimental principle

The dynamic deformation of arc springs mounted in the sealing spring channels can only be detected from the torsion angle data of the primary and secondary flywheels and the torque data. The simplified mechanics model of DMF is shown in Figure 3. T_{INT} represents the transmitted torque of DMF. θ_p and θ_s denote the torsional angle of the primary and secondary flywheel, respectively. What's more, J_p and J_s denote the moments of inertia of the primary and secondary flywheel, and T_L is the load torque acting on the secondary flywheel. The transmitted torque can be expressed in terms of the relative torsion angle $\theta=\theta_p-\theta_s$.

From Figure 3, the equations of motion of DMF can be expressed as

$$\begin{cases} J_p\ddot{\theta}_p = T - T_{INT} \\ J_s\ddot{\theta}_s = T_{INT} + T_L \end{cases} \tag{4}$$

Hence, the transmitted torque can be written as:

$$T_{INT} = \frac{1}{2}\left(T-T_L-J_p\ddot{\theta}_p+J_s\ddot{\theta}_s\right) = \frac{1}{2}\left(T(t)-T_L(t)-J_p\ddot{\theta}_p(t)+J_s\ddot{\theta}_s(t)\right) \tag{5}$$

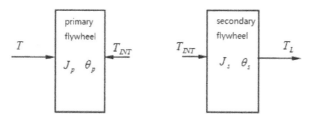

Figure 3.   Schematic diagram of mechanics model of DMF.

The inertia torques of the primary and secondary flywheels can be neglected, when the engine torque does not reciprocate. It can be seen from Equation 5 that the input torque, load torque and the torsion angles of the primary and secondary flywheels should be measured to obtain the dynamic characteristics of DMF. While the input torque reciprocates, the experiment is conducted under excitation. In this case, the inertia torques should be taken into consideration. Using the torsion angles data of the primary and secondary flywheels, the torsional accelerations $\ddot{\theta}_p(t)$ and $\ddot{\theta}_s(t)$ can be calculated by difference operations. Afterwards, using Equation 5, the transmitted torque can be obtained.

The torque from engine may take the form as:

$$T_{eng} = T_0 + \sum_{j=1}^{\infty} \Delta T_j \cdot \cos\left(j\omega_0 t + \psi_j\right) \tag{6}$$

where T_{eng} represents the torque from engine. T_0 denotes the mean torque that is practically needed. ΔT_j and ψ_j denote the oscillation amplitude and the phase angle of number j order. ω_0 means the base frequency, representing the rotational speed. Due to the engine torque, the arc spring will be compressed with oscillation, and the deformation is composed of two parts. Firstly, the arc spring is overall compressed, due to the required torque T_0. Secondly, the arc spring deforming to a certain angle oscillates with small amplitude, due to the reciprocating torque.

Therefore, we put forward two dynamical experiments of DMF. One is the experiment without excitation; the other is the experiment under excitation. From the experiment without excitation, the ability of DMF to transmit engine torque can be investigated. Hence, the dynamic variation of torsional stiffness of DMF can be achieved. From the excited experiment, the dynamic variation of damping of DMF under working conditions can be obtained.

3.2 *Experimental setup*

The dynamical experimental test bench of DMF is shown in Figure 4. DMF is excited by a converter motor, which is controlled by a specified transducer. The specified transducer control the converter motor to generate torsional vibration by adjusting the three phase current of the converter motor. When excitation is needed, the transducer controls the converter motor to produce asymmetric three-phase currents. The excitation magnitude can be modified by changing the magnitude of the asymmetric current. The load imposed on the DMF is achieved by the dynamometer. The torque transferred by the DMF and torsional displacements of the primary and the secondary flywheels are detected by torque sensors and encoders, respectively.

3.3 *Measurements*

When conducting the experiment without excitation, the three-phase current of the converter motor is controlled to be balanced. The working frequency of the converter motor are set by

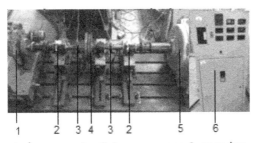

1. dynamometer; 2. torque sensor; 3. encoder;

4. DMF-CS; 5. convertor motor; 6. transducer

Figure 4.   Dynamic experimental setup.

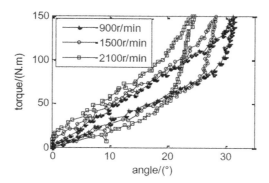

Figure 5. Torque-angle curves without excitation.

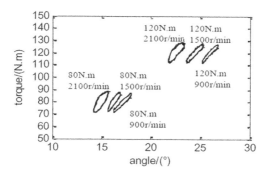

Figure 6. Torque-angle curves with excitation.

the transducer to be 30 Hz, 50 Hz and 70 Hz, which indicates that the rotational speeds of the motor are 900 r/min, 1500 r/min and 2100 r/min, respectively. The load is controlled by the dynamometer to increase linearly from 0 to 150 N·m, then reduce to 0. The relation between the transmitted torques and the relative torsional angles are plotted in Figure 5. When conducting the excited experiment, the three-phase currents of the converter motor are controlled to be asymmetric. The working frequencies of the converter motor also are set to be 30 Hz, 50 Hz and 70 Hz. In addition, the excitation magnitude is set by adjusting the magnitude of the asymmetric current to be 10 N·m. In this case, the loads imposed on the DMF are controlled to be 80 N·m and 120 N·m. The torque-torsional angle curves are plotted in Figure 6.

Hysteretic characteristic is observed in above experiments with or without excitation. The area enclosed by a complete loop denoted the energy dissipated. The curves in Figure 5 and Figure 6 show that the areas enclosed increases as the rotational speed increases, which implies that the damping of DMF increases with the rotational speed. Figure 5 also shows that the stick-slip phenomenon occurs. The arc springs stick to the shell until the external torque is large enough to overcome the static friction. What's more, the stiffness increases with increasing the rotational speed in loading processes. Further, in Figure 6, the leans of the curves become steeper as the rotational speed increases, indicating the equivalent stiffness enlarged with the increase of rotational speed. The experimental results show that DMF characterizes hysteretic and rotational speed-dependent features. In addition, both torsional stiffness and damping increase as the rotational speed increases.

4 CONCLUSIONS

The torsional vibration excitation can be achieved by changing single-phase current of the converter motor. From the experiment with excitation, the relations between the reciprocating torque and torsion angle of DMF are measured, obtaining the dynamical variation of damping of DMF under driving conditions. From the experiment without excitation, the ability to transmit torque under different rotational speeds is investigated, getting the dynamical variation of torsional stiffness of DMF.

Torsional stiffness and damping of DMF characterize speed-dependent, which both increase with increasing the rotational speed. The transmission system goes through the resonant region when engine starts and stops. In this case, the torsional stiffness is low and the damping is large, implying that DMF can effectively reduce the resonant amplitude. However, DMF presents large torsional stiffness and small damping under driving conditions, reducing the damping effect of DMF. Since the magnitude of engine torque fluctuation is rather small at higher rotational speed, DMF can still achieve ideal damping effect.

REFERENCES

Hoon W., Yoon-young K., Haeil J., Gwang-nam L. 2001. Nonlinear Rate-dependent Stick-Slip Phenomena: Modeling and Parameter Estimation. *International Journal of Solids and Structures*, 38, 1415–1431.

Kukhyun A., Jang Moo L., Wonsik Lim et al. 2004. Analysis of a Clutch Damper Using a Discrete Model. *KSME International Journal,* 18(11), 1883–1890.

Lei C., Ming-ran D., Zhengfeng J. 2012. Optimization Method of Performance Parameters of Dual Mass Flywheel. *Transactions of CSICE*, 30(3), 277–283.

Shengtian L. 2004. Influences of a Dual Mass Flywheel Damper on Idling Vibration. *Transactions of the Chinese Society for Agricultural Machinery*, 35(3), 16–19.

Schaper U., Sawodny O., Mahl T., Blessing U. 2009. Modeling and Torque Estimation of an Automotive Dual Mass Flywheel. 2009 American Control Conference, 16(6), 1207–121.

Tao C., Zhenhua L., Chengqian S. 2006. Analysis Method for Elastic Characteristics of Arc Spring. *China Mechanical Engineering*, 17 (1), 493–495.

Walter A., Kiencke U., Jones S. and Winkler T. 2008. Anti-Jerk & Idle Speed Control with Integrated Sub-Harmonic Vibration Compensation for Vehicle with Dual Mass Flywheel. *SAE International Journal of Fuels Lubrication*, 1(1), 1267–1276.

Yuan D., Yuxue P. 2009. Design and Research on Arc Spring in Dual Mass Flywheel. *Journal of Wuhan University of Technology*, 31(5), 114–117.

Zhe L., Jaspal S. 2013. Transmission Torque Converter Arc Spring Damper Dynamic Characteristics for Driveline Torsional Vibration Evaluation. *SAE Int. J. Passeng. Cars-Mech. Syst*, 6(1), 1483.

Manufacturing Engineering and Intelligent Materials – Lu & Abu Bakar (Eds)
© 2015 Taylor & Francis Group, London, ISBN 978-1-138-02832-6

Fuzzy clonal optimization of ventilation system reliability in uranium mine

J.Y. Dai, J.Q. Cui, S.L. Zou & M. Wang
University of South China, Hengyang, Hunan Province, China

ABSTRACT: According to the general characteristics of the uranium mine ventilation system and radioactive pollutants features, reliability model of ventilation system is constructed by applying system reliability and fuzzy programming method subject to fuzzy goals and fuzzy constraints in uranium mine, which can implement iterative optimization of ventilation system reliability by using clonal optimization algorithm. Practical examples show that the fuzzy programming of the system reliability can better improve reliability model of ventilation system and the clonal selection optimization algorithm can better realize optimization control of the ventilation system reliability. This can provide certain reference value for improving the reliability of ventilation system and safety management in uranium mine.

1 INTRODUCTION

Mine ventilation system is a complex system which consists of three major components—ventilation facilities, ventilation power and ventilation network (Lu 2010). Its role is to provide fresh air to the working force, eliminate polluted air, provide guarantees for the safe operation of production systems. There are large amounts of harmful radioactive pollutants such as radon and its progeny in the underground uranium mine and they do harm to safety production. The ventilation system is the most important means of reducing radon in uranium mine (Liang, Zhou & Liu 2006). Therefore, the study of uranium mine ventilation system reliability is particularly important.

How to achieve the optimal reliability of ventilation system in mine? Some experts have carried out reliability evaluation of ventilation system by using analytic hierarchy process, fuzzy comprehensive evaluation method, fault tree analysis and other methods (Liu et al. 2011, Zhang et al. 2012, Niu et al. 2013, Ding & Huang 2013, Duo et al. 2013), which provide a scientific basis for evaluating ventilation system in mine. With the application of fuzzy mathematics, the reliability model of ventilation system can better reflect its intrinsic regularity. Usually, the general fuzzy programming model needs to be converted to a deterministic model to achieve its optimal solution. The ventilation system is a complex system, which can hardly meet the requirements of random fuzzy in uncertainty environment through converting the fuzzy model to a deterministic model (Tang & Wang 2004). However, it can easily overcome the defect of this kind of optimization method and avoid the system error through the introduction of genetic algorithm and neural network algorithm (Lu et al. 2014, Guo et al. 2013). But these algorithms may also have drawbacks such as premature assimilation, slow convergence and easy to fall into local optimum, but immune clonal optimization algorithm combines the excellent properties of relevant algorithms, which overcomes the shortcoming of genetic algorithm and artificial neural network. Using their "immune clone affinity" mechanism can better achieve optimization of system reliability. All of these show a good superiority of immune clonal optimization algorithm (Yao et al. & Shi 2011, Liu et al. 2014). Therefore, this paper uses fuzzy membership function to build the reliability model of ventilation system in uranium mine and uses the immune clonal selection optimization algorithm to implement optimal control, which will provide a decision-making basis for improving reliability of ventilation system.

2 FUZZY CLONAL SELECTION OPTIMIZATION CONTROL OF VENTILATION SYSTEM RELIABILITY IN URANIUM MINE

According to ventilation engineering principle of the system reliability and ventilation accident statistics, combined with the experiences and views of experts and scholars, we can conclude that the reliability of ventilation system consists of reliability of network system, the main ventilator system, radioactive pollutants control system and ventilation management system. The ventilation network system mainly consists of the ventilation mode, ventilation power, ventilation network, ventilation facilities, etc. The main ventilator system consists of equipment installation quality, application quality, mainly fan efficiency, daily maintenance and management, mechanical and electrical equipment and so on. The radioactive contamination control system consists of the concentration of uranium dust, radioactive aerosols, radon and its progeny. Ventilation system management consists of the security management and organizational design, the quality of security personnel, ventilation safety investment, and safety management measures. Taking into account the difference of the relative importance and the uncertainty of systems, fuzzy numbers range is used to determine the relative weights of the systems.

Because its weight coefficients are fuzzy numbers, classical optimization theory is hardly able to achieve the optimum reliability of ventilation system. In order to achieve the purpose, the intelligent algorithms should be applied. For the sake of not changing its problem properties, firstly, we can set membership function to make the coefficient fuzzy, then establish the appropriate integrated model of membership function based on fuzzy judgment decision criteria. Secondly, apply clonal selection mechanism to realize optimization reliability model. Fuzzy clonal selection algorithm process is as follows.

Step 1 Establish fuzzy reliability model of ventilation system in uranium mine;

Step 2 Membership function set. Set membership function of the fuzzy number coefficient according to the nature of fuzzy parameters for reliability model of the ventilation system;

Step 3 Initialization of fuzzy numbers. Initialize the fuzzy goals and fuzzy constraints depending on the different possibility level;

Step 4 The expected value set of the fuzzy objective function. To solve the corresponding expected value range of fuzzy objective function under the condition of meeting the corresponding probability horizontal range and its constraint conditions;

Step 5 Membership function set of the fuzzy goals and fuzzy constraints. Construct its corresponding fuzzy goals and fuzzy constraints membership function according to the fuzzy target range expectations and fuzzy constraint range. Using fuzzy algorithm to integrate their fuzzy goals and fuzzy constraints;

Step 6 Construct the integrated function. According to fuzzy algorithm, integrate fuzzy target integration membership functions and fuzzy constraint integration membership functions;

Step 7 Population initialization. Initialization set of candidate solutions for antibody populations P, P is composed of Memory Population M and the remaining population T;

Step 8 Calculation of antibody affinity. Choose n antibodies from the affinity population to form antibody population Pn;

Step 9 Antibody clonal. Cloning Pn to generate a temporary clonal population C, clonal size is proportional to the antibody-antigen affinity; the higher avidity, the greater the number of antibody clonal;

Step 10 Antibodies mutation. It simulates biological clonal selection variation process, high frequency mutation of the temporary clonal population C can get mutated antibody group C; high frequency mutation is proportional to the antibody-antigen affinity;

Step 11 Antibodies reselect. Recalculate the antibody affinity in the mutated antibody group C, if it is higher than temporary avidity, use the antibody instead of the original antibody and add it to the memory antibodies group M;

Step 12 Antibodies renew. It simulates the process of the biological clonal selection of B cells natural dying and the small proportion of the antibodies must be initialized. In order to ensure the antibody's diversity, we should replace low affinity antibodies with generated new antibodies;

Step 13 stop. Determine whether the termination condition is satisfied, if it is, terminated; otherwise, go to step (Step 8), into the next iteration.

3 CASE STUDY

Take an example of a large uranium mine. It uses forced mechanical ventilation system, but there are still many problems in technology and management—radon and its progeny are low + pass rate in the working faces, which bring more complex radiation and protection problems of uranium mining and increase the difficulty of uranium ventilation project. According to site investigation, expert assessment method, we can use triangular fuzzy membership functions to get the weight fuzzy numbers and reliability fuzzy numbers of the ventilation network, ventilation equipment, pollution control and ventilation management system and its corresponding subsystem in the uranium mine ventilation system (Table 1).

The objective optimization function of ventilation system reliability:

$$\max R_V = \tilde{w}_S \bullet \sum_{i=1}^{3} \tilde{w}_{si} R_S + \tilde{w}_F \bullet \sum_{j=1}^{4} \tilde{w}_{Fj} R_F + \tilde{w}_E \bullet \sum_{k=1}^{3} \tilde{w}_{Ek} R_E + \tilde{w}_M \bullet \sum_{l=1}^{4} \tilde{w}_{Ml} R_M \quad (1)$$

R_V: reliability of ventilation system;
R_S: reliability of ventilation network system;
R_F: reliability of main fan system;
R_R: reliability of radioactive pollutant control system;
R_M: reliability of ventilation management system.

$\widetilde{W}_S$, $\widetilde{W}_F$, $\widetilde{W}_E$, $\widetilde{W}_M$, are the corresponding fuzzy weight range of mine ventilation network mine main fan, radioactive pollutant control system, mine ventilation management.

$\widetilde{W}_{Si}$, $\widetilde{W}_{Fj}$, $\widetilde{W}_{Ek}$, $\widetilde{W}_{Ml}$, are the corresponding fuzzy weight range of the second indicators.

Taking into account the association of the second fuzzy reliability in ventilation system, in order to simplify the problem, we can use series system to achieve ventilation system reliability. Assuming its reliability is affected by the overall reliability of the system, we can use the range number to control, so the constraint equation is Eq.(2):

$$s.t\{R_S * R_F * R_E * R_M \in [0.60, 1.00] \quad (2)$$

Step 1 Establish fuzzy goals and fuzzy constraint equations of ventilation system

Table 1. The second fuzzy weight and fuzzy reliability of ventilation system.

		Weight		Weight	Reliability
Ventilation System Reliability	Ventilation Network Reliability	0.35 0.45 0.55	Ventilation power Ventilation network Ventilation facilities	0.20,0.30,0.40 0.40,0.50,0.60 0.10,0.20,0.30	0.90,0.95,1.0 0.80,0.88,0.96 0.90,0.94,0.98
	Ventilation Equipment Reliability	0.15 0.25 0.35	Equipment quality Installation quality Ventilation efficiency Maintain quality	0.20,0.30,0.40 0.15,0.20,0.25 0.10,0.20,0.30 0.25,0.30,0.40	0.90,0.94,0.98 0.90,0.94,0.98 0.80,0.89,0.98 0.80,0.90,1.00
	Pollutant Control Reliability	0.05 0.1 0.15	Uranium dust Radioactive aerosols Radon and progeny	0.20,0.30,0.40 0.15,0.20,0.25 0.40,0.50,0.60	0.80,0.88,0.96 0.80,0.88,0.96 0.90,0.94,0.98
	Ventilation Safety Management	0.1 0.2 0.3	Security organization Quality personnel Ventilation safety input Safety measures	0.15,0.20,0.25 0.15,0.25,0.35 0.10,0.20,0.30 0.25,0.35,0.45	0.90,0.94,0.98 0.80,0.89,0.98 0.80,0.88,0.96 0.80,0.89,0.98

Because the various coefficients of ventilation system reliability model are triangular fuzzy numbers, so depending on the sets A of triangular fuzzy number, we set its cut set α to establish the fuzzy goals and fuzzy reliability constraint equations of ventilation systems on the different possibilities level.

$$\max R_V = \left(\widehat{W}_S\right)_R^a \bullet \sum_{i=1}^{3} \left(\tilde{w}_{si}\right)_R^a R_S + \left(\tilde{w}_F\right)_R^a \bullet \sum_{j=1}^{4} \left(\tilde{w}_{Fj}\right)_R^a R_F$$

$$+ \left(\tilde{w}_E\right)_R^a \bullet \sum_{k=1}^{3} \left(\tilde{w}_{Ek}\right)_R^a R_E + \left(\tilde{w}_M\right)_R^a \bullet \sum_{l=1}^{4} \left(\tilde{w}_{Ml}\right)_R^a R_M \tag{3}$$

$$s.t \begin{cases} R_s * R_F * R_E * R_M \in [R_{\min}, R_{\max}] \\ w_i \in \left[(\tilde{w}_i)_L^\alpha, (\tilde{w}_i)_R^\alpha\right] \\ w_{jk} \in \left[(\tilde{w}_{jk})_L^\alpha, (\tilde{w}_{jk})_R^\alpha\right] \\ R_{jk} \in \left[(\tilde{R}_{jk})_L^\alpha, (\tilde{R}_{jk})_R^\alpha\right] \end{cases} \tag{4}$$

Step 2 Set expected value range of the fuzzy objective function
According to the theory of fuzzy parameter programming, expectation range of ventilation system reliability fuzzy goal can be obtained by Eqs. (5–8):

$$f_a^- = G(x) = \max_{x \in F_1} \left\{ \left(\widehat{W}_S\right)_L^a \bullet \sum_{i=1}^{3} \left(\tilde{w}_{si}\right)_L^a R_S + \left(\tilde{w}_F\right)_L^a \bullet \sum_{j=1}^{4} \left(\tilde{w}_{Fj}\right)_L^a R_F + \left(\tilde{w}_E\right)_L^a \right.$$

$$\left. \bullet \sum_{k=1}^{3} \left(\tilde{w}_{Ek}\right)_L^a R_E + \left(\tilde{w}_M\right)_L^a \bullet \sum_{l=1}^{4} \left(\tilde{w}_{Ml}\right)_L^a R_M \right\} \tag{5}$$

$$f_a^+ = G(x) = \max_{x \in F_2} \left\{ \left(\widehat{W}_S\right)_R^a \bullet \sum_{i=1}^{3} \left(\tilde{w}_{si}\right)_R^a R_S + \left(\tilde{w}_F\right)_R^a \bullet \sum_{j=1}^{4} \left(\tilde{w}_{Fj}\right)_R^a R_F + \left(\tilde{w}_E\right)_R^a \right.$$

$$\left. \bullet \sum_{k=1}^{3} \left(\tilde{w}_{Ek}\right)_R^a R_E + \left(\tilde{w}_M\right)_R^a \bullet \sum_{l=1}^{4} \left(\tilde{w}_{Ml}\right)_R^a R_M \right\} \tag{6}$$

$$F_1 = \left\{ x \mid s.t, R_{ijk} \geq 0; \right\} \tag{7}$$

$$F_2 = \left\{ f_a^+ \succ f_a^-, x \in F_1 \right\} \tag{8}$$

We can assume that the α cut sets is 0.5 in the primary and secondary index fuzzy trigonometric functions, Applying the clonal selection optimization algorithm, we can obtain that the expected value range of the fuzzy objective function of ventilation system reliability is [0.6838, 1.6980].

Step 3 The membership function set and its optimization solution of fuzzy goals and fuzzy constraints.

According to the expected value range of fuzzy objective functions and fuzzy constraint functions, we can set membership function model which can construct the fuzzy goals:

$$\mu_G(x) = \begin{cases} 0 & G(x) \leq 0.6838 \\ \left(G(x) - f_a^-\right) / \left(f_a^+ - f_a^-\right), & 0.6838 \prec G(x) \leq 1.6980 \\ 1 & G(x) \succ 1.6980 \end{cases} \tag{9}$$

According to the complex characteristics of the uranium mine ventilation system, its expected value range of fuzzy constraint function is [0.60, 1.0], so its fuzzy constraint membership function model is Eq.(10):

$$\mu_C(x) = \begin{cases} 0 & C(x) \leq 0.60 \\ \left(C(x) - 0.60\right) / \left(1.00 - 0.60\right), & 0.60 \prec C(x) \leq 1.00 \\ 1 & C(x) \succ 1.00 \end{cases} \tag{10}$$

Table 2. The reliability of ventilation system under different possibilities optimal level.

	First index	Second index	α Cut sets 0.2	α Cut sets 0.5	α Cut sets 0.8
Ventilation System Reliability Optimization	Ventilation Network Reliability	Ventilation power	0.9876	0.9879	1
		Ventilation network	0.96	0.9598	0.96
		Ventilation facilities	0.9317	0.9742	0.9235
	Ventilation Equipment Reliability	Equipment quality	0.9615	0.977	0.9797
		Installation quality	0.9218	0.9309	0.934
		Ventilation efficiency	0.8706	0.8039	0.8115
		Maintain quality	0.8082	0.9079	0.9725
	Pollutant Control Reliability	Uranium dust	0.8119	0.8	0.8
		Radioactive aerosols	0.8195	0.8603	0.8129
		Radon and its progeny	0.9233	0.9163	0.9183
	Ventilation Safety Management	Security organization	0.9043	0.9129	0.9168
		Quality personnel	0.9576	0.8965	0.9262
		Ventilation safety input	0.8287	0.8213	0.8298
		Safety measures	0.9583	0.9073	0.8963
Optimal ventilation system reliability			0.9183	0.9104	0.8963

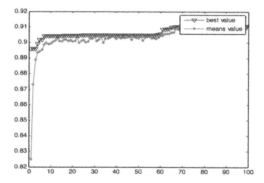

Figure 1. The optimal reliability iterative optimization schematic when the possible level is 0.50.

Simplified into consideration, assume that the importance of fuzzy goals is equal to fuzzy constraints in the uranium mine ventilation system reliability. Fuzzy membership function of the ventilation system reliability is expressed as follows:

$$\mu_{gc}(x) = \max \min \{\mu_G(x), \mu_c(x)\} \tag{11}$$

According to the rule of the maximum membership degree, we can use clonal selection optimization algorithm to solve its iterative optimization. Applying the primary and secondary fuzzy reliability weight coefficient of ventilation system on the different probability level, the optimum reliability of system under different probability level can be, respectively, obtained (Table 2, Fig. 1), which can provide decision basis for ventilation system upgrade.

4 CONCLUSION

Based on the characteristics of uranium mine ventilation system and radioactive pollutants, we build the reliability model of fuzzy goals and fuzzy constraints in uranium ventilation system by applying system reliability and fuzzy programming method. And then, applying cloning optimization algorithm to carry out iterative optimization reliability of the uranium mine ventilation system, this will provide a reference basis for uranium mine ventilation system modification. Practical examples show that the application of system reliability and fuzzy programming can better improve the reliability model of ventilation system, the application of clonal selection optimization algorithm can better realize the optimization control of ventilation system reliability. It provides a good application prospect for improving the uranium mine ventilation system reliability and realizing uranium mine ventilation transformation and safety management.

ACKNOWLEDGEMENTS

This paper is funded by the National Natural Science Foundation of China. (No. 51174116).

REFERENCES

Ding, Houcheng. Huang, Xinjie, 2013. Coal Mine ventilation reliability evaluation based on AHP-FCE. *Journal of Natural Disasters.* 07:153–159.

Duo, Yili. Hai, Jun. Chen, Yang. 2013. Ventilation System Reliability Analysis based on FMEA and FTA. *Coal mine safety 04*:177–179.

Guo, Yinan. Wang, Chun. Yang, Jichao. 2013. Mine ventilation network based on cultural particle swarm optimization algorithm. *Journal of Southeast University (Natural Sciences)* S1:48–53.

Liang, Zheng. Zhou, Xinghuo. Liu, Changrong. 2006. Research on Uranium mine ventilation and radon reducing. *Journal of Safety Science and Technology* 02:53–56.

Liu, Ruochen. Ma, Chenlin. He, Fei. Ma, Wenping. Jiao, Licheng. 2014. Reference direction based immune clone algorithm for many-objective optimization. *Frontiers of Computer Science.*

Lu, Guobin. Cheng, Peng. Zhang, Junwu. 2014. BP neural network reliability evaluation of ventilation systems. *Journal of Liaoning Technical University (Natural Sciences)* 01:23–27.

Liu, Zheng. Hu Hanhua. Cui Tiantian. Han Yujian. 2011. Safety Evaluation of Mine Ventilation System Based on Improved Fuzzy AHP Mining Research and Development 03:81–84.

Lu, Tao. 2010. Mine ventilation system reliability evaluation method and its development trend. *Coal Mining 04*:1–4.

Niu Xiaozheng, Lv Lixing. Lin Jifei. Li Jilong. 2013. Mine ventilation system evaluation method based on fuzzy mathematics method. *Mining and Metallurgy 07*:19–24.

Shi, Xuhua. 2011. Artificial immune network based on multi Agent system and its application research, East China University Of Science and Technology.

Tang, Jiafu. Wang, Dingwei. 2004. Understanding of Fuzzy Optimization: Theories and Methods. *Journal of System Science and Complexity.* 17(1):117~136.

Yao, Xulong. Hu, Nailian. Zhou, Lihui. Li, Yong. 2011. Ore matching of underground mine based on Immune clonal selection optimization algorithm. *Journal of Beijing university of science and technology* 05:526–531.

Zhang, Sen. Chen, Kaiyan. Guo, Yipeng. Zhang, Bao. 2012. Application of Fuzzy Analytic Hierarchy Process in rujigou mine ventilation system evaluation. *Coal Mine Safety 01*:183–186.

Manufacturing Engineering and Intelligent Materials – Lu & Abu Bakar (Eds)
© 2015 Taylor & Francis Group, London, ISBN 978-1-138-02832-6

Efficient design approach for multimedia classroom control system applying PLC technique

Y.C. Chen, W.G. Li, Q.L. Yang & L.L. Huang
Faculty of Mechanical and Electrical Engineering, Kunming University of Science and Technology,
Kunming, Yunnan, P.R. China

ABSTRACT: According to the basic control requirement of multimedia classroom, this paper presents an efficient design approach applying PLC technique. The control objects of multimedia classroom system include the projector, the screen, the light, the door, and the curtain. The detailed control approaches and the control ladder diagrams are designed. The control principles and functions of all control objects are also analyzed in detail, and the designs of the control programs are mainly analyzed. The proposed design approach is validated by our simulated experiment results.

1 INTRODUCTION

The traditional multimedia classroom relies heavily on the operation staff, which also helps solving any incident during the class time for the teaching staffs. Multimedia classroom staffs often help teachers to turn on the control system before class and to turn off the control system after class, they also help to deal with the emergency which takes place during the class time, and finally, they need to lock the door while all the classes are over. These works increase the load of classroom staff, and reduce the work efficiency. With more and more application of multimedia platform in classroom, many problems have also been appeared. In this paper, a design approach for control system of multimedia classroom is presented to solve above problems and the proposed approach design the control system of multimedia classroom by using the PLC (programmable Logic Controller) technique and the configuration software, which will reduce many complicated workloads for classroom staffs.

At present, many researchers have proposed a lots of control approaches about multimedia classroom control system applying PLC technique. Li etc. presented the design of intelligent lighting control system for classroom (Li et al. 2007), the control system design of classroom light based on MCU is proposed by Li and Zhao (Li & Zhao 2004), design of step motor control system based on PLC was proposed by Li and Li (Li & Li 2008), experimental system of controlling stepping motor based on PLC was proposed by He etc. (He et al. 2008).

2 DESIGN OF CONTROL LADDER DIAGRAM FOR MULTIMEDIA CLASSROOM

2.1 *Control function of multimedia classroom*

The main control functions of multimedia classroom include: (1) The power-on and the power-off; (2) The running and the stopping of the projector; (3) The falling and the rising of the screen; (4) The running and the stopping of the sound; (5) The opening and the closing of door; (6) The running and the stopping of the lamp; (7) The opening and the closing of curtain.

2.2 *Design of control ladder for multimedia classroom*

1. Design of power control
   The power of multimedia classroom is directly connected with standard AC power, in this paper, we use the "SET" command to keep to be connected to power, and to protect the multimedia classroom equipment. In some emergency situations, we can use this command to shut off or turn on the power at any time.

2. Design of projector control
   Projector control rules: the projector is turned on while the class is on, and is turned off after class. A normally-opened contact "M1" is used to turn on the projector, and a normally-closed contact "M2" is used to turn off the projector at any time.

3. Design of screen control
   The falling of screen involves the control of forward rotation and reverse rotation of motor, position control of screen, time-required control. Screen position control uses the timer "T250" and "DO" data transmission control, the falling of screen requires 32 seconds. The descriptions of the procedures are as follows: (1) The motor is controlled by a normally-closed timer contact "T250", the motor stops rotating automatically after 32 seconds and the screen stop falling. (2) Screen position is controlled by a "DO" timer, it beams "T250" data to "DO", and then the screen is falling down. (3) Selection of the falling time of screen. Through the study of multimedia classroom, and considers of the screen standard specification, the "32 seconds" is regard as the required time in this design of screen control. (4) Selection of normally-closed contact.

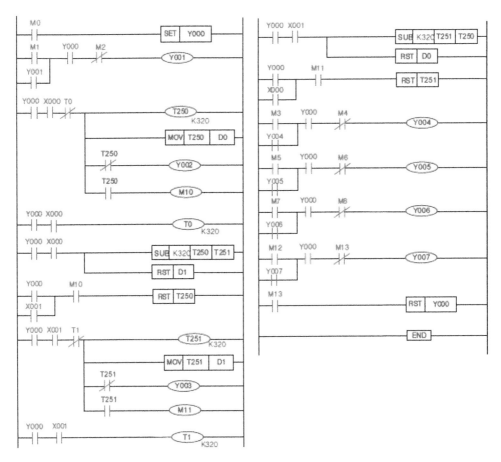

Figure 1.   Ladder diagram of control system for multimedia classroom.

Because the timer "T250" will begin a new timing after 32 seconds, we add a non-cumulative timer "T0" to guarantee the power supply, the timer "T250" will be disconnected after 32 seconds and can not continue timing. The control of screen rising is basically the same as the screen falling. The overall control of the screen requires the falling and the rising of screen do not affect each other, and they can interact their position with each other. (Wang 2004).

4. Design of sound control

Excellent audio effect of multimedia classroom can bring active atmosphere to the teaching, and improve the quality of teaching. Modern multimedia classrooms require multimedia devices have the advantages of simple operation, ensuring teachers to focus on teaching more time, and making the classroom sound be controlled by PLC. The ideas of sound control are as follows: "M3" turn on the sound in classroom, and "M4" turn off the sound in classroom.

5. Design of door control

We can control the opening and closing of the door on the Kingview interface, which is convenient for the teachers and students who are late since some other reasons, the control principle is the same as above control ideas, "M5" controls the opening of the door, and "M6" controls the closing of the door.

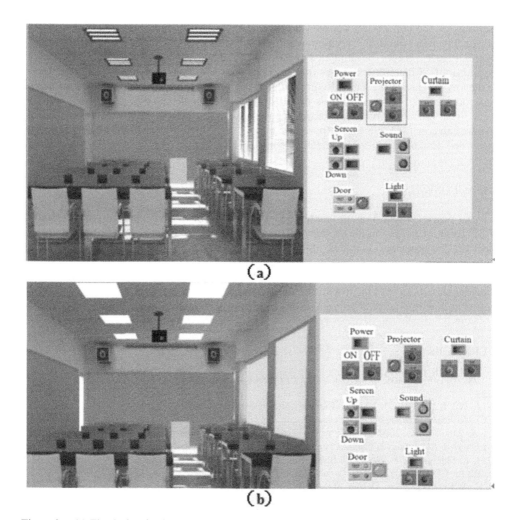

Figure 2.   (a) The devices in classroom are turned off. (b) The devices in classroom are turned on.

6. Design of lamp control

We need to turn on and turn off the lamps in the classroom by the requirement of display of multimedia projector, and turn off the lamps in order to ensure the projection quality while the projector is working; We need to turn on the classroom lamps in order to see the words on the blackboard clearly while the teachers need to lecture on the blackboard. So the convenience of the classroom lamp control is essential in multimedia teaching system. The control ideas are as follows: "M7" controls the lamp to be turned on; "M8" controls the lamp to be turned off.

7. Design of curtain control

Effect of the opening of curtains is similar to the opening of lamps, and is one of the factors that affecting the quality of the multimedia classroom teaching. The control ideas are as follows: "M12" controls the closing of curtain; "M13" controls the opening of curtain.

Concludes all the contents above mentioned, the ladder diagram of control system for multimedia classroom is shown in Figure 1.

3 EXPERIMENT

In order to validate the correctness and stability of the design by the application of PLC, the settings of the connection between the PLC and computer need to be firstly configured, which includes baud rate, communication mode, data bit, stop bit and communication ports etc.. Figure 2 shows the simulated experiment results of control for sound, lamps, curtains, and doors. Figure 2(a) denotes these devices are turned off, and Figure 2(b) denotes these devices are turned on. From Figure 2(b), we can obtain the lamp is turned on (the top of classroom is bright), the curtain is pulled down (the right of classroom), and the door is opened (the northwest of classroom is opened); According to the simulated experiment results, we can conclude that the proposed design approach is implemented successfully, and the proposed design approach is completely correct. (Han 2011).

4 CONCLUSIONS

This paper proposed a design approach of control system for multimedia classroom applying PLC technique. The simulated experiment results show the proposed approach is able to meet the requirements of system design, and the proposed approach is convenient for teachers operating the control system. Through the communication between the monitoring computer and automatic control unit in PLC, we can realize the control and management of multimedia classroom system. It will save the workload of staff, and will reduce the cost of operation and maintenance. I believe this system will be widely used in the future.

REFERENCES

Han, X.X. 2011. *From basic to the practice of PLC and Kingview*, Beijing: China Mechanical Industry Press.
He, Y. & Wu, S.J. & Guo, X.Y. 2008. Experimental System of Controlling Stepping Motor Based on PLC. *Mechanical Management and development in Chinese*. 23(5): 72–74.
Li, L. & Zhao, G.K. 2004. The Control System Design of Classroom Light Based on MCU. *Machinery & Electronics in Chinese*. (6): 69–70.
Li, Q.H. & Shang, J.L. & Zhang, G.B. & Yang, H.X. 2007. Design of Intelligent Lighting Control System for Classroom. *Computer Measurement & Control in Chinese*. 15(8): 1011–1013.
Li, W.H. & Li, G.Q. 2008. Design of Step Motor Control System Based on PLC. *Instrumentation Technology in Chinese*. (9): 23–25.
Wang, T.Y. 2004. *Principle and application of programmable controller*, Beijing: China National Defense Industry Press.

Manufacturing Engineering and Intelligent Materials – Lu & Abu Bakar (Eds)
© *2015 Taylor & Francis Group, London, ISBN 978-1-138-02832-6*

The application of active turbocharging control technology on range-extender of electric vehicle

D. Wang, D.C. Qian & C.X. Song

State Key Laboratory of Automotive Simulation and Control, Jilin University, Changchun, China

ABSTRACT: The active turbocharging control technology was applied to a range extender to improve its efficiency. For the research on control method and energy-saving mechanism of active turbocharging control technology, the traditional and active turbocharger control method were compared by simulation and bench test. The test and simulation results showed that the active control method could realize a wider range of boost pressure ratio and intake pressure, deceased pumping loss at part load and brake specific fuel consumption, which eventually improved the efficiency of range extender.

1   INTRODUCTION

Extended-range electric vehicle is an ideal transition solution from hybrid vehicles to pure electric vehicles (Joao et al. 2012). The configuration of extended-range electric vehicle includes large capacity battery and range extender composed of IC engine and generator, which is generally complex, and the curb weight is generally large. Therefore, the lightweight of range extender, especially the downsizing design of IC engine is particularly important to extended-range electric vehicle (Michael et al. 2011).

Recently, turbocharging technology is the major technological solution of IC engine downsizing, which directly improves the power and emission performance, and reduces fuel consumption. For a turbocharging system except for Variable Geometry Turbochargers (VGT), the boost pressure is controlled by waste gate Jay (2008). In recent years there has been active turbocharging control system, which uses electric or vacuum actuator to control wastegate. Exhaust bypass valve opening degree is no longer a linear correlation with the intake pressure, thereby allowing boost pressure ratio is continuously adjustable over a wide range.

In this paper, the active turbocharging control technology was applied to a small displacement gasoline engine in a range extender. The IC engine of range extender requires high efficiency, and its response characteristic is not so important. Thereby, in this paper, the operating point of the range extender was optimized, and its efficiency and energy saving potential were analysed by simulation, with the pressure ratio and throttle position as load regulation means.

2   PRINCIPLES OF TURBOCHARGING CONTROL SYSTEM

For a traditional turbocharging system with fixed geometry, the boost pressure is controlled by wastegate. The wastegate is a valve that opens the turbine housing to relieve excess exhaust backpressure, triggered by a boost pressure actuator which is activated by boost pressure fed from the compressor. Because the preload of boost pressure actuator is fixed, the maximum boost pressure is also a fixed value. Because of actuator damping, control system hysteresis and the length of intake pipe, the action of wastegate lags behind exhaust pressure fluctuations, and there is a phase difference between the fluctuations of intake manifold pressure and exhaust pressure Corky (1997). The reasons above lead to the fluctuation of a turbocharged engine.

In recent years, there has been some application of active turbocharging closed-loop control system, which is based on electric or vacuum actuator and shown as Figure 1. The obvious

Figure 1. Turbocharger with active control system.

feature of active turbocharging system is that an independent power source is added in the traditional structure to drive the wastegate actuator instead of the boost intake air. So the wastegate actuator and intake pressure is no longer physically associated, and a greater range of target boost pressure can be achieved. In the control strategy, the engine speed, throttle opening degree and some other parameters are collected by ECU to look-up table to obtain the target pressure ratio, and compare it with the actual boost pressure. Then ECU determines the vacuum degree of vacuum actuator to control the opening degree of turbine bypass valve. When the actual boost pressure is higher than target boost pressure, the vacuum degree of vacuum actuator is increased to increase the opening degree of bypass valve allowing more exhaust to bypass turbine. And when less than target, the vacuum degree is reduced to decrease the valve opening degree, so that more exhaust go through turbine and the boost pressure is higher. That is the way to realize closed-loop control of boost pressure.

3 MATHEMATICAL MODELS OF TRADITIONAL AND ACTIVE TURBOCHARGING SYSTEM

3.1 *Traditional control methods*

Traditional wastegate mechanism is composed of diaphragm cylinder, return spring, push rod, valve and other components, which is a typical force-balance pneumatic proportional controller Wenjia (2013). The structure of controller is shown as Figure 2. The lower chamber of diaphragm cylinder and pipeline after compressor are connected with a pipe. The absolute pressure after compressor acts on the membrane surface. A return spring with stiffness coefficient k is installed in the upper chamber, and the upper chamber is connected with atmosphere. Diaphragm is connected to valve with a push rod. The pressure inside valve is equal to absolute pressure before turbine. The pressure outside valve is equal to exhaust pressure after turbine, approximately equal to the atmospheric pressure. The force balance equation is:

$$\begin{cases} \left(p_{in} - p_{atm}\right)A_0 + \left(p_{ex} - p_{atm}\right)A_1 = m\ddot{x} + b\dot{x} + k\left(x + x_0\right), \\ \qquad\qquad \left(\left(p_{in} - p_{atm}\right)A_0 + \left(p_{ex} - p_{atm}\right)A_1 > kx_0\right) \\ x = 0, \\ \qquad\qquad \left(\left(p_{in} - p_{atm}\right)A_0 + \left(p_{ex} - p_{atm}\right)A_1 \le kx_0\right) \end{cases} \tag{1}$$

where, x is valve displacement, the k is stiffness coefficient of return spring, the p_{in} is the absolute pressure after compressor, the p_{atm} is atmospheric pressure, the A_0 is diaphragm area, the A_1 is valve surface area, the p_{ex} is absolute pressure before turbine, the m is quality of motion body, the b is viscous friction coefficient. The kx_0 is the preload of valve.

According to Eq.1, the traditional control mechanism is a normally closed agency, and the valve is closed by the preload. The range of boost pressure is determined by the stiffness

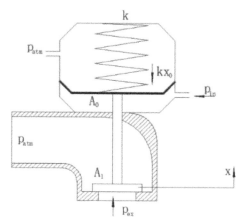

Figure 2. The structure of a traditional controlled wastegate.

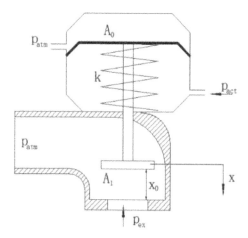

Figure 3. The structure of a active controlled wastegate.

coefficient of return spring k and its preload kx_0, and is limited by the absolute pressure after compressor p_{in} and absolute pressure before turbine p_{ex}.

3.2 *Active control method*

Active turbocharging control device is composed of vacuum source, proportional valve, diaphragm cylinder, return spring, push rod, valve and other components, which is an electrically controlled servo vacuum actuator. The structure is shown as Figure 3. Vacuum source provides vacuum power, which after the adjustment of the proportional valve forms a certain degree of vacuum to act in lower chamber. A return spring with stiffness coefficient k is installed in the upper chamber, and the upper chamber is also connected with atmosphere. Diaphragm is connected to valve with a push rod. The pressure inside valve is equal to absolute pressure before turbine, the pressure outside valve is equal to exhaust pressure after turbine, approximately equal to the atmospheric pressure. The force balance equation of the control device is (Ogata 2013):

$$\begin{cases} \left(p_{atm} - p_{act}\right)A_0 = -m\ddot{x} - b\dot{x} - k(x - x_0) + \left(p_{ex} - p_{atm}\right)A_1 \\ x \leq x_0 \end{cases} \qquad (2)$$

where, x is valve displacement, the k is stiffness coefficient of return spring, the p_{act} is the absolute pressure in lower chamber, the p_{atm} is atmospheric pressure, the A_0 is diaphragm area, the A_1 is valve surface area, p_{ex} is absolute pressure before turbine, the m is quality of motion body, the b is viscous friction coefficient, x_0 is the maximum valve opening position, kx_0 is preload of valve.

According to Eq. 2, Active control mechanism is a normally open agency, and the valve is closed when the vacuum power is larger than preload of return spring kx_0. Because the absolute pressure in lower chamber p_{act} can be adjusted by proportional valve, the range of boost pressure is expanded greatly, and is no longer dependent on the return spring stiffness k and return spring preload kx_0.

3.3 Simulation model

The target range extender in this paper is composed by internal combustion engine, generator, shafting and other parts, the basic performance parameters of each part is shown in Table 1. The CFD software GT-suite and Matlab/Simulink were used to build the simulation model of traditional and active control system. In GT-suite environment, the model of gasoline

Table 1. Parameters of the range-extender.

Engine	
Displacement (mL)	599
Cylinder	4 Inline
Bore X stroke (mm)	67×42.5
Peak power (kW)	70
Turbocharger	
Manufacturer	Hitachi
Type	HT07-4 A
A/R	In 0.35
	Ex 0.45
Trim	In 54
	Ex 65
Generator	
Peak power (kW)	60
Continuous power (kW)	90
Voltage (V)	300

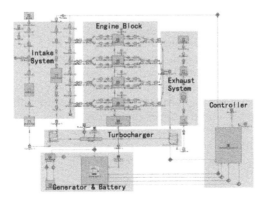

Figure 4. Part of the simulation model.

engine, turbocharger, shaft and generator was established. In Matlab/Simulink environment, the mathematical model of turbo control systems was established. The two softwares were connected by "harness". The simulation model is shown as Figure 4.

4 TESTING AND SIMULATION

The test bench of the IC engine used on the target range extender with active control system is shown as Figure 5. As the bench test results, the boost pressure ranges of traditional and active control methods at the widely-opened-throttle condition is shown as Figure 6. The active control method can realize a wider range of boost pressure ratio and intake pressure.

In the analysis of brake specific fuel consumption and range extender efficiency at different engine speed, the tradition and active control methods were compared at two target generation power of 15 kw and 30 kw. The brake specific fuel consumption of the IC engine is shown as Figures 7 and 8. The efficiency of range extender is shown as Figures 9 and 10. The calculation results show that the minimum brake specific fuel consumption is reduced by 5% to 10%, and the range extender efficiency is increased by one to two in percentage.

For further study of energy saving mechanism of active control method, the PMEP of IC engine at three conditions of 15 kW and 30 kW were simulated. The calculation result is shown as Figure 11 and Figure 12. It is proved that the pumping loss at part load is decreased using active control method, which is the main reason of the improvement of engine efficiency.

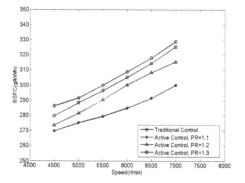

Figure 5. Photograph of engine benchmark testing.

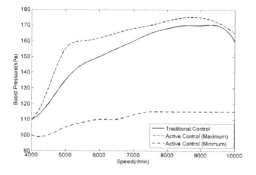

Figure 6. Boost pressure curves under traditional and active control at WOT.

Figure 7. BSFC curves at 15 kW generating power (PR, boost pressure ratio).

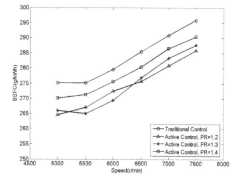

Figure 8. BSFC curves at 30 kW generating power.

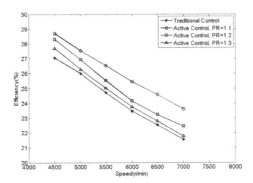

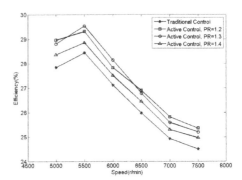

Figure 9. Range-extender efficiency curves at 15 kW generating power.

Figure 10. Range-extender efficiency curves at 30 kW generating power.

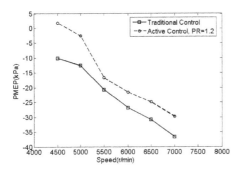

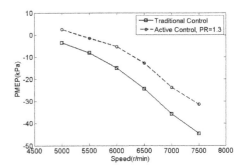

Figure 11. PMEP curves at 15 kW generating power.

Figure 12. PMEP curves at 30 kW generating power.

5 CONCLUSIONS

According to range extender's actual needs of the internal combustion engine, the active turbocharger control method was applied in the design of range extender. The adjustment range of boost pressure was studied through bench test, with the boost pressure ratio and throttle opening degree as load regulation means. The operating point of the range extender was optimized by numerical Computation and Simulation. The test and simulation results showed that when using active control technology, the minimum brake specific fuel consumption was reduced by 5% to 10%, and the range extender efficiency is increased by 1% to 2%. And the active control method could realize a wider range of boost pressure ratio and intake pressure, deceased pumping loss at part load and brake specific fuel consumption, which eventually improved the efficiency of range extender.

REFERENCES

Corky, B. 1997. *Maximum Boost: Designing, Testing and Installing Turbocharger Systems*. Cambridge: Bentley (Robert) Inc.

Jay, M. 2008. *Turbo: Real World High-Performance Turbocharger Systems*. North Branch: CarTech Inc.

Joao, R. et al. 2012. Analysis of Four-Stroke, Wankel, and Micro-Turbine Based Range Extenders for Electric Vehicles. *Energy Conversion and Management* 2012(58): 120–133.

Michael, A.M. et al. 2011. The GM "Voltec" 4ET50 Multi-Mode Electric Transaxle. *SAE International Journal of Engines* 2011(4):1102–1114.

Ogata, K. 2013. *Modern Control Engineering*. Beijing: Publishing House of Electronics Industry.

Wenjia, L. 2013. *Modelling and Control of Turbocharged Gasoline Engine Intake System*. Jilin University.

Manufacturing Engineering and Intelligent Materials – Lu & Abu Bakar (Eds)
© *2015 Taylor & Francis Group, London, ISBN 978-1-138-02832-6*

In-wheel-motor vehicle ABS control strategy based on regenerative braking torque adjustment

S.X. Song, W.C. Zhao & Q.N. Wang
State Key Laboratory of Automotive Simulation and Control, Jilin University, China

ABSTRACT: In this paper, the author analyzes the Electro-mechanical Braking system of In-wheel-motor vehicle. A new ABS control strategy based on In-wheel-motor regenerative braking torque adjustment is proposed, and a vehicle speed estimate method is built. The simulation and analysis for the proposed control strategy are done in the AMESim Simulink co-simulation platform. The result shows that the proposed control strategy is suitable for In-wheel-motor vehicle. It can ensure the braking stability of EV in a wide range of road adhesion coefficient.

1 INTRODUCTION

With the advent of 21st century, the increasing depletion and the higher price of oil resources have become a topic of concern worldwide (Gao et al. 1999). At the same time, the rapid development of transportation also leads to serious environmental pollution day after day. The data released by Chinese Research Institute shows that the vehicle emissions of carbon monoxide accounted for more than 80% with regard to air pollution in Beijing. Therefore, it is imminent to solve the problem of environmental pollution. In this situation, the electric car has attracted more attention in recent years.

Compared with the fossil fuel vehicles, not only does the electric vehicle have lower cost, but also has the advantage of zero emissions. One of the main technical features of electric vehicle is that its driving motor can provide braking torque when the vehicle is braking, so as to realize the function of braking energy recovery. The collected braking energy can be converted into electrical energy, then stored in the energy storage system which contributes to extend the driving range as well as reduce the energy consumption (EHSANI et al. 1999).

In-wheel-motor vehicle has more technical advantages than other forms of electric vehicles. The In-wheel-motor vehicle can distribute the torque of four wheels independently and apply independently regenerative braking torque control. With the enhancing ability of the In-wheel-motor, it becomes possible to achieve the anti-lock brake control by adjusting the braking torque of the In-wheel-motor. In this paper, the author focused on the ABS control of the In-wheel-motors and a new ABS control strategy based on adjusting motor regenerative braking torque is proposed.

2 VEHICLE SPEED ESTIMATE

Vehicle speed is an important reference variable adopted by numerous vehicle stability control system. Therefore, the vehicle state estimation problem is of great practical and theoretical application value. Research institutes around the world have been committed to the research on reference speed estimation technology. In recent years, with the popularity of high-precision micro vehicle acceleration sensor, there are more technical means to carry on vehicle speed estimation. Integrating four wheel speed signals with vehicle body acceleration sensor signal for the vehicle speed estimation becomes a trend of development. In signal processing and observer built respect, scholars from various countries commonly used Kalman filter algorithm, Rhomberg observer, robust observer, sliding mode observer, fuzzy observer, and so on, which has obtained a series of research findings (Yu & Gao 2009). For the In-wheel-motor used in EV, the wheel rotary speed signal can be obtained by the resolver sensor. The sampling frequency and the signal quality of the resolver are much higher than traditional car wheel speed sensors, and the accuracy of the signal is much higher than traditional hall sensors. Taking into account this feature, the Kalman filter is used to build velocity observer so as to estimate the slip ratio of each wheel of In-wheel-motor vehicle. The system state space model of vehicle speed estimation is shown as follows (Guo et al. 2011).

$$x_k = Ax_{k-1} + w_k \tag{1}$$

$$z_k = H_k x_k + v_k \tag{2}$$

where, x_k, x_{k-1} is the state vector at time of k, $k-1$, A is the matrix of system state, w_k is the process noise, z_k is the observation, H_k is the matrix of observation model, v_k is the observation noise;

The mean value of process noise in the speed estimation model can be presented as $E(w_k) = e_w$, the covariance is Q_k. The mean value of observation noise in the speed estimation model can be presented as $E(V_k) = e_v$, the covariance is R_k. The state vector can be presented as $x_k = [u_k \ a_k]^T$, where a_k is the acceleration, u_k is the vehicle velocity.

The system state transition matrix is shown as follows:

$$A = \begin{bmatrix} 1 & \Box t \\ 0 & 1 \end{bmatrix}$$

System noise vector:

$$w_k = \begin{bmatrix} 0 \\ w_k \end{bmatrix}$$

Wheel speed measuring vector:

$$H_k = [1 \quad 0]$$

The initial value of physical quantities is shown as follow:

$$\begin{cases} x_0 = [u_k \quad 0]^T \\ p_0 = \begin{bmatrix} 1 & 0 \\ 0 & 1 \end{bmatrix} \\ Q_0 = \begin{bmatrix} 0 & 0 \\ 0 & W_k^2 \end{bmatrix} \\ e_v = 0 \\ R_0 = 0 \end{cases}$$

Calculation steps of vehicle speed estimation based on adaptive Kalman filter algorithm are expressed as follows:

Predicted state estimate:

$$\hat{x}_{k|k-1} = A\hat{x}_{k-1|k-1} \tag{3}$$

Predicted estimate covariance:

$$P_{k|k-1} = AP_{k-1|k-1}A^T + Q_k \tag{4}$$

314

Measurement residual:

$$\tilde{y}_k = z_k - H_k\,\hat{x}_{k|k-1} \tag{5}$$

Innovation covariance:

$$S_k = H_k P_{k|k-1} H_k^T + R_k \tag{6}$$

Optimal Kalman gain:

$$K_k = P_{k|k-1} H_k^T S_k^{-1} \tag{7}$$

Updated state estimate:

$$\hat{x}_{k|k} = \hat{x}_{k|k-1} + K_k \tilde{y}_k \tag{8}$$

Updated estimate covariance:

$$P_{k|k} = (I - K_k H_k) P_{k|k-1} \tag{9}$$

In the updated estimate, the filter optimizing the predicted value in the predicted process by using the observation value of the measurement process, in order to make the estimate of the value more accurate.

3 ABS CONTROL STRATEGY OF IN-WHEEL-MOTOR VEHICLE

In-wheel-motor-vehicle has more technical advantages than other forms of electric vehicles. It can distribute the torque of four wheels independently and apply regenerative braking torque control.

In the In-wheel-motor vehicle structure, there are two braking components, which are regenerative braking system and the hydraulic braking system, consisting of an Electro-mechanical Braking system. In-wheel-motor-vehicle braking system control strategy would be more complicated and achieve better performance (Zhao et al. 2013). ABS in an In-wheel-

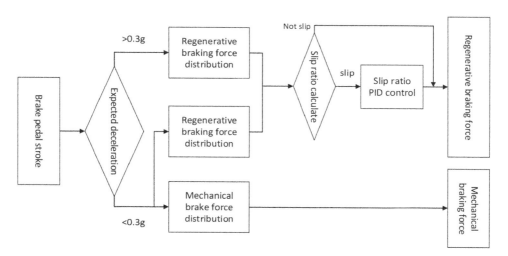

Figure 1. ABS control strategy of the in-wheel-motor vehicle.

motor-vehicle can be achieved by only adjusting braking torques of In-wheel-motors, which has precise torque control ability, fast response, high precision adjustment, but need to consider the coordinate control problems with the hydraulic braking system. The ABS control strategy of the In-wheel-motor vehicle is shown in the diagram (Fig. 1).

The ABS control process is expressed as follows: when the driver brakes, firstly, the system runs according to a conventional case, at the same time the system estimates the slip ratio of each wheel; when the slip ratio is higher than 0.2, the PID controller is enabled to adjust the slip ratio (Lin et al. 2013), then outputs the regenerative brake torque signal to keep the vehicle stable.

4 SIMULATION AND RESULTS

In this paper, AMESim platform is used to build In-wheel-motor vehicle, driving environment and driver model, control strategy of electro-mechanical braking system is built in Simulink, and use AMESim Simulink co-simulation analysis, respectively, in the high-adhesion road (adhesion coefficient is 0.8) and low adhesion road (adhesion coefficient is 0.2). Specific simulation parameters are shown in Table 1.

Simulation results are shown in Figure 2 to Figure 5, sub-graph shows the mechanical braking and regenerative braking force distribution curve of the wheels, sub-graph b is the comparison curve between estimating sliding ratio of the wheel and the actual sliding ratio. Figures 3 and 4 indicate on the left front wheel and right rear wheel simulation results of the test vehicle in the initial speed of 60 km/h when braking on road with 0.2 adhesion coefficient. Figure 5 and Figure 6 shows the left front wheel and right rear wheel simulation results of the test vehicle in the initial speed of 60 km/h when braking on 0.8 road adhesion coefficient road.

On the low adhesion coefficient road, braking at low intensity, the wheel locking tendency occurred, so the mechanical brake system was not operated in the braking condition.

Table 1. Simulation parameter.

Complete vehicle shipping mass (kg)	1400
Wheelbase (mm)	2700
Track (mm)	1486
Distance from Cg to front shaft (mm)	1350
Height of center of mass (mm)	375
Wheel radius (m)	0.315
Peak power of the motor (kw)	80
Peak torque of the motor (Nm)	800
Battery character	320V 80 Ah

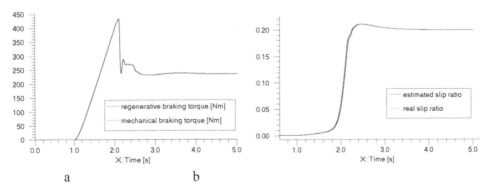

a   b

Figure 2. Braking torque and slip ratio of LF wheel ($\varphi = 0.2$).

316

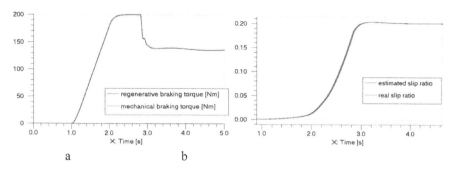

a   b

Figure 3.   Braking torque and slip ratio of RR wheel ($\varphi = 0.2$).

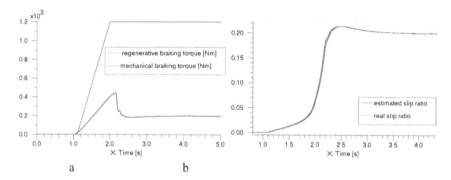

a   b

Figure 4.   Braking torque and slip ratio of LF wheel ($\varphi = 0.8$).

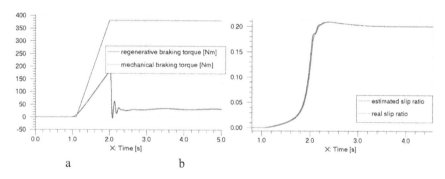

a   b

Figure 5.   Braking torque and slip ratio of RR wheel ($\varphi = 0.8$).

The whole braking process is completed only by using the regenerative braking system. When the wheel slip trend occurs, ABS controls the size of the regenerative braking force by PID controller to keep the wheel slip ratio in a reasonable range. On high adhesion coefficient road, the vehicle performed a high intensity braking, and the mechanical braking system and regenerative braking system operated together. When wheel lock trend appears, ABS system maintained the wheel slip ratio in a reasonable range, only by controlling the size of the regenerative braking torque.

5 CONCLUSION

In this paper, the ABS control system of In-wheel-motor vehicle has been studied, and a new control strategy based on motor torque adjustment is proposed. In the AMESim vehicle system, a dynamic model is established, and the simulation analysis is done. The results show:

1. The resolver sensor on the In-wheel-motor combining with Kalman filter algorithm can achieve an ideal reference speed estimation;
2. ABS control can be achieved by only adjusting the braking torque of In-wheel-motors;
3. By adjusting the motor torque, the ABS function can be achieved in a wide range of road adhesion coefficient when the vehicle brakes. During the whole control process, braking force changes more smoothly with better comfortable feelings.

REFERENCES

Ehsani, M., Gao, Y. & Butler, K.L. 1999. Application of Electrically Peaking Hybrid (ELPH) Propulsion System to a Full-size. *Vehicular Technology* 48(6): 1779–1787.

Gao, Y., Chen, L. & EHSANI, M. 1999. Investigation of the effectiveness of regenerative braking for EV and HEV. *SAE* 1999–01: 2910–2915.

Guo, H., Wu, Z., Xin, B. & Chen, H. 2011. Vehicle Velocities Estimation Based on mixed EKF. *2011 Chinese Control and Decision Conference (CCDC)* 2030–2035.

Lin, H., Song, C. & Zheng, Z. 2013. Simulation and Parametric Analysis of ABS for EV with Electric Brake. *Advanced Materials Research* 655: 1469–1473.

Yu, Z. & Gao, X. 2009. Review of Vehicle State Estimation Problem under Driving Situation. *Journal of Mechanical Engineering* 45(5): 20–33.

Zhao, W., Gu, X. & Wang, C. 2013. Optimization Design of Electro-hydraulic Brake System with In-wheel Electric Vehicle. *Journal of Nanjing University of Aeronautics & Astronautics* 45(6): 871–874.

Manufacturing Engineering and Intelligent Materials – Lu & Abu Bakar (Eds)
© 2015 Taylor & Francis Group, London, ISBN 978-1-138-02832-6

Research on control strategy of In-wheel-motor vehicle electro-mechanical braking system

S.X. Song, S.Q. Fan & Q.N. Wang
State Key Laboratory of Automotive Simulation and Control, Jilin University, China

ABSTRACT: Based on the analysis of the braking process of In-wheel-motor vehicle, an electro-mechanical hybrid braking system control strategy is proposed for In-wheel-motor vehicle, and a solution for the realization of this control strategy is designed. A simulation model is established in AMESim and Simulink platform, and the simulation and analysis are conducted under NEDC circulations. The simulation results show that the control strategy proposed has an advantage in terms of energy recovery, and the control strategy makes the actual braking force distribution of the In-wheel-motor vehicle close to the ideal braking force distribution curve (I-curve), which improves the brake stability of In-wheel-motor vehicle substantially. The solution proposed is an ideal electro-mechanical hybrid braking system solution of In-wheel-motor vehicle.

1 INTRODUCTION

With the development of automotive industry, the increasing air pollution and oil shortage brought by the traditional internal combustion engine attract more and more attention (Gao et al. 1999). The appearance of Electric Vehicle (EV) can solve the double problems of energy conservation and environmental protection effectively, which is considered to be an ideal substitute of conventional vehicle (EHSANI et al. 1999). An EV has a lot of advantages such as sustainable energy sources, zero emissions and low noises. Meanwhile, it can achieve the function of braking energy recovery through its own driving motor.

Braking energy recovery means part of the mechanical energy can be converted into electrical energy and stored in the energy storage system such as power battery or super capacitor through the motor generating when the vehicle is braking and the energy can be used while driving, in order to extend the driving range of EV, and reduce the workload of brake and extend the life of mechanical braking system at the same time.

The electro-mechanical hybrid braking system of EV is composited by the regenerative braking system and the traditional hydraulic braking system. In the implementation of the many advantages above, the complexity of the system is increased and higher requirements are put forward to the control strategy of the system. In this paper, the electro-mechanical hybrid braking system of In-wheel-motor vehicle is analyzed, and a new braking force distribution strategy and implementation method of the strategy are proposed. Also, the strategy is simulated and analyzed.

2 VEHICLE BRAKING FORCE ANALYSIS

The vehicle is regarded as a rigid body, when the car is braking along the standard road surface, the force situation is shown as follows.

In Figure 1, L is the wheelbase, a is the distance between the mass center of the vehicle to the center line of the front axle, b is the distance between the mass center of the vehicle to the

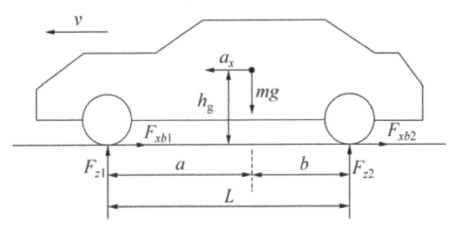

Figure 1. Vehicle braking force.

center line of the rear axle, $L = a + b$. Hg is the height of the vehicle mass center. F_{xb1}, F_{xb2} are the total braking force of the front and rear wheels, for an EV, there are (Lian et al. 2013)

$$\begin{cases} F_{xb1} = F_{xbr1} + _{Fxbm1} \\ F_{xb2} = F_{xbr2} + F_{xbm2} \end{cases} \tag{1}$$

In formula (1): F_{xbr1}, F_{xbr2} are the regenerative braking force of the front and rear axle, respectively, F_{xbm1}, F_{xbm2} are the mechanical braking force of the front and rear axle, respectively. As depicted in Figure 1, take moments of the ground contact point of the front and the rear wheels, respectively, then get the formula below.

$$\begin{cases} F_{Z1}L = mgb + m\dfrac{du}{dt}h_g \\ F_{Z2}L = mga - m\dfrac{du}{dt}h_g \end{cases} \tag{2}$$

In formula (2), F_{z1} is the ground normal reaction force to the front wheels; F_{z2} is the ground normal reaction force to the rear wheels; mg is the vehicle gravity; (du/dt) is the vehicle acceleration.

Set $(du/dt = zg)$, z is called braking intensity, it can be obtained as below.

$$\begin{cases} F_{z1} = mg\left(b + zh_g\right)/L \\ F_{z2} = mg\left(a - zh_g\right)/L \end{cases} \tag{3}$$

It can be shown after the derivation

$$F_{xb2} = \frac{1}{2}\left[\frac{G}{h_g}\sqrt{b^2 + \frac{4h_g L}{mg}F_{xb1}} - \left(\frac{mgb}{h_g} + 2F_{xb1} \right) \right] \tag{4}$$

Formula (4) is the ideal braking force distribution curve of front and rear wheels (I-curve).

3 THE BRAKING FORCE DISTRIBUTION CONTROL STRATEGY

There are some electro-mechanical hybrid braking force distribution control strategy, such as the parallel braking force distribution, the hybrid parallel braking force distribution and the ideal braking force distribution. In traditional EV constructions, since most of them are single axle driven vehicles, regenerative braking torques can't be applied to the front and rear axles at the same time; therefore, the ideal braking force distribution will not be realized (Wang et al. 2012). The In-wheel-motor vehicle has the characteristic of four-wheel independent drive, which provides the basis to achieve the ideal braking distribution strategy. According to the technical features of In-wheel-motor vehicle, the established braking force distribution strategy of electro-mechanical hybrid braking system is described as follows.

The driver-expected deceleration value can be calculated based on the opening travel of the brake pedal applied by the driver. If the driver-expect deceleration is less than 0.3 g, which is judged as low intensity of braking, the braking force will be provided by the In-wheel motors singly, in order to achieve the maximization of the regenerative braking energy recovery, the distribution method is as follows.

$$F_{xbe2} = \frac{1}{2}\left[\frac{G}{h_g}\sqrt{b^2 + \frac{4h_gL}{G}F_{xbe1}} - \left(\frac{Gb}{h_g} + 2F_{xbe1}\right)\right] \tag{5}$$

In the formula (5), F_{xbe1}, F_{xbe2} are the regenerative braking force of the front and rear wheels provided by the In-wheel-motors; G is the vehicle gravity; h_g is the height of the vehicle mass center from the ground; b is the distance between the mass center of the vehicle to the center line of the rear axle; L is the wheelbase of the vehicle.

When the driver-expect deceleration is more than 0.3 g, the total braking torque of the front and rear axles can be calculated according to I-curve and whether the regenerative braking torque can satisfy the demand of brake or not will be judged. Considering the protection of the In-wheel-motor system and the power battery system, the maximum braking torque of the motor and the maximum charging power of the battery are set as the constraint quantity of regenerative braking torque. The distribution strategy of braking torque under the middle and high intensity of braking is as follows.

$$\begin{cases} F_{xbe\_pre1} = \min\left(F_{xbe\_cal1}, \dfrac{T_{\max}}{r}\right) \\ F_{xbe\_pre2} = \min\left(F_{xbe\_cal2}, \dfrac{T_{\max}}{r}\right) \end{cases} \tag{6}$$

$$\begin{cases} F_{xbe1} = F_{xbe\_pre1}, \; (F_{xbe\_pre1} + F_{xbe\_pre2} \leq \dfrac{P_{chg\_max}}{\eta_{chg}v}) \\ F_{xbe2} = F_{xbe\_pre2} \\ F_{xbe1} = \dfrac{F_{xbe\_pre1}}{F_{xbe\_pre1} + F_{xbe\_pre2}}\dfrac{P_{chg\_max}}{\eta_{chg}v} \\ F_{xbe2} = \dfrac{F_{xbe\_pre2}}{F_{xbe\_pre1} + F_{xbe\_pre2}}\dfrac{P_{chg\_max}}{\eta_{chg}v}, \; (F_{xbe1} + F_{xbe1} > \dfrac{P_{chg-max}}{\eta_{chg}v}) \end{cases} \tag{7}$$

$$\begin{cases} F_{xbm1} = F_{xb1} - F_{xbe1} \\ F_{xbm2} = F_{xb2} - F_{xbe2} \end{cases} \tag{8}$$

In the formula (6), F_{xbe_pre1}, F_{xbe_pre2}, are the original-set value of regenerative braking force of the front and rear axle, respectively; F_{xbe_cal1}, F_{xbe_cal2} are the brake force demand of the front and rear axle calculated according to I-curve; $T_{\max}$ is the maximum braking torque which can be provided by the motor drive system in a transient state; r is the wheel rolling radius; P_{chg_max} is the instant charging power limitation of the battery; η_{chg} is the charging efficiency; v is the speed of the vehicle.

This control strategy requires that the brake operating mechanism wouldn't trigger the hydraulic braking system in the initial brake stage, thereby maximizing the braking energy

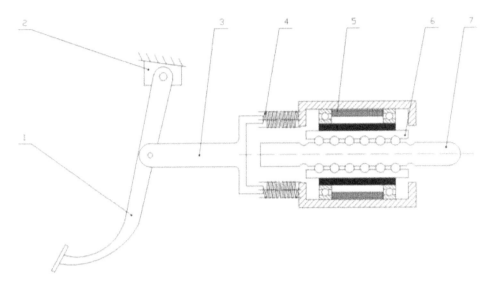

Figure 2. A new type of brake operating mechanism.

Table 1. Simulation parameters.

Total mass (kg)	1400
Wheelbase (mm)	2700
Longitudinal position of front shaft from vehicle center of gravity (mm)	1386
Height of the vehicle center of gravity (mm)	375
Rolling radius of the tire (m)	0.29
Peak power of In-Wheel-motor (kw)	40
Peak power of In-Wheel-motor (Nm)	450
Battery parameters	320 V 80 Ah

recovery under low braking intensity (Zhang et al. 2008). Obviously, traditional automotive brake operating mechanism cannot achieve this feature (Yu et al. 2008). In this paper, a new type of brake operating mechanism is chosen, which has a longish spare travel and the foot feeling can be simulated in the spare travel, the mechanism also has a motor assistance which can adjust the size of the hydraulic braking torque. Its structure is shown in Figure 2.

In Figure 2, component 1 is braking pedal, component 2 is pedal travel sensor, component 3 is pedal pusher, component 4 is return spring, component 5 is assistant electric motor, component 6 is ball screw nut, and component 7 is ball screw.

4 SIMULATION AND ANALYSIS

The model of the In-wheel-motor vehicle, the driving environment and the driver is built in AMESim platform, while the model of the vehicle's electro-mechanical hybrid braking system control strategy is built in Simulink. The co-simulation is conducted under NEDC conditions. The simulation parameters are shown in Table 1.

The simulation results are shown in Figure 3 to Figure 5. As shown in Figure 3, most of the braking conditions in a NEDC condition are low braking intensity, which are consistent with the actual urban driving conditions. Under most of low braking intensity conditions, braking torque is provided by In-wheel-motors and the mechanical brake will participate in the work when largish brake torque is demanded.

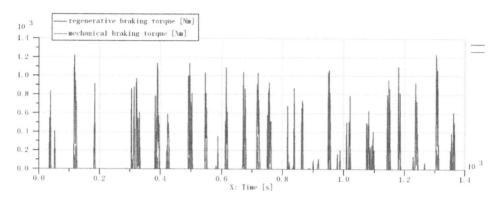

Figure 3. Braking torque distribution of electro-mechanical braking system.

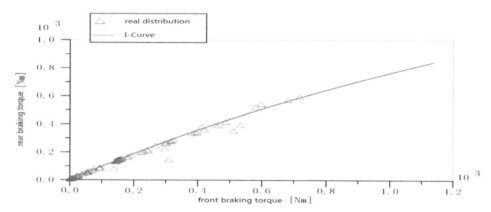

Figure 4. The actual distribution of the front and rear braking torque.

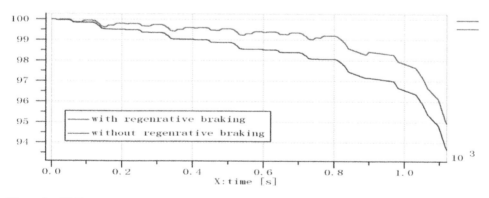

Figure 5. SOC curve.

It is obvious in Figure 4 that under the new braking force distribution strategy, the braking force distribution ratio of the front and rear wheels is close to the ideal braking force distribution curve, even slightly below the curve, which can improve the stability of the EV effectively when the vehicle is braking.

Figure 5 shows the curve of SOC's range with time range during the whole NEDC cycle with the regenerative braking function is opened or closed. The results show that the braking

323

energy recovery effect is obvious in terms of energy saving, an NEDC cycle can save energy 29.5% more than no braking energy recycling.

5 CONCLUSIONS

In this paper, an electro-mechanical hybrid braking system control strategy is proposed based on a new type of brake operating mechanism. The advantage of the control strategy in terms of energy recovery is proved through the simulation of an In-wheel-motor vehicle model carrying on this control strategy. Furthermore, the control strategy makes the actual braking force distribution close to I-curve, which improves the brake stability of EV substantially. This is an ideal electro-mechanical hybrid braking system solution of In-wheel-motor vehicle.

REFERENCES

Ehsani M., Gao, Y. & Butler, K.L. 1999. Application of Electrically Peaking Hybrid (ELPH) Propulsion System to a Full-size. *Vehicular Technology* 48(6): 1779–1787.

Gao, Y., Chen, L. & EHSANI, M. 1999. Investigation of the effectiveness of regenerative braking for EV and HEV. *SAE* 1999–01: 2910–2915.

Lian, Y., Tian, Y., Hu, L. & Yin, C. 2013. A new braking force distribution strategy for electric vehicle based on regenerative braking strength continuity. *J. Cent. South Univ* 20: 3481–3489.

Wang, M., Sun, Z., Zhuo, G. & Cheng, P. 2012. Maximum Braking Energy Recovery of Electric Vehicles and Its Influencing Factors. *Journal of TongJi University (Natural Science)* 40(4): 583–588.

Yu, Z., Feng, Y. & Xiong, L. 2008. Review on Vehicle Dynamics Control of Distributed Drive Electric Vehicle. *Journal of Mechanical Engineering* 49(8): 106–114.

Zhang, W., Pan, J. & Liang, W. 2008. Research on Regenerative Braking of Permanent Magnetic Synchronous Motor Servo System. *Micro-motor* 41(3): 4–6.

Manufacturing Engineering and Intelligent Materials – Lu & Abu Bakar (Eds)
© *2015 Taylor & Francis Group, London, ISBN 978-1-138-02832-6*

Research of Traction Control System based on fuel and ignition control

D. Wang, S.Q. Fan & C.X. Song

State Key Laboratory of Automotive Simulation and Control, Jilin University, Changchun, China

ABSTRACT: In order to develop a Traction Control System (TCS) for vehicles with mechanical throttle engine, injection & ignition control method has been used. This method can limit the engine driving torque through spark retard, spark cut or fuel cut, so as to get rid of the dependency on electronic throttle or ABS in traditional TCS. The research designed the algorithm of the traction control system and integrated it into engine management system. Specially, a "Threshold and PID" control method and a launch control sub-program was built for low speed condition, when it is difficult to observe the wheel slip. Finally, a formula racecar was chosen as the testing platform. The road test proved that the control method has a satisfactory effect.

1 INTRODUCTION

Traction Control System (TCS) plays an important role in vehicle active safety systems, Sohel (2003). In order to prevent the excessive slip of driving wheel and improve the vehicle dynamic performance, TCS detects the slip ratio of driving wheel and keeps it in an acceptable range by getting the drive axis' output torque under control (Deping & Wencheng 1998).

The specific methods of traction control System to control the drive axis' output torque are shown as below (Jawad et al. 2000):

(A) For vehicles with electric throttle or Drive-by-Wire, TCS controller can change the throttle position in order to limit the engine torque; (B) for vehicles with Anti-Brake System, TCS controller can apply some brake force on driving wheel in order to decrease the drive torque; (C) TCS controller can control the injection pulse width or ignition timing in order to control the engine torque.

In the passenger car area, Method A and B are more common at present. And the researches of traction control system are mainly focused on the two methods above as well. However, for vehicles without the electric throttle or Anti-Brake System, such as small and medium-sized trucks or buses, low-cost minivans, construction vehicles, and rule-limited racecars, traction control could only be implemented by Method C.

2 FUNDAMENTALS OF TRACTION CONTROL SYSTEM

The composition of the traction control system based on injection and ignition control will be relatively simple, which is composed of four wheel speed sensors, throttle position sensor, Engine Management System (ECU), injectors and ignition coils, shown as Figure 1.

ECU detects the speed of the four wheels through wheel speed sensors, which can be used to calculate the driving wheel slip ratio, and determine the driving state. When it is necessary to limit the drive torque, ECU will control the ignition coils to retard ignition timing, or control the injectors to cut fuel injection. Besides, ECU can estimate the driver's acceleration intent by throttle position sensor, and decide the target slip ratio of driving wheel.

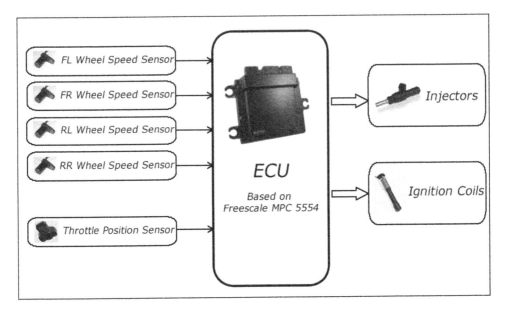

Figure 1.   System fundamental.

3   ALGORITHM DEVELOPMENT

3.1   *Traction control main-program*

The traditional algorithm of traction control system is Incremental-PI method, which is simple to develop, easy to calibration, and considered to be the most classic way (Dafeng et al. 2005). Since there's no derivative calculation part, it can only play a role when driving wheel slip ratio exceeds the target range, but cannot predict the possibility of excessive driving wheel slip by the trend of wheel slip ratio. And it has certain limitations on the low-adhesion road or aggressive driving conditions.

This paper uses a "Threshold & PID" control method based on target slip ratio. The program flow chart is shown as Figure 2.

The input error is the difference between the driving wheel slip ratio S_r and the target slip ratio S_T:

$$e = S_r - S_T = \frac{V_{drive} - V_{\mathrm{Re}f}}{V_{\mathrm{Re}f}} \times 100\% - S_T \qquad (1)$$

The output is the Engine-Torque Reduced Ratio:

$$R_{ctrl} = \begin{cases} 0,\ S_r \leq 0; \\ K_p T_d \dfrac{de}{dt},\ 0 < S_r < S_T \\ K_p \left[e + \dfrac{1}{T_i} \int e\,dt + T_d \dfrac{de}{dt} \right],\ S_r > S_T \end{cases} \qquad (2)$$

where, V_{drive} is the Driving wheel Speed, $V_{\mathrm{Re}f}$ is the Vehicle Reference Speed, K_p, T_i, T_d are the proportion constant, derivative time constant and integration time constant of a PID control. $V_{\mathrm{Re}f}$ is equal to the maximum one of drive wheel speed, and V_{drive} is calculated as the average value of driving wheel speed. If driving wheel slip ratio $S_r \leq 0$, traction control system won't intervene the engine. When $0 < S_r \leq S_T$, the system will only run derivative calculation part but

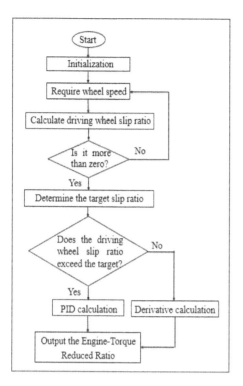

Figure 2.   Main-program flow chart.

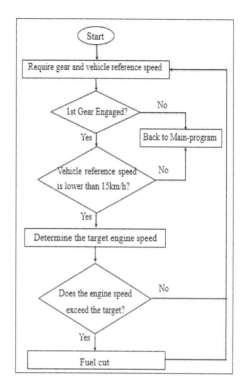

Figure 3.   Sub-program flow chart.

proportion or integration calculation part. As the driving wheel slip ratio exceeding the target value, $S_r > S_T$, a full-PID calculation will be engaged. With this control algorithm, TCS can detect the driving wheel slip ratio trend in time domain and begin to control the torque before an excessive slip, which will achieve a safer and smoother traction control effect.

3.2   Launch control sub-program

When a vehicle starts in the low-adhesion road or launches from a standing state in a high acceleration, the driving wheel angular acceleration will be large enough to cause a seriously excessive slip. In this moment, the vehicle reference speed is almost zero. The driven wheel speed is too low to measure accurately, as well as the reference vehicle speed. This will cause a horrible error in slip ratio calculation as well as the torque control.

This paper designed a sub-program based on engine speed limitation for the launch condition. The sub-program flow chart is shown as Figure 3. Only meet a) using the 1st gear and b) vehicle reference speed is lower than the threshold of 15 km/h, this sub-program will be triggered. If one of above conditions doesn't meet, system will quit the sub-program and turn to the traction control main-program.

The Launch control Sub-program uses the vehicle reference speed as input value, calculates the target engine speed under a given vehicle speed, and finally controls the driving wheel speed.

Target Engine Speed (RPM):

$$n_{t\arg et} = \begin{cases} n_{Launch}, V_{\mathrm{Re}f} < 5 \\ \dfrac{1000V_{\mathrm{Re}f} \cdot i_1 \cdot i_f (1+S_T)}{120\pi R_r}, 5 \leq V_{\mathrm{Re}f} \leq 5 \end{cases} \tag{3}$$

where, n_{Launch} is launch engine speed, typically it's a speed higher than idle; i_1 is 1st gear ratio; i_f is final gear ratio; R_r is rolling radius of wheel.

4 ENGINE TORQUE CONTROL

There are many ways to control the engine brake torque including reducing electric throttle opening degree, decreasing injection pulse width, retarding ignition timing or intermittent fuel cutting. Since this paper aims for the vehicles without electric throttle, and gasoline engine must use the mixture around $\lambda = 1$ (Banish. 2007), we choose retarding ignition timing and intermittent fuel cutting as the way to torque-control.

There is not a linear relationship between the engine brake torque and the ignition angle, neither is the fuel-cut level. So it's necessary to take a dynamometer test with the target engine. Test results are shown as Figure 4 and Figure 5.

Along with the ignition timing retard, engine brake torque is gradually reduced. When retard angle reached 20 degree, the ratio of torque reducing by ignition R_{retard} will range from 10% to 15% in different engine speed.

Along with the fuel-cut level increasing, engine brake torque is significantly reduced. The ratio of torque reducing by fuel-cut R_{cut} will range from 40% to 50% with 25% fuel-cut, and may reach 100% when fuel-cut level get 50%.

Taking the NVH performance and the driving comfort into account (Delagrammatikas et al. 2011), it should be priority to only use ignition timing retard when the Engine-Torque Reduced Ratio R_{Ctrl} doesn't exceed the ratio of torque reducing by ignition R_{retard} at 20 degree retard. That means:

$$R_{retard} = R_{Ctrl} \qquad (4)$$

However, too much ignition timing retard always means high exhaust gas temperature, which will damage catalytic converter or exhaust pipe (Matt & Jerry. 2010). Therefore, if

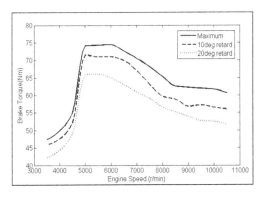

Figure 4. Engine torque curves under different ignition timing.

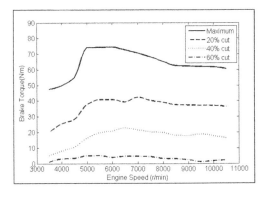

Figure 5. Engine torque curves under different fuel-cut level.

328

R_{retard} at 20 degree retard cannot fulfill R_{Ctrl}, ignition timing retard and intermittent fuel cutting will work together. The ratio of torque reducing by fuel-cut R_{cut} will be:

$$R_{cut} = R_{Ctrl} - R_{retard}(Max) \tag{5}$$

In addition, the maximum value of R_{Ctrl} is set to 100%. So the engine won't produce drag torque, which will make a negative impact on the adhesion condition between tire and road.

5 ROAD TESTING AND RESULT

The testing vehicle platform, Gspeed2013, a Formula Student Car developed by Jilin University, is shown as Figure 6. Its technical specification is shown as Table 1. The controller is based on Freescale MPC5554 CPU and takes engine and TCS in control. A Honeywell GT101 Hall speed sensor and a 24-tooth encoder disk were mounted on every wheel.

5.1 Launch and acceleration testing

Two rounds of launch and acceleration test were taken on a dry road and a wet puddle. Each round of test consisted of three different vehicle conditions: without traction control system, with traction control system, with traction control system and launch control sub-program.

Figure 6. Testing vehicle—gspeed 2013 formula student car.

Table 1. Testing vehicle specification.

Vehicle layout	Mid-engine and rear-wheel-drive
Engine	0.6 L Turbocharged gasoline engine
Peak power (kW)	70
Peak torque (Nm)	75
Total mass (kg)	290
Front weight (kg)	138
Rear weight (kg)	152
Wheelbase (mm)	1600
Track (mm)	Front:1250 Rear:1220
Wheel diameter (mm)	510

During the test, the vehicle would launch from standing state and accelerate to 100 km/h. The testing result is shown as Table 2, in which "W/O TCS" means without traction control system, "W/TCS" means with traction control system, "W/TCS + LC" means with traction control system and launch control sub-program. Figures 7–9 are driving wheel speed and vehicle reference speed curves of the three conditions on a dry road, while Figures 10–12 show the curves on a wet puddle.

Through the test, it's easy to find that traction control system can effectively reduce the driving wheel excessive slip during launch and acceleration. Under the effect of launch control sub-program, the slip ratio of driving wheel can be limited in a more satisfactory range at a low speed below 15 km/h.

5.2 Aggressive driving test

The aggressive driving test was carried out on the venue as Figure 13. The lap time is shown as Table 3. Figure 14 is the distribution statistics comparison of driving wheel slip ratio, without and with traction control system.

Test result shows that with traction control system the driving wheel slip ratio samples are more concentrated in the range of 6% to 15%, and the lap time is significantly shorter.

Table 2.   0–100 km/h acceleration time.

	W/O TCS	W/TCS	W/TCS + LC
Dry	4.29 s	3.65 s	3.43 s
Wet	4.82 s	4.62 s	4.45 s

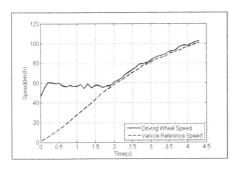

Figure 7.   Dry road acceleration speed curve without TCS.

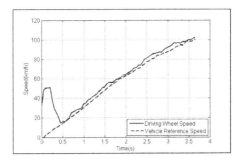

Figure 8.   Dry road acceleration speed curve with TCS.

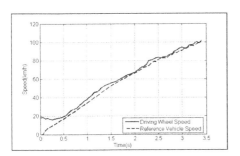

Figure 9.   Dry road acceleration speed curve with TCS and launch control sub-program.

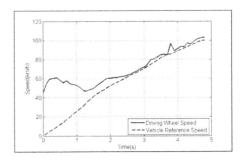

Figure 10.   Wet puddle acceleration speed curve without TCS.

330

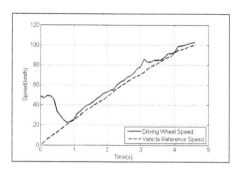

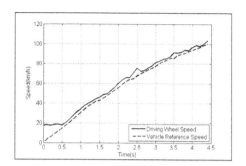

Figure 11. Wet puddle acceleration speed curve with TCS.

Figure 12. Wet puddle acceleration speed curve with TCS and launch control sub-program.

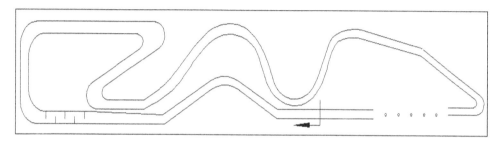

Figure 13. Aggressive driving test venue.

Table 3. Aggressive driving test lap time.

	W/O TCS	W/TCS
Lap 1	49.22 s	47.69 s
Lap 2	45.28 s	43.85 s
Lap 3	44.36 s	42.99 s

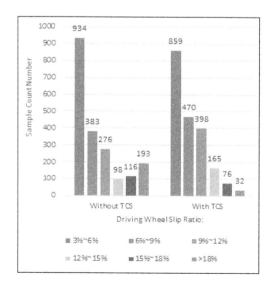

Figure 14. Distribution statistics comparison of driving wheel slip ratio.

6 CONCLUSIONS

This research applied injection and ignition control in traction control system, developed a "Threshold & PID" control method based on target slip ratio, designed a sub-program based on engine speed limitation for the launch condition, and also carried out an engine dynamometer test in order to get the relationships between engine brake torque, ignition timing and fuel-cut level, and proposed distribution plan between ignition timing retard and intermittent fuel cutting.

The conclusion of road test is as below:

1. Injection and ignition based traction control system can limit the driving wheel slip ratio in a very satisfactory range;
2. Traction control system can effectively improve the dynamic performance and driving stability;
3. A launch control program based on engine speed limitation can further improve the control effect of a traction control system in low-speed launch condition.

REFERENCES

Banish, G. 2007. *Engine Management Advanced Tuning*. North Branch: CarTech Inc.

Dafeng, S. et al. 2005. Target controller for traction control system based on vehicle rapid development system. *Journal of Jilin University (Engineering and Technology Edition)* 35(1).

Delagrammatikas et al. 2011. The Traction Control System of the 2011 Cooper Union FSAE Vehicle. *SAE Paper, SAE International*: 2011-01-1108.

Deping, W. & Wencheng, L. 1998. Control methods for automobile traction control system. *World Automobile* 1998(9): 15–18.

Jawad, B. et al. 2000. Traction Control Applications in Engine Control. *SAE Paper*: 2000-01-3464.

Matt C. & Jerry H. 2010. *Performance Fuel Injection Systems*. New York: Penguin Group (USA) Inc.

Sohel Anwar. 2003. Brake-based vehicle traction control via generalized predictive algorithm. *SAE Paper*: 2003-01-0323.

Manufacturing Engineering and Intelligent Materials – Lu & Abu Bakar (Eds)
© *2015 Taylor & Francis Group, London, ISBN 978-1-138-02832-6*

Design and analysis of hydraulic cylinder fail-safety mechanism

M.H. Ho
Department of Mechanical Engineering, Taoyuan Innovation Institute of Technology, Jhongli, Taiwan, R.O.C.

P.N. Wang
Department of Material and Fiber, Taoyuan Innovation Institute of Technology, Jhongli, Taiwan, R.O.C.

ABSTRACT: In this study, the hydraulic cylinder with internal non-slip safety locking device was designed and analyzed. From the market demand and fail-safety requirement, the hydraulic cylinder prototype specifications are developed. The dimensions and mechanisms of hydraulic cylinder model were initially designed to make sure parts combination as requirement. The stress and deformation of designed model with material properties were analyzed by Solid work COSMOS module. The finite element analysis results showed that the clamp clip cutting seams helps to uniform rod and clamp clip stress distribution. The bottom end and front end of the clamp clip with 16 cuttings have the best results in terms of stress and deformation. But the bottom end and front end of the clamp clip with 8 cuttings for rod holding effect would be better for reducing machining costs and increasing durability. It would be better to reduce the stress concentration phenomenon by rounded treatment of the cut clip end.

1   INTRODUCTION

Hydraulic machinery has features of large force output and stability loading. But when the power source is not expecting shutdown, how to prevent expected movement of the hydraulic cylinder is a problem. It is necessary to have a fail-safe device to protect personnel and machinery security. Plant safety in Europe, the use of a hydraulic cylinder failure safety device, based on the consideration of the plant not only being environmentally safe, but also on behalf of the importance of the safety of the plant's staff. The fail-safe devices were widely used in metal forming, die casting, civil construction nodal forward and machining tools holders. However, if the system is unpredictable, it would result in safety accidents, so that the staff would be injured or killed at work. Currently, industrial hydraulic cylinder production technology has made great progress, the product stability and control technology has been in a considerable level.

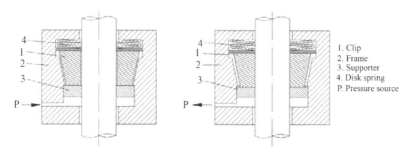

1. Clip
2. Frame
3. Supporter
4. Disk spring
P. Pressure source

Figure 1.   The design concept of hydraulic cylinder fail-safety mechanism.

About the design of machine tools holder, Anon (1970) had reviewed several typical hydraulic and mechanical clamping mechanisms for mold and hydraulic platform for rapid transfer of the mold holding system. (Burkov et al. 1980) had developed the tools holder in machine tool. The machine tool used the disc-type spring and double-acting hydraulic cylinder as gripping and releasing the spindle and tool. When supplied a compressive force to the disc spring, the mechanism would release the clamped spindle sleeve, and let the tool detached. These devices can also be used for milling or other processing machine or NC machine. Yankovskii (1983) developed a new type of pneumatic cylinder rotary coupling section with fail-safe protection. This device is simple, inexpensive and minimal friction surface and it has connector to be connected at both ends of the cylinder to determine the work piece clamping or release. (Monakhov et al. 1976) developed a mechanical arm with pieces grip device for lathe chuck. The innovative grip device can clamp the work piece diameter from 80 mm to 250 mm. Steinberger (1985) had designed a lathe clamping device which can keep clamp force at the speed is 8000 rpm and with a 160 mm diameter of work piece. When the clamp cylinder pressure is removed, the integrated sleeve can be maintaining the clamping force. Cappella (1995) used a mathematical model to simulate and the predicate the power of the molding machine clamping mechanism. The high flexible modulus was composed of small hydraulic cylinder and the clamping system. Few years ago, the industry for the production and study of the hydraulic cylinder failure safety device are rare. But, some suppliers like JAKOB have sale hydro-mechanical spring clamping systems for machine tool. The detail design and analysis about clamp clip isn't studied. In this research, the hydraulic cylinder fail-safety mechanism is designed and analyzed to find out suitable number of cutting seams and to realize the stress distribution of rod under clamp force.

Design concept: The hydro-mechanical spring clamping systems work through interaction of mechanical and hydraulic systems. The clamping force is applied mechanically through a pre-loaded disk spring packet. There are two working types are provided as spring clamping or spring pressure cylinders. The hydraulic pressure is only required for the release stroke during which the rod is moved. The Figure 1 shows the layout of hydraulic cylinder failsafety mechanism. When the hydraulic pressure P applied into the cylinder, it will push supporter 3 and let the clip 1 to press disk springs 4 in compressed condition, resulting in the clip being separated from the cylinder rod. When pressure source P loses, the disk spring 4 will push cone type clip 1 to clamp rod until the pressure P applied again.

2 METHODS

In the study, firstly, the parts and selecting disc spring were planned from the market requirements and specifications. The parts models were built in Solidwork CAD module. The finite element analysis software used was Solidwork COSMOS module. The boundary conditions with reference to the mechanism actual action were set in order to analyze the assemble model closer to reality.

2.1 *Mechanisms design*

Figure 2 is hydraulic cylinder assembly diagram. The analysis parts are included cylinder rod, holder, clip, supporter and frame. The materials used in analysis are assumed homogeneous and isotropic. Based on the consideration of how to make the clamping force evenly in the cylinder rod, so the design of the clamping clip sleeve have three design parameters: the slot

Figure 2. Cylinder parts layout.

divide position, the number of cutting slots, and the slots cutting length. The clip models of the cutting numbers have 4 types: uncut, 4-slots, 8-slots and 16-slots. It has consideration of the geometry of the clip's cutting depth to maintain the part integrity and to avoid too weakening. The slot designs for clip are the width of 2 mm and the length of 200 or 250 mm. The cut slot length of clip sleeve is more than half of 300 mm; this purpose is to make the clip has a little strength but not too weak. Table 1 shows clip detail design and slot types for analysis.

2.2 Analysis tools

The computer hardware used in this study was **IBM** compatible PC. The analysis software used was Solidwork 2009 and **COSMOS** FEA module. The analysis models and materials properties are shown in Table 2. The brass clip is used as material because of its resilience and flexiblility. The other parts used are AISI 1035 as analysis material.

Table 1.   The analysis of clip cutting type models and results.

Clip cutting type				Results
Model name	Bottom side	Front side	Cutting length (mm)	Maximum stress (MPa)
N0W0	0	0	0	339
N4W4-200	4	4	200	342
N4W4-250	4	4	250	347
N8W8-200	8	8	200	491
N8W8-250	8	8	250	350
N16W16-200	16	16	200	392
N16W16-250	16	16	250	282

Table 2.   The material properties for analysis.

Designation	Properties			
Grade	Young modulus (GPa)	Yield stress (MPa)	Tensile strength (MPa)	Poisson ratio
AISI 1035	205	283	585	0.29
Brass	110	276	552	0.33

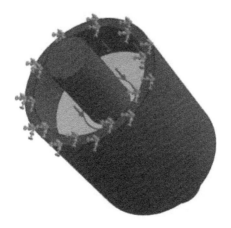

Figure 3.   Boundary conditions.

2.3 *Boundary conditions*

The boundary conditions are shown in Figure 3. The cylinder frame upper side surface remains fixed and a pressure force thrusts 125,000 kg in the holder's upper surface. This force caused holder to push the clip sleeve to generate clamping force on rod.

3 RESULTS AND DISCUSSIONS

To make clear the hydraulic rod clamp and release action, the design of the clamp clip and the rod contact stress distribution must be uniform. However, since the geometry of squeezed clip sleeve is cone type, the rod under the hydraulic clamping force will be not uniform. For example, in Figure 4(a), the uncut clip sleeve applying clamping stress to rod, only the front and the rear end of rod have holding force, but the force is very small at intermediate section.

Let both ends of clip hold the rod by uniform elastic force, it will be easy to lose clip sleeve and rod. To understand the deformation and stress distribution of clip and rod, the models with Table 1 conditions were analysed. This design method can reduce the clip and rod to produce airtight phenomenon which caused the clip sleeve and the rod to not successfully release. The rods with clamped stress distribution are shown in Figure 4. Figure 4 shows the stress distribution of clamped rods under different clip types. These results show that the more cut seams, the more uniform clamp stress distribution. The longer cut seams also have assisted the stress distribution. Figure 4(d) N16W16-250 has best clamp stress distribution. From Figure 4(c), the rod under 8 cutting seam has reached the uniform stress effect. The clip sleeve is made of brass due to the production of large clamping surface and also to avoid rod damage after repeated clamp.

Table 1 shows the stress analysis results under different cutting models. The maximum stresses varied from 282 to 491 MPa. Because the parts which were designed had no lead angle or rounded corners, so obviously the maximum stress occurred in the clip contact end. In future design, the models will be modified.

The uncut clip sleeve has high stress distribution at cone tip area which is shown in Figure 5(a) and also the rod has high stress at contact area in Figure 4(a). It would be affected by the supporter block at clip end, and the deformation was constrained in clip tip. But the clip with cutting 4, 8, 16 slots in the front and end let stress concentration be reduced as shown in Figure 5(b–d); and the clip sleeve has an effect for uniform stress distribution.

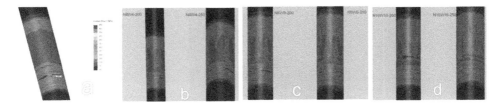

Figure 4. The rod stress distribution of N0W0 (a), N4W4 (b), N8W8 (c) and N16W16 (d).

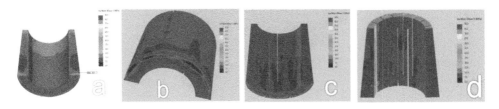

Figure 5. The clip stress distribution of N0W0 (a), N4W4 (b), N8W8 (c) and N16W16 (d).

The clip with 2 mm slot width provides sufficient deformation space for clip under loading stage. Excluding the effect of stress concentration, the overall safety factor is of about at least two times. Because too many cutting slots will reduce the strength of clip and make the clip incapable for long-term use. From the consideration of analysis results and machining cost, it is recommended that the cutting number of 8 slots to achieve the desired requirement.

4 CONCLUSIONS

–   This mechanism guarantees the greatest reliability because the clamping force is maintained fully independent of the oil pressure or leak-losses.
–   The analysis results show that the design of clip with cutting slots can help make stress distribution uniform.
–   The best stress distribution is clip with 16 slots at front and end. But considering manufacturing cost and durability, the best is clip with 8 slots at front and end.
–   The cut slots stress concentration phenomenon can be reduced by rounded treatment.

REFERENCES

Anon, 1970. Developments in hydraulic clamping systems–1. *Plastic Design Process* 10(9): 26–31.
Burkov, V.A. & Tikhon, A.O. 1980. New Spring/Hydraulic Devices for Mechanized Tool-Clamping and Release. *Machines & Tooling* 51(5): 25–28.
Cappella, A.W. 1995. Predicting clamp force on toggle type molding machines by using a math model. *Anual Technical Conference—ANTEC, Conference Proceedings* 3: 4238–4241.
Monakhov, G.A. & Melet'ev, G.A. 1976. Gripping Devices of Industrial Robots. *Russian Engineering Journal* 56(9): 3–5.
Steinberger, J. 1985. Clamping Devices for Modern Lathes. *Werkstatt und Betrieb* 118(5): 281–284.
Yankovskii, M.A. 1983. Safety Coupling for Air-Delivery to Rotating Pneumatic Cylinders. *Soviet Engineering Researc* 3(7): 87–88.

Manufacturing Engineering and Intelligent Materials – Lu & Abu Bakar (Eds)
© *2015 Taylor & Francis Group, London, ISBN 978-1-138-02832-6*

Development of a hierarchical stability controller of 4WD electric vehicle with in-wheel motors

F. Xiao, C.X. Song & S.K. Li
State Key Laboratory of Automotive Simulation and Control, Jilin University, China

ABSTRACT: According to the characteristic that each wheel torque of 4 Wheel Drive (4WD) electric vehicle is independently controllable, the control allocation method with hierarchical structure to optimize the distribution of motor torque can improve the handling stability of the vehicle. The controller is composed of an upper controller and a lower distributor, of which the upper controller can command the desired yaw moment based on the measurements from sensors, and the lower controller is used to ensure that the desired value of yaw moment commanded by the upper controller is indeed obtained from the torque allocated to each wheel. The whole vehicle and steering system models are established for simulation and demonstration through the application of MATLAB and Simulink. The simulation results show that the control allocation method with hierarchical structure can effectively improve the handling stability of the vehicle.

1   INTRODUCTION

Under the double pressures of energy saving and environment protection, the various forms of electric vehicle has become the focus of research in current automotive industry. The 4 WD electric vehicle with in-wheel motors has become a more promising one (Satoshi, 2010). With the promotion of electric vehicles, their stability control problem has become increasingly important (Takuya et al. 2006, Esmailzadeh et al. 2003). Compared with the traditional vehicles, the torque distribution method among the four wheels for the 4 WD electric vehicles is comparatively flexible, which may improve the stability control of the vehicle dynamics (Shin-ichiro & Yoichi 2001, Yoichi 2004). In order to improve the stability of the vehicle, this paper presents a hierarchical control method, which comprises an upper controller and a lower controller. The upper controller is used to generate the generalized longitudinal force and yaw moment required to achieve vehicle motion according to the signals such as steering wheel angle and position of acceleration/brake pedal. The longitudinal force is regulated by taking the deviation of demand speed and actual speed as the input variable, and the yaw moment is calculated by using linear quadratic regulator after the comparison between reference state and actual state. The lower controller was designed to distribute properly the torque on each motor according to the generalized longitudinal force and yaw moment. In this paper, a 4 WD electric vehicle model was established, and a simulation test was performed to verify the effectiveness of the hierarchical vehicle stability control algorithm.

2   STRUCTURE OF HIERARCHICAL INTEGRATED CONTROL

In this paper, a hierarchical control structure is used by the integrated control algorithm as shown in Figure 1, the controller of which uses upper and lower structures. Firstly, an expected value reference model is established in the upper controller, expected vehicle

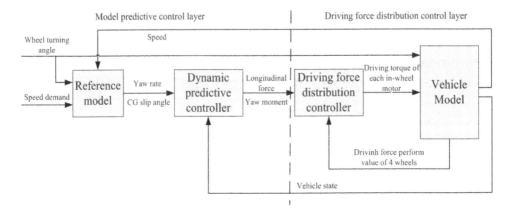

Figure 1. Hierarchical control structure.

travelling parameters under ideal conditions such as yaw rate and side-slip angle can be calculated based on the current steering wheel angle and speed information. Then, we compare the yaw rate and side-slip angle with the feedback we got from the vehicle model. If the actual values have large difference from the expected values, the model predictive controller can calculate the total yaw moment, namely, additional yaw moment, which is needed to take the vehicle back to a stable state, and according to the current speed and the target speed, the total longitudinal force demand in the current vehicle state can be figured out through a particular control algorithm. Based on the total longitudinal force demand and additional yaw moment, the torque of each wheel can be figured out by the lower controller according to certain rules.

2.1 2-DOF reference model

Vehicle 2-DOF planar model has been widely used as an approximate description model of dynamic characteristics (Motoki & Masao 2001), the state-space expression of the planar model in chassis coordinates is:

$$\dot{x} = Ax + E\delta_f \tag{1}$$

where

$$\dot{x} = [\beta, \gamma]^T$$

$$A = \begin{bmatrix} -\dfrac{2(C_f + C_r)}{mu} & -1 - \dfrac{2(C_f l_f - C_r l_r)}{mu^2} \\ \dfrac{2(C_f l_f - C_r l_r)}{I_z} & \dfrac{2(C_f l_f^2 + C_r l_r^2)}{I_z u} \end{bmatrix},$$

$$E = \begin{bmatrix} \dfrac{2C_f}{mu} \\ \dfrac{2C_f l_f}{I_z} \end{bmatrix},$$

In Eq. 1, β represents the vehicle side-slip angle, γ represents the entire vehicle yaw rate, u represents the speed, m represents the vehicle mass, l_f, l_r represent the distances between CG and front, rear axle, C_f, C_r represent the cornering stiffness of front, rear axle, I_z represents the rotational inertia of the vehicle around the z axis.

The desired yaw rate for the vehicle can therefore be obtained as follows:

$$\gamma_{des} = \frac{u/(l_f + 1_r)}{1 + \dfrac{m}{L^2}\left(\dfrac{l_f}{C_r} - \dfrac{1_r}{C_f}\right)u^2}\delta_f \tag{2}$$

the desired slip angle can be described as Eq.(3).

$$\beta_{des} = \frac{\delta_f}{l_f + 1_r + \dfrac{mV^2\left(\dfrac{1_r}{C_f} - \dfrac{l_f}{C_r}\right)}{2(l_f + 1_r)}}\left(1_r - \frac{l_f}{2C_r(l_f + 1_r)}mV^2\right) \tag{3}$$

2.2 Dynamic predictive controller

The desired yaw rate and the desired slip angle cannot always be obtained. It is not safe to try and obtain the above desired yaw rate if the friction coefficient of the road is unable to provide tire forces to support a high yaw rate. Hence, the desired yaw rate must be bounded by a function of the tire-road friction coefficient. The objective of the dynamic predictive controller is to determine the desired yaw torque for the vehicle so as to track the target yaw rate and target slip angle from the reference model; dynamic predictive controller can calculate the generalized force of longitudinal force and yaw moment that is needed to return the vehicle to a stable state by comparing the error between the expected parameters value and the actual vehicle feedback value in various states (Zhu et al. 2008).

3 ESTABLISHMENT OF VEHICLE MODEL

This paper established the 15-DOF 4 WD EV model for the control stability simulation through the application of MATLAB and Simulink. The vehicle model includes 6-DOF body model, steering system model, suspension model, tire model, and so on [8]. Since this paper focusses on verifying the effectiveness of the hierarchical vehicle stability control algorithm, the model just simplifies the complex motor model as the torque output based on the motor external characteristics. The model established in this paper can reflect the dynamic

Table 1. Part of the vehicle simulation parameters table.

Parameter	Value	Unit
Vehicle mass m	1240	kg
Distance between CG and front axle l_f	1157	mm
Distance between CG and rear axle l_r	1453	mm
Inertia moment I_z	1662	kg·m²
Front track B_f	1463	mm
Rear track B_r	1463	mm
Height of CG h_g	510	mm
Peak torque of motor	300	Nm
Rolling radius	307	mm

Where ca = interface adhesion; δ = friction angle at interface; and $k1$ = shear stiffness number.

characteristics of the vehicle well and the effectiveness of control algorithm can be evaluated easily. Table 1 shows part of the vehicle model parameters.

4 SIMULATION AND ANALYSIS

In order to demonstrate the effectiveness of above control method, the simulation under lane-change operating condition was carried on. The vehicle kept an even speed of 80 km/h on a road whose adhesion coefficient is 1, as is shown in Figure 2, the angle input of front wheel changes as sine wave with a cycle of 2 s. The simulation results are shown in Figure 3 to Figure 5.

Analysis of simulation results: Through the comparative analysis on the actual value of vehicle yaw rate and CG slip angle in the simulation process with the target value of the reference model, we know actual yaw rate follows the reference yaw rate very well. As we can see from the β-γ phase diagram, the control-imposed vehicle encloses smaller area in the β-γ phase diagram, which indicates that the stability margin has been improved after the control. Through the comparative analysis on the equal distribution and the optimizing distribution of the longitudinal force, we know that the control method to distribute each wheel torque optimally can provide greater adhesion for wheels with large vertical load, so has the steering stability been improved when the vehicle is close to adhesion limit.

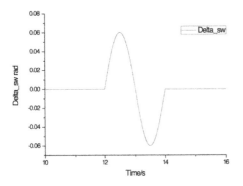

Figure 2. Wheel turning angle.

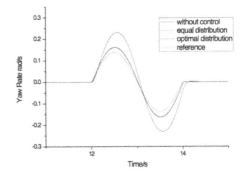

Figure 3. Yaw rate.

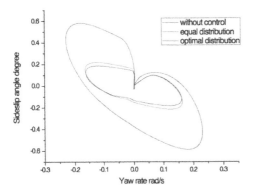

Figure 4. Vehicle slip angle.

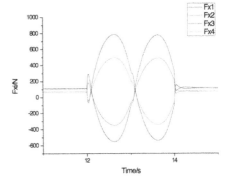

Figure 5. Distribution results of longitudinal force on each electric wheel.

5 CONCLUSION

This paper established a hierarchical control algorithm to achieve the control of yaw moment applied to 4 WD electric vehicle based on characteristics of vehicle with in-wheel motors. The algorithm can optimize the distribution on each wheel and make full use of the adhesion of tire. The simulation results show that the proposed algorithm achieves an effective distribution of longitudinal force on four wheels, reduces the yaw rate and the lateral acceleration, and improves the handling stability of the vehicle.

REFERENCES

Esmailzadeh, E., Goodarzi, A. & Vossoughi, G.R., 2003. Optimal yaw moment control law for improved vehicle handling, *Mechatronics* 2003, 13(7): 659 –675, 2003.
Motoki, S.& Masao, N., 2001, Yaw-moment control of electric vehicle for improving handling and stability, JSAE Review 22 (2001) 473–480, doi: 10.1016/S0389-4304(01)00130-8.
Satoshi, M. 2010. Innovation by in-wheel-motor drive unit//AVEC10, 2010.
Shin-ichiro, S. & Yoichi, H., 2001. Advanced motion control of electric vehicle with fast minor feedback loops: basic experiments using the 4-wheel motored EV "UOT Electric March II". JSAE Review (S0389-4304), 2001, 22(4): 527–536.
Takuya, M., Pongsathorn, R. & Masao, M., 2006. Direct Yaw-moment Control Adapted to Driver Behavior Recognition // SICE-ICASE Int Joint Conf, 2006: 534–539.
Yoichi, H., 2004, Future Vehicle Driven by Electricity and Control—Research on Four-Wheel-Motored "UOT Electric March II". IEEE Transactions on Industrial Electronics (S0278–0046), 2004, 51(5): 954–962.
Zhu, S., Jiang, W., Yu, Z. & Zhang, L., 2008, Stability Control of Four In-Wheel-Motor Drive Electric Vehicle by Control Allocation," Journal of System Simulation, 2008, 20(17): 4840–4846.

Manufacturing Engineering and Intelligent Materials – Lu & Abu Bakar (Eds)
© *2015 Taylor & Francis Group, London, ISBN 978-1-138-02832-6*

Stability control of 4WD electric vehicle with in-wheel motors based on an advanced fuzzy controller

F. Xiao, C.X. Song & S.K. Li
State Key Laboratory of Automotive Simulation and Control, Jilin University, China

ABSTRACT: Each wheel torque can be controlled independently, so 4 Wheel Drive (4WD) electric vehicle can control the vehicle stability through controlling the motor driving to generate yaw moment, which is different from the conventional vehicles. 4WD EVs have potential applications in control engineering. In this paper, a vehicle co-simulation platform is constructed through the application of AMEsim and Simulink. Additionally, an advanced fuzzy controller is designed to determine the corrective yaw moment that is required to stabilize the vehicle, and applies a virtual yaw moment around the vertical axis of the vehicle so as to compensate for deviations between CG slip angle and yaw rate. Several maneuvers are simulated to demonstrate the performance and effectiveness of the proposed controller, and the simulation results show that the proposed stability control system can effectively improve the handling stability of the vehicle.

1 INTRODUCTION

In order to solve the problem of energy emergency and environmental pollution, it is becoming a hotspot to develop a new generation of electric vehicles. In automobile research and development, an all-wheel drive way based on in-wheel motors is regarded as a structure of a breakthrough (Yoichi 2004). When a vehicle drives at fast speed on a slippery road or some certain roads, its turning is prone to under-steer characteristics and its aligning is prone to over-steer characteristics, which will make the yaw force of the vehicle lateral close to adhesion limit or reach saturation, even the dynamics stability can be lost. While in direct yaw moment control, additional yaw moment is generated by adjusting the distribution of longitudinal force and lateral force on four wheels, which makes the vehicle move according to the desired trajectory, and the vehicle driving stability can be kept (Takuya et al. 2006). When compared with a traditional ICE vehicle, we know an electric vehicle with in-wheel motors has characteristics that its motors have quick response and each wheel torque can be controlled precisely and independently, etc., which make it easier to implement Direct Yaw moment Control (DYC) of the vehicle (Shin-ichiro & Yoichi 2001, Motoki & Masao 2001). In this paper, optimal distribution rules are made and the controller is designed based on fuzzy control algorithm (Esmailzadeh et al. 2003), vehicle slip angle errors and yaw rate errors are compensated according to vehicle stability control logic. Fuzzy controllers are generally well-suited for complex nonlinear dynamic control problems, and can act as supervisory modules (Farzad et al. 2004). Based on MATLAB and Simulink, a co-simulation model is built in this paper and thus the effectiveness of the stability control algorithm can be validated.

2 VEHICLE DYNAMIC MODEL

In this paper, a vehicle 15-DOF dynamic model is established with MATLAB and Simulink in the light of the characteristics of 4 WD vehicles, including vehicle body submodel, steering system submodel, tire model, suspension model, etc. The complexity of motor control makes it hard to establish an accurate model of motor control system, so a simpler motor control model has been chosen in the paper.

3 STRUCTURE OF FUZZY CONTROLLER

The driver model can calculate the wheel turning angle δ according to the expected position x_d and y_d, thereby the vehicle slip angle β_d and the expected yaw rate γ_d can be calculated. Error values e_β and e_γ are equal to the results that the actual slip angle and yaw rate feedback from vehicle model minus their expected values from the desired value estimator. In order to overcome the influence of nonlinear factors such as tire as well as to improve the accuracy of vehicle stability control, a fuzzy logic controller is designed in this paper. Figure 2 is the block diagram of fuzzy controller. The control input signals of the controller are the vehicle slip angle error e_β and the yaw rate error e_γ. The fuzzy controller calculates the desired torque of each wheel that error compensation needed based on the input value. The yaw moment is provided by the regenerative braking torque of control motor so as to improve the efficiency of energy recovery.

3.1 *Driver model*

A driver model is used to trace the desired path for the closed-loop simulation. Figure 1 shows a schematic diagram of the driver model. The driver model manipulates the steering angle to compensate the error between the estimated position and the desired position. The estimated position x^* and y^* can be calculated from the following equations.

$$x^* = x + \left(V_x \cos\varphi - V_y \sin\varphi\right) \cdot \frac{L}{V} \tag{1}$$

$$y^* = y + \left(V_x \sin\varphi + V_y \cos\varphi\right) \cdot \frac{L}{V} \tag{2}$$

Table 1. Part of the vehicle simulation parameters table.

Parameter	Value	Unit
Vehicle mass m	1240	kg
Distance between CG and front axle l_f	1157	mm
Distance between CG and rear axle l_r	1453	mm
Inertia moment I_z	1662	kg·m²
Height of CG hg h_g	510	mm
Peak torque of motor	300	Nm
Rolling radius	307	mm

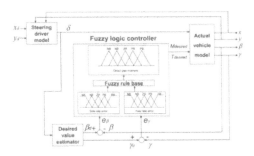

Figure 1. Block diagram of fuzzy controller.

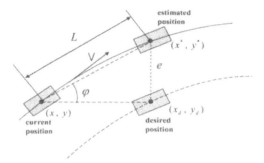

Figure 2. Steering by the driver model.

$$e = \sqrt{(x_d - x^*)^2 + (y_d - y^*)^2} \tag{3}$$

$$\delta = PID(s) \cdot e \tag{4}$$

In formulas, x^* represents the estimated longitudinal displacement, y^* represents the estimated lateral displacement, x_d represents the expected longitudinal displacement, y_d represents the expected lateral displacement, δ represents turning Angle, φ represents vehicle heading, e represents the displacement error between estimated position and expected position, L represents the predictive distance, and PID represents gain coefficients.

3.2 Desired value estimator

Vehicle 2-DOF planar model has been widely used as an approximate description model of dynamic characteristics [6], the state-space expression of the planar model in chassis coordinates is:

$$\dot{x} = Ax + E\delta_f \tag{5}$$

where

$$\dot{x} = [\beta, \gamma]^T$$

$$A = \begin{bmatrix} -\dfrac{2(C_f + C_r)}{mu} & -1 - \dfrac{2(C_f l_f - C_r l_r)}{mu^2} \\ \dfrac{2(C_f l_f - C_r l_r)}{I_z} & \dfrac{2(C_f l_f^2 + C_r l_r^2)}{I_z u} \end{bmatrix}, \quad E = \begin{bmatrix} \dfrac{2C_f}{mu} \\ \dfrac{2C_f l_f}{I_z} \end{bmatrix}$$

In Eq. 1, β represents the vehicle side-slip angle, γ represents the entire vehicle yaw rate, u represents the speed, m represents the vehicle mass, l_f, l_r represent the distances between CG and front, rear axle, C_f, C_r represent the cornering stiffness of front, rear axle, I_z represents the rotational inertia of the vehicle around the z axis.

The desired yaw rate for the vehicle can therefore be obtained as follows:

$$\gamma_{des} = \frac{u/(l_f + l_r)}{1 + \dfrac{m}{L^2}\left(\dfrac{l_f}{C_r} - \dfrac{l_r}{C_f}\right)u^2} \delta_f \tag{6}$$

The desired slip angle can be described as Eq.(3).

$$\beta_{des} = \frac{\delta_f}{l_f + l_r + \dfrac{mV^2\left(\dfrac{1_r}{C_f} - \dfrac{l_f}{C_r}\right)}{2(l_f + 1_r)}}\left(1_r - \frac{l_f}{2C_r(l_f + 1_r)}mV^2\right) \qquad (7)$$

4 SIMULATION AND ANALYSIS

In order to demonstrate the effectiveness of the designed vehicle stability control system, the simulation under lane-change operating condition was carried on. The vehicle kept an even speed of 80 *km/h* on a road whose adhesion coefficient is 0.2, as is shown in Figure 3, the angle input of front wheel changes as sine wave with a cycle of 2*s*. The simulation results are shown in Figure 4 to Figure 7.

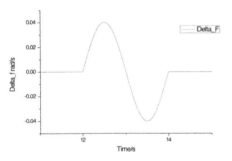

Figure 3.   Wheel turning angle changes.

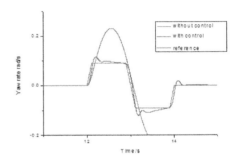

Figure 4.   Yaw rate as sine wave.

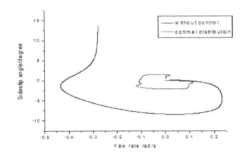

Figure 5.   Body slip angle.

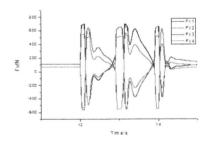

Figure 6.   Distribution of longitudinal force on each wheel.

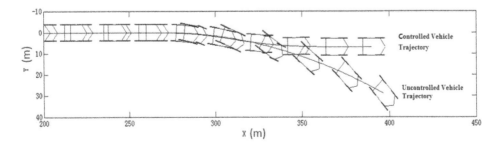

Figure 7.   Vehicle trajectory.

Analysis of the results: as we can see from Figure 4, due to the limitation of adhesion coefficient of road surface, the reference yaw rate appears to be a constant value; the reference yaw rate is able to be followed by controlling the vehicle, while the ideal vehicle trajectory cannot be followed by the vehicle without control. From the β-γ phase diagram, we know the control-imposed vehicle enclose smaller area in the β-γ phase diagram, which indicates that the stability margin has been improved after the control.

5 CONCLUSION

In this paper, the stability control of 4 WD electric vehicle with in-wheel motors is taken by the regenerative braking torques of motors. Also, a controller is designed for calculating the needed yaw moment to compensate for deviations between yaw rate and vehicle slip angle. A co-simulation platform based on MATLAB and Simulink is built, thus the feasibility of the vehicle stability control algorithm can be validated under the operating condition of a low adhesion road. The simulation results show that the control algorithm can rationally distribute the torque on 4 wheels when the vehicle lost its stability, so as to make it feasible that the vehicle trajectory can be followed, therefore the handling stability of the vehicle can be improved.

REFERENCES

Esmailzadeh, E., Goodarzi, A. & Vossoughi, G.R., 2003. Optimal yaw moment control law for improved vehicle handling, Mechatronics 2003, 13(7): 659 –675, 2003.
Farzad, T., Shahrokh, F. & Reza K., 2004 Fuzzy Logic Direct Yaw—Moment control System for All-Wheel—Drive Electric Vehicles, Vehicle System Dynamics, 41(3): 203–221, 2004, doi: 10.1076/vesd. 41.3.203.26510.
Motoki, S. & Masao, N.,2001, Yaw-moment control of electric vehicle for improving handling and stability, JSAE Review 22 (2001) 473–480, doi: 10.1016/S0389-4304(01)00130–8.
Shin-ichiro, S. & Yoichi, H., 2001. Advanced motion control of electric vehicle with fast minor feedback loops: basic experiments using the 4-wheel motored EV "UOT Electric March II". JSAE Review (S0389-4304), 2001, 22(4): 527–536.
Takuya, M., Pongsathorn, R. & Masao, M., 2006. Direct Yaw-moment Control Adapted to Driver Behavior Recognition//SICE-ICASE Int Joint Conf, 2006: 534–539.
Yoichi, H., 2004, Future Vehicle Driven by Electricity and Control—Research on Four-Wheel-Motored "UOT Electric March II". *IEEE Transactions on Industrial Electronics* (S0278-0046), 2004, 51(5): 954–962.

Manufacturing Engineering and Intelligent Materials – Lu & Abu Bakar (Eds)
© 2015 Taylor & Francis Group, London, ISBN 978-1-138-02832-6

A new-style local-scale reconfigurable precise turn-milling machine for microminiature

Z.X. Li, X. Jin, Z.J. Zhang & W.W. Lv
School of Mechanical Engineering, Beijing Institute of Technology, Beijing, China

ABSTRACT: Turn-milling technology has proved to be the mainstream for precise and microminiature machining. Local-scale RMT efficiently meets the demands of complete machining. Based on turn-milling technology and the idea of local-scale RMT, a new-style local-scale reconfigurable precise turn-milling machine for microstructure is introduced.

Keywords: local-scale RMT; microminiature; turn-milling technology

1 INTRODUCTION

As the important composition of microminiature manufacturing technology, microminiature machining technology is the important developing orientation of advanced machining technology and the researching frontiers of multi-subject crossing. It has gained more attention of the scholars and the research organizations (Workshop 2000, Sun et al. 2004, Jia 2004, Liu 2005). According to the recognized saying, miniature parts are defined as those whose size range is between ten microns to ten millimetres (0.01 mm~10 mm)., It has also been named the microscale/mesoscale parts in the world Workshop report (2000).

Micro miniature technology is divided into two fields, one is the technique of silicon micromachining and LIGA, but its 3D ability is inferior, and it needs super-clean condition, restricted line width and powerful ray source. So, the technique of silicon micromachining and LIGA does not match the machining of the high bearing structural parts. The other field is the micromachining technology based on traditional machining technology, named precise and microminiature machining technology. Its goal is to solve the problems which are how to bear capacity of micro-structural components and microminiaturize the micro-manufacture systems. Beijing Institute of Technology has first applied turn-milling technology to micro-manufacture. With a series of academic and experimental researches, many productions have come out. Based on the micromachining demand for the turn-milling technology, and with the combination with local-scale RMT theory, a new-style local-scale reconfigurable precise turn-milling machine for micro-structural parts developed by BIT is introduced.

2 PRECISE AND MICROMINIATURE HIGH ROTATING SPEED TURN-MILLING TECHNOLOGY

2.1 *Development of precise and microminiature machining technique*

Internationally, the technical development of precise and microminiature machining technology mainly has the following characteristics (Zhang et al. 2007):

1. High rotating speed NC microminiature machining technology, mainly solve the problem of cutting speed of miniature parts.

2. Reconfigurable multiple process combines machining equipment and technology. Reconfigurable machine technology developed after the reconfigurable machining technology which is a completely new concept. This new technology requires the machine must be rearranged or replaced quickly and reliably. Reconfiguration will allow adding, removing, or modifying machine structure to adjust production capacity in response to changing market demands or technologies.
3. Multiple process combines complete machining technology. Complete machining means to finish all the machining process by clamping on one or maybe two machines, so it's called comprehensive processing technology or combined machining. Now, complete machining is the developing tendency of microminiature machining.

2.2 *Turn-milling technology is the mainstream for precise and microminiature machining*

Turn-milling is a relatively new concept in machining technology, wherein both the workpiece and the tool are given a rotary movement simultaneously. It's not the simple combination of turning and milling, but is a new cutting theory and technology in the situation that the NC technology has developed greatly. With high rotating speed turn-milling, high surface quality and shape precision can be achieved. In recent years, many researches have been carried on turn-milling technology. And many scholars are of the viewpoint that turn-milling technology is fit for the machining of large shaft and featheredged parts. The turn-milling machining equipment for large- and medium-sized parts have come out. Beijing Institute of Technology (BIT) advanced the high rotating speed precise and microminiature turn-milling technology, which proved to be the mainstream technology of precise and micro-manufacture in practice and in theories. BIT also developed the high rotating speed NC turn-milling machine, and now the improved machine CXKM25-III I has come out (see in Fig. 1)

Turn-milling technology can be applied to precise and microminiature machining for the following reasons:

1. With the integration of microstructures, many gyration and small structures are distributed in a single part. The machining often needs the combination of many processes such as turning, milling, drilling, broaching and grinding etc. However, times of clamping lead to badly losing of precision. Because of the special structure, some parts can only be clamped one or two times. So the conflict of complex microstructure, repetitious clamping and precision assurance comes out. Fortunately, the turn-milling technology solves the conflict to a tee for its advanced concept "once clamped, complete machining".

Figure 1. The high rotating speed NC turn-milling machine developed by BIT for precise and microminiature.

2. General machining process can't complete the precise machining of microminiature parts because it can't assure the necessary cutting speed. With high rotating speed milling, the turn-milling technology can gain normal even high cutting speed when the microminiature parts are rotating in low speed. So the precision can be achieved.

Many academic and experimental researches have been carried out on precise and microminiature turn-milling technology, and many productions are gained. Professor Zhang zhi-jing made minute calculations for cutting speed, cutting force value and orientation by theory analysis and experimental sum up, and proved the feasibility and advantage of precise and microminiature turn-milling.

3. LOCAL-SCALE RECONFIGURABILITY OF PRECISE AND MICROMINIATURE TURN-MILLING MACHINE

"Once clamped, complete machining" is the core idea of turn-milling technology. In order to meet the demand of complete machining process and to answer the changing market demands rapidly, the turn-milling machine should be local-scale reconfigurable.

RMT (Reconfigurable Machine Tool) was brought forward by University of Michigan in 1996 (Xu & Wang 2007). A Reconfigurable Machine Tool is designed to process a given family of machining features and is constructed from a set of standard modules. An RMT provides a cost-effective solution to mass customization and high-speed capability. There is no known systematic method or a scientific basis for designing RMTs (Moon 2007). RMT is based on full-scale reconfigurable theory (Clayton 2002). Considering the practical manufacture, not all the machines need full-scale reconfiguration. Thus, local-scale RMT emerges (Zhang et al.).

3.1 *Definition of local-scale RMT*

Because of the small dimensions and low stiffness in mesoscale machining, a new manufacturing paradigm should be applied to completely machine the parts without changing machine tools or even fixtures, which lead to emergence of the local-scale RMT. Different from the definition of RMT given by Koren, a local-scale RMT is designed to realize the complete machining function with the ability to change or update several indicated modules (Zhang et al.).

Local-scale RMT developed after RMT, and it has the following characteristics (Zhou 2009):

1. Some of the machine modules have standard interface and can be reconfigured with the machining demands.
2. Some of the machine structures can be rearranged or replaced with the changing of machining objects.
3. The machine can alter its machining ability or function without redundancy.
4. The machine control system should have independent modularization control interface.

3.2 *Local-scale reconfigurability of precise and microminiature turn-milling machine*

Based on the machine main body and combined with machining demands of micromimiature parts, the local-scale reconfigurable modules of the precise and microminiature turn-milling machine (Fig. 2) include tool post module, NC rotary table module, milling unit module and automatic tool changing module.

Modules and the machine main body or modules and modules are joined with standard mechanism interface. And the modules are joined with the machine control system independently. The tool post module uses the existing product for the turning of microstructure. The NC rotary table module can accomplish precise 90° rotation, and can accomplish both radial and axial milling with the combination of the milling unit. The milling unit is the key module of the machine. By replacing the milling motor spindle with grinding motor spindle, it can also grind workpieces. In the automatic tool changing module, there

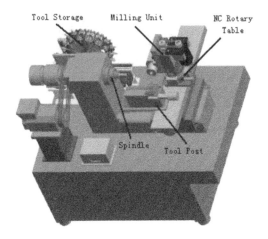

Figure 2.   The structure of precise and microminiature turn-milling machine.

is a double-row discal tool storage, and its unique structure assures the precision and celerity of automatic tool changing.

The above modules are local-scale reconfigurable, and so the machine function is greatly extended. With the mechanism interface holding the line, the modules can be reconfigured expediently.

3.3   *Precision ensuring of the module interface*

Each module has certain precision, thus the mechanism interface has decisive effect on the motion precision of the reconfigured machine. The mechanism interface set of the machine needs high repeatable precision and good interface rigidity. The over-positioning technique can ensure the reliability of mechanism interface, increase the interface area, and heighten the rigidity and precision. Unreasonable positioning design leads to the interference, and even leads to the phenomenon that the element can't be assembled or positioned. So, the over-positioning structure should be designed reasonably and it's vital. With the increasing of the number of the positioning elements and the heightening of positioning precision, the rigid joint difficulty of the reconfigurable mechanism interface increases also. To solve the problem, double-state positioning[9] emerges. It means, in rigid over-positioning mechanism interface joint, to improve the assembly characteristic with elastic elements and without losing positioning precision. Different from the rigid over-positioning mechanism interface, double-state mechanism interface ensure the positioning precision and repeatable precision by assembly. The elastic elements compensate the abrasion. Double-state positioning idea has been applied in the local-scale reconfigurable high rotating speed NC turn-milling machine.

4   APPLICATION OF THE LOCAL-SCALE RECONFIGURABLE HIGH ROTATING SPEED TURN-MILLING MACHINE

Many experiments of machining process have been carried out on the machine. Many micro-structures that can't be machined by turning has been machined, such as Micro slender shaft (length/diameter: 8.0 mm/0.5 mm), featheredged part (Fig. 3) etc.

Also, many complex microminiature parts (Fig. 4) have been machined by complete machining on the machine for some research organizations.

All the data and information gained from the experiments give technique assistance to the microminiaturization of mechanical parts in the field of aeronautics and weapon production.

Figure 3. Featheredged part (Diameter: 2.5 mm, featheredge: 0.3 mm, length: 5 mm).

Figure 4. Example of complete machining.

5 CONCLUSION

BIT advanced the high rotating speed precise and microminiature turn-milling technology, and also developed the high rotating speed NC turn-milling machine. The technology has provided a new machining process for precise and microminiature parts. Local-scale RMT was brought forward and has been applied to the turn-milling machine. Now further research is being carried on local-scale high rotating speed precise and microminiature turn-milling technology and machine.

ACKNOWLEDGEMENTS

This research is supported by National Science and Technology Major Project 2012ZX04010061).

REFERENCES

Clayton B. 2002. CoE Creates World's First Full-Scale Reconfigurable Machine Tool [R]. Michigan Engineering. July 25.
Jia chun-de. 2004. Research on Turn-milling Technology (In Chinese). PhD Dissertation. Beijing: Beijing Institute of Technology.

Liu ke-fei. 2005. Research on Precise and Micro-miniature Turn-milling (In Chinese). PhD Dissertation. Beijing: Beijing Institute of Technology.

Moon, Y. 2000. Reconfigurable Machine Tool Design: Theory and Application. PhD Dissertation, The University of Michigan.

Sun ya-zhou, Liang ying-chun, Cheng kai. 2004. Micro-scale and Meso-scale mechanical manufacturing (In Chinese). *Chinese Journal of Mechanical Engineering [J]*, 40(5): 1~6.

Workshop report. 2000. Workshop on Micro/Meso-Mechanical Manufacturing. Evanston, Illinois, USA, Northwestern University.

Xu hong, Wang qing-ming. 2005. Reconfigurable Machine Tool Design (In Chinese) [J]. *China Mechanical Engineering*, 16(7): 588~593.

Zhang zhi-jing, Jin xin, Zhou min. 2007. Precise and Microminiature Manufacturing Theory, Technology and Its Appliance (In Chinese). *Chinese Journal of Mechanical Engineering*, 43(1): 49~61.

Zhang zhi-jing, Zhou min, Yuan wei. Local-Scale RMT for Mesoscale Turn-Milling Technology. International Conference on Reconfigurable Mechanisms and Robots ASME/IEEE.

Zhou min. 2009. Key Technology of Local-scale Reconfigurable Micro Turn-milling. PhD Dissertation. Beijing: Beijing Institute of Technology.

Manufacturing Engineering and Intelligent Materials – Lu & Abu Bakar (Eds)
© 2015 Taylor & Francis Group, London, ISBN 978-1-138-02832-6

SBR for coal chemical wastewater treatment and bacteria ecological analysis

L. Wang, Y.J. Sun, S.D. Lu, X.H. Zhao & A.Z. Ding
College of Water Sciences, Beijing Normal University, Beijing, China

ABSTRACT: Coal chemical wastewater is very difficult to address because it contains many complex toxic substances. This study is based on a coal chemical wastewater treatment plant in Shaanxi Yulin. The main part of the wastewater treatment plant is SBR (sequencing batch reactor activated sludge process) which has the benefits of operational simplicity and flexibility. Activated sludge samples were analyzed by molecular ecological methods. The SBR efficiency for the removal of organic matter is very high with the COD removal efficiency of 97%, while the NH3-N removal efficiency is approximately 99% at the reactors' steady state. The activated sludge samples in SBR reactors have been analyzed by ecological methods. Raw water from the SBR reactors was used as the unique substrate to incubate degrading bacteria in the sludge, simulating the actual situation. The colonies were successfully grown and 16S rDNA Polymerase Chain Reactions (PCR) were constructed to examine microbial diversity. The sequencing results showed that most of the coal chemical degradation bacteria is belonged to Protebacteria and Actinobacteria. The dominant species were Arthrobacter arilaitensis Re117, Stenotrophomonas maltophilia K279a, and Acinetobacter oleivorans DR1. The results are helpful for the training of function bacterium agents in the coming research and provide a theoretical reference for coal chemical industry wastewater treatment.

Keywords: coal chemical wastewater; SBR; ecological analysis; 16S rDNA PCR; sequencing

1 INTRODUCTION

Coal is one of the main fossil energies in our country, and coal consumption accounted for 2/3 of the total energy consumption at all levels (Zhao Qiang et al. 2012). The coal chemical industry based on coal as raw material, transforms the coal into gas, liquid and solid fuel and produces various chemical products, through a series of chemical reactions (Ma Zhong-xue et al. 2007). Coal chemical wastewater is often in high concentrations of gas washing water with the following main characteristics: complex composition, containing a large amount of solid suspended particles, volatile phenol, polycyclic aromatic hydrocarbons, pyrrole, furan, imidazole, naphthalene, heterocyclic compounds containing nitrogen, oxygen, sulfur, cyanide, oil, ammonia nitrogen and sulfur as well as other toxic and harmful substances. Its COD and chromaticity values are very high (Shi Guang-Mei. 1993). If the coal chemical wastewater is untreated or improperly treated and discharged, it can cause serious pollution of water. Accordingly, using economic and reasonable treatment technology, guaranteeing the coal chemical wastewater discharging standard or reuse has great practical significance (Han Hong-Jun et al. 2010; Yu Hai et al. 2014).

At present, coal chemical wastewater is mainly treated by biochemical methods. The Sequencing Batch Reactor (SBR) is perhaps the most promising and viable of proposed activated sludge modifications made for the removal of organic carbon and nutrients. In a relatively short period, it has become an increasingly popular treatment for domestic and industrial wastewaters and is known as an effective biological treatment system due to its simplicity and flexibility of operation (Heloísa Fernandes et al. 2013).

Although coal chemical wastewater treatment technology has been well developed, the understanding of the diversity of microbial populations and their functional relevance is still very limited, which largely restricts the improving of the wastewater treatment system (Wu Yang et al. 2008). This study is based on a coal chemical wastewater treatment plant in Shaanxi Yulin, the main part of the wastewater treatment plant is SBR. This study researched the SBR reactors in the field, and analyzed activated sludge samples by molecular ecological methods, finding dominant bacteria. The study used raw water to cultivate bacteria colonies, then sent them to sequence instead of DNA Sequencing and Clone, the purpose is discovering a effective method to culture bacteria and helpful for the training of function bacterium agents in the coming research. Using molecular ecological techniques to study the structure and function of activated sludge, can help people understand the structure of the function and role of bacteria in the degradation of harmful substances. These techniques can also reveal the relationship between structure and function of bacteria, so as to optimize coal chemical wastewater activated sludge treatment system and provide a theoretical basis for the treatment of coal chemical wastewater (Hyeok Choi et al. 2006; Binbin Liu et al. 2006).

2 PROJECT OVERVIEW

2.1 Devices introduction

The wastewater treatment plant provides services for China National Coal Group Corp's Methanol To Olefin (MTO) project. Its tasks are treating, reusing and recycling the wastewater, as well as achieving wasterwater "Zero Liquid Discharge". The plant covers an area of about 117676 m² with processing capacity of 650 m³/h. The field mainly includes domestic sewage pools and pump rooms, grilles, industrial sewage pools and pump rooms, accident water pools, dosing, high-efficiency clarifiers, SBR pools, effluent pools, sludge concentration pools, Biological Aerated Filters (BAF), blower rooms, monitoring water pools, sludge dewatering systems.

2.2 Technological process

The coal chemical wastewater treatment process is "Adjusting tank—clarification—SBR-BAF". The specific process is shown in Figure 1.

2.3 Fluent and effluent water quality

This research analyzed the influent water in SBR reactors, and discovered that the influent water mainly includes methanol, ethanol, propanol, acetic acid, as well as small amounts of cyanide, phenols and oils. The influent water quality is shown in Table 1.

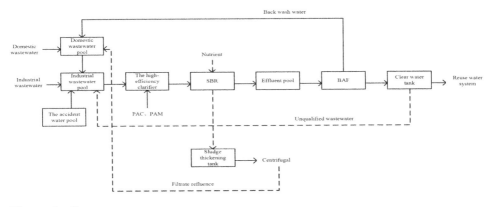

Figure 1. Sewage treatment process.

3 MATERIALS AND METHODS

3.1 *Major materials and equipments*

Materials used in this research were sampling barrels, a microscope (Olympus), funnels, tubes, a −20°C refrigerator, a −4° Crefrigerator, NaCl, agar, disposable dishes, high-pressure steam sterilization pot, a 37°C incubator, a PCR instrument, ddH$_2$O, mix and primers.

3.2 *Learning and sampling in the plant*

We studied the SBR reactors and continuously took activated sludge samples. The SBR reactor's reaction stages are shown in Table 2. The SBR reactors' influent COD and NH3-N as well as removal efficiency are shown in Figure 2 and Figure 3.

The SBR reactors' activated sludge samples were selected in a period which the effluent indicators were moving from an unsteady state to a steady state. The samples were consistently taken at the aeration reaction stage. One part of the mud mixture was filtered, placed

Table 1. Inflow water indicators.

Substances	COD (mg/L)	BOD (mg/L)	NH3-N (mg/L)	SS (mg/L)	pH	CN- (mg/L)	Sulfide
Values	≤1300	≤600	≤300	≤100	6~9	≤0.25	≤1.1

Table 2. SBR reaction stages.

Reaction stages	Filling	Aeration reaction	Stirring	Settling	Drawing
Time (h)	1	8	1	1	1

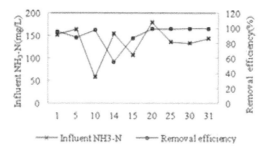

Figure 2. Influent COD and removal efficiency in SBR reactors.

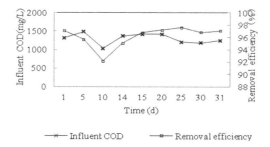

Time (d)

Figure 3. Influent NH$_3$-N and removal efficiency in SBR reactors.

into 15 mL centrifuge tubes and stored in a −20°C refrigerator, another part of the mud mixture was served in 50 mL centrifuge tubes and stored in a 4°C refrigerator for subsequent experiments.

3.3 Colonies incubated and 16S rDNA Polymerase Chain Reaction (PCR)

The solid medium was made by SBR reactor influent water mixing with agar and NaCl (100 mL water, 1.5 g agar, 1 g NaCl). The solid medium was high-pressure steam sterilized, then poured plates. Mud mixture samples at the effluent indicators' steady state were diluted and painted on the plates. The dilution used in this study was 10^{-1}, 10^{-2}, 10^{-3}, 10^{-4}. The plates were incubated in a 37°C incubator for 72 h, until the colonies grew up. The appropriate plates were seleted with numerable colonies, every colony on them was separately cultivated.

16S rDNA Polymerase Chain Reactions (PCR) were constructed to examine microbial diversity. We chose 60 colonies to perform 16S rDNA PCR. The primers used for the reactions were 27F (5′-AGA GTT TGA TCC TGG CTC AG-3′) and 1492R (5′-GGC TAC CTT GTT ACG ACT T-3′). The PCR system was 50 µL, including ddH$_2$O 19 µL, mix 25 µL, each primer 2 µL, DNA template 2 µL. The PCR products were evaluated by agarose (1%, w/v) gel electrophoresis. The electrophoresis results were satisfactory, then the PCR products were sequenced by a biotechnology company. All different sequences were believed to be Operational Taxonomic Units (OTUs). The sequences were compared with all accessible sequences in the databases using the BLAST server at NCBI (National Centre of Biotechnology Information) (Liu, W.T. et al. 1997; Yujiao Sun et al. 2010).

4 RESULTS AND DISCUSSION

4.1 Colonies incubated results

After 72h in the 37°C incubator, the colonies grew up, the separated colonies were also cultivated, as shown in Figure 4.

To find a way to cultivate the same colonies with the wastewater treatment plant SBR reactors, the influent water in the SBR reactors was made into medium without any additional nutrients. After that, the mud mixture samples were painted on the medium, just simulating the actual situation. The colonies incubated could represent the actual bacteria in SBR reactors theoretically. From Figure 4 we can see that every colony has its own shape, then we used 16S rDNA PCR to examine microbial diversity.

4.2 16S rDNA PCR and sequencing results

The electrophoresis results were satisfactory, and PCR products were sequenced. Sequencing results are shown in Table 3.

As seen in Table 3, approximately 37.25% of the colonies are similar to Arthrobacter arilaitensis Re117, and the similarity is very high. Arthrobacter arilaitensis is one of the major bacterial species found at the surface of cheeses, especially in smear-ripened cheeses, where it contributes

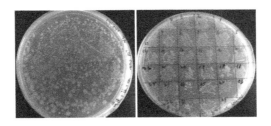

Figure 4.   Colonies incubated result and separately incubated result.

Table 3. Sequencing results.

OTUs	OTU numbers	Percentage of total content	Approximate strains	NCBI reference sequence	Homologous similarity	Organism
1#	19	37.25%	*Arthrobacter arilaitensis Re117*	NC_014550	100%	G⁺, *Actinobacteridae;*
2#	5	9.80%	*Stenotrophomon-as maltophilia K279a*	NC_010943	99%	G⁻, *Proteobacteria*
3#	7	13.73%	*Acinetobacter oleivorans DR1*	NC_014259	99%	G⁻, *Proteobacteria*
4#	1	1.96%	*Acinetobacter baumannii ATCC 17978*	NC_009085	99%	G⁻, *Proteobacteria*
5#	2	3.92%	*Pseudomonas mendocina ymp*	NC_009439	97%	G⁻, *Proteobacteria*
6#	2	3.92%	*Paracoccus denitrificans PD1222*	NC_008687	100%	G⁻, *Proteobacteria*
7#	1	1.96%	*Exiquobacterium SP.ATIb*	NC_012673	100%	G⁺, *Firmicutes*
8#	1	1.96%	*Rhodobacter capsulatus SB 1003*	NC_014034	99%	G⁻, *Proteobacteria*
9#	2	3.92%	*Vibrio harveyi ATCC BAA-1116*	NC_009783	100%	G⁻, *Proteobacteria*
10#	2	3.92%	*Aliivibrio salmonicida LFI1238*	NC_011312	100%	G⁻, *Proteobacteria*
11#	2	3.92%	*Serratia liquefaciens ATCC 27592*	NC_021742	98%	G⁻, *Proteobacteria*
12#	1	1.96%	*Acinetobacter SP.ADP1*	NC_005966	100%	G⁻, *Proteobacteriar*
13#	1	1.96%	*Bacillus toyonensis BCT-7112*	NC_022781	99%	G⁺, *Firmicutes*
14#	3	5.88%	*Pseudomonas stutzeri A1501*	NC_009434	99%	G⁻, *Proteobacteria*
15#	1	1.96%	*Cronobacter turicensis Z3032*	NC_013282	99%	G⁻, *Proteobacteria*
16#	1	1.96%	*Pseudomonas fluorescens Pfo-1*	NC_007492	99%	G⁻, *Proteobacteria*

to the typical colour, flavour and texture properties of the final product. A. arilaitensis Re117 is well-equipped with enzymes required for the catabolism of major carbon substrates present at cheese surfaces such as fatty acids, amino acids and lactic acid (Monnet C et al. 2010). Approximately 13.73% of the colonies are similar to Acinetobacter oleivorans DR1. The genus Acinetobacter is ubiquitous in soil, aquatic, and sediment environments and includes pathogenic strains, such as A. baumannii. Many Acinetobacter species isolated from various environments have biotechnological potential since they are capable of degrading a variety of pollutants. Acinetobacter sp. strain DR1 has been identified as a diesel degrader. Here we report the complete genome sequence of Acinetobacter sp. DR1 isolated from the soil of a rice paddy (Jung J et al. 2010). About 9.80% of the colonies are similar to Stenotrophomonas maltophilia K279a. Stenotrophomonas maltophilia is a nosocomial opportunistic pathogen of the Xanthomonadaceae. The organism has been isolated from both clinical and soil environments in addition to the sputum of cystic fibrosis patients and the immunocompromised. Whilst relatively distant phylogenetically, the closest sequenced relatives of S. maltophilia are the plant pathogenic xanthomonads (Crossman LC et al. 2008). Approximately 5.88% of the colonies are similar to Pseudomonas stutzeri A1501. Pseudomonas stutzeri A1501 genome contains genes involved in broad utilization of carbon sources, nitrogen fixation, denitrification,

Table 4. Contents of different main species in samples.

Microbial classification	Percentage of samples
δ-Proteobacteria	3.92%
γ-Proteobacteria	54.91%
Actinobacteria	37.25%
Firmicutes	3.92%

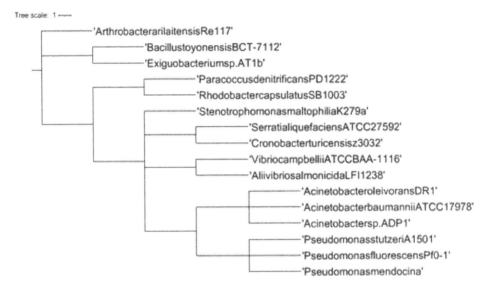

Figure 5. Phylogenetic tree of samples.

degradation of aromatic compounds, biosynthesis of polyhydroxybutyrate, multiple pathways of protection against environmental stress, and other functions that presumably give A1501 an advantage in root colonization (Yan Y et al. 2008). There are still a number of colonies are similar to other strains, such as Acinetobacter baumannii ATCC 17978, Acinetobacter SP.ADP1, Exiquobacterium SP.ATIb, Serratia liquefaciens ATCC 27592 and so on.

The contents of the different main species in the samples are summarized in Table 4. The Proteobacteria is the most frequent species. Proteobacteria represent a major group of bacteria and they include many notable genera. Some are free-living (nonparasitic), and include many of the bacteria responsible for nitrogen fixation. Some Actinobacteria are major contributors to biological buffering of soils and have roles in organic matter decomposition (Ningthoujam et al. 2009).

All of the samples were classified into 16 different groups, as shown in Figure 5.

5 CONCLUSIONS

1. The SBR removal efficiency for organic matter is very high in the chemical industry wastewater treatment plant. At the reactors' steady state, the removal efficiency for COD is approximately 97% and the removal efficiency for NH_3-N is about 99%. SBR is a promising and viable method for the removal of organic carbon and nutrients.
2. The bacteria in the SBR reactors mainly consists Proteobacteria, Actinobacteria and Firmicutes. The Proteobacteria is the most common species, about 58.83% of the samples belong to Proteobacteria, about 37.25% of the samples belong to Actinobacteria, about 3.92% of the samples belong to Firmicutes.

3. Approximately 37.25% of the colonies are similar to Arthrobacter arilaitensis Re117, 13.73% of the colonies are similar to Acinetobacter oleivorans DR1, and 5.88% of the colonies are similar to Pseudomonas stutzeri A1501. These bacteria colonies have effects on organic matter removal, like COD, NH3-N, P, and the other less content species play supporting roles in removal of organic matter. The results are helpful for training function bacterium agents in the coming research and provide a theoretical reference for coal chemical industry wastewater treatment.

ACKNOWLEDGMENTS

This work was supported by the National Natural Science Foundation of China (Grant No. 51178048, No. 51378064).

REFERENCES

Binbin Liu, Feng Zhang, Xiaoxi Feng, et al. (2006) Thauera andAzoarcus as functionally important genera in a denitrifying quinoline-removal bioreactoras revealed by microbial community structure comparison. *FEMS Microbiol Ecol*, 55: 274–286.

Crossman L.C, Gould V.C, Dow J.M, et al. (2008) The complete genome, comparative and functional analysis of Stenotrophomonas maltophilia reveals an organism heavily shielded by drug resistance determinants. *Genome Biol*, 9(4): R74.1–R74.13.

Han Hong-Jun, Li Hui-Qiang, Du Mao-An, et al. (2010) A/O/ammonia gasification wastewater treatment process. *Water supply and drainage in China*, 26(6): 75–77.

Heloísa Fernandes, Mariele K. Jungles, Heike Hoffmann, et al. (2013) Full-scale Sequencing Batch Reactor (SBR) for domestic wastewater: Performance and diversity of microbial communities. *Bioresource Technology*, 132: 262–268.

Hyeok Choi, Kai Zhang, Dionysios D. Dionysiou, et al. (2006) Effect of activated sludge properties and membrane operation conditions on fouling characteristics in membrane bioreactors. *Chemosphere*, 63: 1699–1708.

Jung J, Baek J.H, Park W. (2010) Complete genome sequence of the diesel-degrading Acinetobacter sp. strain DR1. *J Bacteriol*, 192(18): 4794–4795.

Liu W.T, Marsh T.L, Cheng H, et al. (1997) Characterization of microbial diversity by determining terminal restriction fragment length polymorphisms of genes encoding 16S rRNA. *Appl. Environ. Microbiol.*, 63: 4516–4522.

Ma Zhong-xue, Yang jun. (2007) The development of the coal chemical technology and new chemical technology [J]. *Journal of petroleum and chemical industry in gansu province*, 12(4): 1–5.

Monnet C, Loux V, Gibrat J.F, et al. (2010) The arthrobacter arilaitensis Re117 genome sequence reveals its genetic adaptation to the surface of cheese. PLoS One, 5(11): 1–14.

Ningthoujam, Debananda S., Tamreihao, et al. (2009) Antagonistic activities of local actinomycete isolates against rice fungal pathogens. *Afr. J. Microbiol. Res.*, 3 (11): 737–742.

Shi Guang-Mei. (1993) Coal gasification wastewater characteristics of water quality analysis [J]. *Journal of Harbin engineering college*, 26(2): 69–70.

Wu Yang, Xie Qing, Long Yong-Bo, et al. (2008) Microbial total DNA extracted from activated sludge method. *Journal of Anhui Agri. Sci*, 36(11): 4452–4454.

Yu Hai, Sun Ji-Tao, Tang Feng. (2014) The new coal chemical industry wastewater treatment technology is reviewed. Industrial water and wastewater, 4(3): 1–5.

Yujiao Sun, Jiane Zuo, Longtao Cui, et al. (2010) Diversity of microbes and potential exoelectrogenic bacteria on anode surface in microbial fuel cells. *J. Gen. Appl. Microbiol.*, 56:19–29.

Yan Y, Yang J, Dou Y, et al. (2008) Nitrogen fixation island and rhizosphere competence traits in the genome of root-associated Pseudomonas stutzeri A1501. *Proc Natl Acad* Sci USA, 105(21): 7564–7569.

Zhao Qiang, Sun Ti-Chang, Li Xue-Mei, et al. (2012) Present situation and development direction of coal gasification wastewater treatment process. *Industrial water and wastewater*, 43(4): 1–6.

Manufacturing Engineering and Intelligent Materials – Lu & Abu Bakar (Eds)
© 2015 Taylor & Francis Group, London, ISBN 978-1-138-02832-6

Study on rare earth hydrogenation reaction thermodynamics by computer simulation

W.L. Luo, X.H. Zeng, A.D. Xie & T.Y. Yu
College of Electronic and Information Engineering, Jinggangshan University, Jian, Jiangxi, China

ABSTRACT: Through establishing the thermodynamics computational model of rare earth hydrogenation reaction, the thermodynamic characteristics of several light rare earth of La, Ce, Pr, Nd, Sm hydrogen absorption reaction, in the 213.15 K to 373.15 K temperature range, have been obtained using computer simulation method. The results show that La, Ce have better properties of H2 absorption than Pr, Nd and Sm, so La, Ce are the main study objects of hydrogen storage material of rare earth.

1 INTRODUCTION

Hydrogen is a kind of green energy material (Zhang 2014, Sun 2012), and its high density storage is one of the key technologies for the development and application of hydrogen energy (Zhu 2011). Among current hydrogen storage materials, rare earth hydrogen storage materials have the best performance of absorption and deabsorption of hydrogen with low pressure and high density in the room temperature (Lin 2014, Lin & Li et al. 2013), so they have been the most widely used in nickel-metal hydride batteries and fuel cells (Zhang et al. 2012). The market environment of electric vehicles demands higher properties of nickel-metal hydride batteries and fuel cells (Zhu et al. 2011), so the development of rare earth hydrogen storage material with high performance becomes a major issue in the field of clean energy.

To study the thermodynamics properties of rare earth hydrogenation reaction is one of the important ways to search for a new type of rare earth alloy for hydrogen storage. According to quantum mechanics, thermodynamics and statistical physics, and chemical reaction theory, through establishing a thermodynamics theoretical computational model of rare earth hydrogenation reaction, this project has obtained the thermodynamics characteristics of some light rare earth of La, Ce, Pr, Nd, Sm hydrogenation reaction by method of computer simulation in the 213.15 K to 373.15 K temperature range. The differences of hydrogen affinity of these rare earth metals could point the way to explore new rare earth alloy for hydrogen storage.

2 COMPUTATIONAL MODEL

According to quantum mechanics, thermodynamics and statistical physics, and chemical reaction theory, we have established the following thermodynamics computational model of rare earth hydrogenation reaction.

2.1 Calculations of enthalpy H and entropy S for rare earth hydrogenation reactions

The rare earth R (R = La, Ce, Pr, Nd, Sm) hydrogenation reactions are a kind of gas-solid two-phase reactions, and there is

$$R(s) + H_2(g) \rightarrow RH_2(s). \tag{1}$$

For gas phase reactant H_2, its internal energy $E = En + Ee$, entropy $S = Sn + Se$, wherein the nuclear motion energy En and electron motion energy Ee, nuclear motion entropy Sn and electron motion entropy Se can be calculated by computation of single molecule by quantum mechanics and then by statistical thermodynamics theories. Further, according to the thermodynamic function, $H = E + PV = E + RT$, the H value can be obtained.

For solid product RH_2, its volume V change in a limited temperature range is very small, so there is $H = E + PV \approx E$, i.e., its H value is approximately equal to the E value. Nuclear movement in solid RH_2 is mainly the vibration, and no obvious translation and rotation, so with the approximation of nuclear vibration instead of the entire nuclear motion, then the nuclear kinetic energy En in solid RH_2 is approximately equal to nuclear vibration energy Ev in gas RH_2, therefore its nuclear motion enthalpy $Hn = Ev$. Similarly, the nuclear and the electronic entropy S of RH_2 are approximately equal to its gas-phase nuclear vibration entropy Sv and electronic entropy Se, i.e. $S = Sv + Se$.

As for the rare earth metal R, its H and S value at different temperatures can be calculated by the relations between H, S and T derived from thermodynamic theory (Robert & Melvin 2003). That is,

$$H_T - H_{298} = aT + \frac{1}{2}(b \times 10^{-3})T^2 + \frac{1}{3}(c \times 10^{-6})T^3 - \frac{d \times 10^5}{T} - A \tag{2}$$

$$S_T = a\ln T + (b \times 10^{-3})T + \frac{1}{2}(c \times 10^{-6})T^2 - \left[\frac{d \times 10^5}{2T^2}\right] - B \tag{3}$$

2.2 Calculations of enthalpy change ΔH, entropy change ΔS, free energy change ΔG inrare earth hydrogenation reactions

From the rare earth hydrogenation reaction type (1), there is

$$\Delta H = \Delta H_n + \Delta H_e = H(RH_2) - H(R) - H(H_2) + \Delta H_e \tag{4}$$

The nuclear movement enthalpy change ΔH_n can be obtained by $H(RH_2) - H(R) - H(H_2)$ from above calculations, and the electron enthalpy change ΔH_e can be obtained by $\Delta H_e = D_0(H_2) - D_0(RH_2)$, where $D_0(H_2)$, $D_0(RH_2)$ are the chemical dissociation energies of gas phase H_2 and RH_2.

From reaction (1), similarly, the entropy change ΔS and free energy change ΔG can be calculated as

$$\Delta S = S(RH_2) - S(R) - S(H_2) \tag{5}$$
$$\Delta G = \Delta H - T\Delta S. \tag{6}$$

3 RESULTS AND DISCUSSIONS

According to the above computational model of rare earth hydrogenation reaction, the ΔH, ΔS, and ΔG have been computed and shown in Figure 1 to Figure 3.

Figure 1 shows that enthalpy changes ΔH in reactions of rare earth of La, Ce, Pr, Nd with H_2 are negative, which suggests these reactions of absorbing H_2 are exothermic, and heat released by La, Ce absorbing H_2 is greater than heat released by Pr, Nd. In other words, the absorption amount of H_2 by La, Ce is more than that by Pr, Nd under the same temperature. However, the enthalpy changes of Sm absorbing H_2 is positive, which is an endothermic reaction, and shows that Sm absorbing H_2 cannot be carried out spontaneously in our computing temperature range.

Figure 2 shows the entropy changes ΔSof rare earth of La, Ce, Pr, Nd, Sm absorbing H2 are all negative, which expresses that the reaction system order degree increases after rare earth absorbing H_2, while reaction system transforms from gas-solid phase to mainly solid

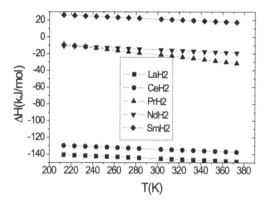

Figure 1.   The relationship between ΔH and temperature T for reactions of rare earth hydrogenation.

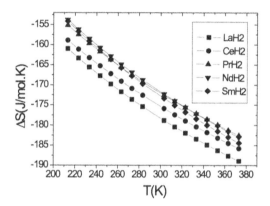

Figure 2.   The relationship between ΔS and temperature T for reactions of rare earth hydrogenation.

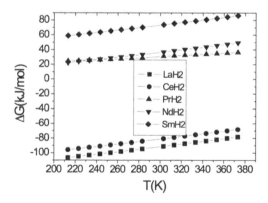

Figure 3.   The relationship between ΔG and temperature T for reactions of rare earth hydrogenation.

phase. With the increase of temperature, the entropy changes have a tendency to decrease, which indicates the absorption reaction will weaken gradually. Under the same temperature, entropy reductions of La, Ce are maximum, which indicates that La, Ce absorb H_2 more easily than Pr, Nd, Sm.

Figure 3 shows the $\Delta G < -60$ kJ/mol for reactions of La, Ce absorbing H2, which manifests that in the temperature range of our calculation, La, Ce are capable of spontaneous absorption of H_2. But the $\Delta G > 0$ for reactions of Pr, Nd, Sm absorbing H2, shows no spontaneous absorption of H_2, and appropriate activation energy must be provided for the reactions. Therefore, La, Ce have better properties to absorb H_2 than Pr, Nd, Sm. That ΔG for reactions of La, Ce, Pr, Nd, Sm absorbing H_2 increase with temperature, shows the increasing temperature is not conducive to the absorption reaction.

4 CONCLUSIONS

According to the quantum mechanics, thermodynamics and statistical physics, and chemical reaction theory, through the establishment of the thermodynamics computational model of rare earth hydrogenation reaction, and by the method of computer simulation, this work have obtained thermodynamic characteristics of several light rare earth of La, Ce, Pr, Nd, Sm hydrogen absorption reaction in the 213.15 K to 373.15 K temperature range. Results show that the hydrogenation reactions of La, Ce, Pr, Nd are exothermic with heat release order of La, Ce, Pr, Nd, but hydrogenation reaction of Sm is endothermic. La and Ce can spontaneously absorb H_2, but Pr, Nd and Sm cannot spontaneously absorb H_2. Under the same temperature, the hydrogen absorption capacities of La, Ce are much stronger than that of Pr, Nd and Sm, with hydrogen absorption strength followed by La, Ce, Pr Nd, and Sm. Therefore, La, Ce are the focus materials of rare earth hydrogen storage. Although Pr, Nd, Sm have relatively poor abilityto absorb H_2, but may also have other traits. The application of rare earth hydrogen storage materials are also worthy of further research and exploration.

ACKNOWLEDGEMENTS

This research is completed under the financial aid of the Jiangxi province Science-technology Support Plan Project (2010BGA00900).

REFERENCES

Li J.L., Wang Y., Zhang X. 2013. Research progress of rare earth magnesium based hydrogen storage alloy. *Metal Functional Materials* 20(4):21–25.
Lin H. 2013 Present development situation of rare earth hydrogen storage alloy material. *Research on Metal Materials* 39(4):30–34.
Lin H. 2014 Development status and prospects of rare earth functional materials. Research on metal materials 40(1):26–31.
Robert C.W., Melvin J.A. 2003 CRC Handbook of Chemistry and Physics. Florida: CRC Press, D-45–47.
Sun X. 2012 Development status and prospect of hydrogen energy. *Technology and Market* 19(4):261–262.
Zhang, C. 2014 New progress in research and application of world hydrogen energy technology. *Petroleum and Petrochemical Energy* 25(8):56–59.
Zhang L.H., Zhang L.J. 2012 Industry status of rare earth hydrogen storage alloy and Ni-MH battery in nearly two years. *Rare Earth Information* 28(1):16–19.
Zhu D., Zen X.b., Ren H.J. 2011 Study on the properties of Ni-MH battery for hybrid vehicle. *Journal of Hefei University of Technology (NATURAL SCIENCE EDITION)* 34(12):1792–1794.
Zhu L. 2011 Discussion on technical problems in the development of hydrogen energy. *Computers and Applied Chemistry* 28(8):1079–1081.

Author index

Printed and bound by CPI Group (UK) Ltd, Croydon, CR0 4YY

24/10/2024

01778293-0007